Fiber Optic Chemical Sensors and Biosensors

Volume II

Editor

Otto S. Wolfbeis, Ph.D.
Professor
Institute of Organic Chemistry
Karl-Franzens University
Graz, Austria

CRC Press
Boca Raton Ann Arbor Boston London

Library of Congress Cataloging-in-Publication Data

Fiber optic chemical sensors and biosensors / Wolfbeis, Otto S.

    p.   cm.
    Includes bibliographical references
    ISBN 0-8493-5508-7
    ISBN 0-8493-5509-5
    1. Fiber optics. 2. Sensors--chemical. 3. Spectroscopy.
    I. Otto S. Wolfbeis
    QD972.P962P65 1991   91-19435
    612'.10444--dc21

Direct all inquiries to CRC Press, Inc., 2000 Corporate Blvd., N. W., Boca Raton, Florida, 33431.

© 1991 by CRC Press, Inc.

International Standard Book Number 0-8493-5508-7
International Standard Book Number 0-8493-5509-5

Library of Congress Card Number 91-19435
Printed in the United States

PREFACE

The development of optical and fiber optic sensors for chemical and clinical parameters is a tremendously fast growing area. These devices have been named "optrodes" and, later, "optodes" by Lübbers in 1975. Particularly in combination with fiber optic waveguides (an offshot of the communication industry), optrodes offer quite new possibilities for remote and *in vivo* sensing, and for inexpensive disposable probes. After many years of worldwide activities it appeared to be timely to publish a book on the subject. It is intended to cover the various aspects of optical chemical sensors including spectroscopy, waveguide theory, physical and analytical chemistry, biochemistry and biophysics, medicine, opto-electronic components, and material sciences.

In view of the interdisciplinary character of the optrode technology, it is obviously necessary that experts from various fields contribute to such a monograph. Because I am aware of the fact that, in these days, nobody wants to read a two-volume book, it is built up in a "modular structure" so that the reader can select certain chapters when interested in a particular subject. Many chapters are self consistent. To get information, for instance, on the state of the art of fluorescence-based pH sensors and their configurations, it would suffice, at first, to read the chapters on Spectroscopic Methods, Sensing Schemes, and pH Sensors. Inevitably, however, there will be some overlap.

The book is organized in the following manner; in the first sections, fundamental aspects are treated (Chapters 1–7), followed by a description of specific sensors for important chemical species (Chapters 8–11) and their applications (Chapters 12–14). Optical methods for temperature measurement, which is an essential part of all analytical procedures requiring some level of precision, is treated in an Chapter 15. The final sections cover bioanalytical methods (Chapters 16–17) and their applications (Chapter 18). Luminescence-based probes for chemoreception which, in the editors opinion, hold great promise for the future, are treated in Chapter 21.

The editor would like to express his sincere appreciation to the authors for their effort and diligence in preparing their manuscripts. I'd like to thank my family which has allowed this book to become part of it for quite a while.

I'm aware of the fact that the present book is neither complete nor perfect. It could just as well have been organized in quite a different way. However, if we would have gone for perfection, it would not have been completed within the next decade.

Otto S. Wolfbeis
May 1991

THE EDITOR

Otto S. Wolfbeis, Ph.D., is Professor of Chemistry and Head of the Analytical Division of the Institute of Organic Chemistry at the Karl Franzens University in Graz, Austria.

Dr. Wolfbeis obtained his Ph. D. degree in 1972 from KFU Graz. He served as a post-doctoral fellow at the Max-Planck Institute for Radiation Chemistry in Mülheim, Germany from 1972–1974 and at the Technical University of Berlin in 1977. He has been at KFU since 1975. It was in 1987 that he assumed his present position.

Dr. Wolfbeis is a member of the Austrian, Swiss, and German Chemical Societies, the European Photochemistry Association (EPA), and the Society of Photoinstrumentation Engineers (SPIE). He heads the Austrian Working Group on Chemical Sensors, acts on the advisory boards of several analytical journals, and is active in several European commissions. He has organized or co-organized several conferences on topics related to optical spectroscopy and optical methods of analysis.

He received the Sandoz Prize in 1981, the Feigl Prize for microanalysis in 1986, and was awarded the highly reputed Heinrich-Emanuel Merck Prize for Analytics in 1988.

Dr. Wolfbeis has authored more than 200 papers on optical sensors, fluorescence spectros-copy of plant natural products such as coumarins, flavons, and alkaloids, and has authored a monograph on the fluorescence of organic natural products. Other fields of research include three-dimensional fluorescence spectroscopy of biological liquids and synthetic and spectro-scopic work on fluorescent probes and indicators. He holds a number of patents related to fluorescence technology. He has given approximately 20 invited lectures at international meetings and numerous quest lectures at Universities and Institutes. His current major re-search interests are in optical sensors and biosensors.

DEDICATION

To Tina, Claudia, Gudrun, and Barbara, and my many friends in the marvelous land between California and the New York Island.

CONTRIBUTORS

Mark A. Arnold
Department of Chemistry
The University of Iowa
Iowa City, Iowa

Floreal Blanc, CEA
Centre d'Etudes Nucleaires
Boite Postale 6
Fontenay, France

Loic J. Blum
Laboratoire de Genie Enzymatique
CNRS-Universite Lyon 1
Batement 308
Villeurbanne, France

Gilbert Boisde, CEA
Centre d'Etudes Nucleaires
Boite Postale 6
Fontenay, France

R. Stephan Brown
Department of Chemistry
Erindale College
University of Toronto
Mississauga, Ontario, Canada

Pierre R. Coulet
Laboratoire de Genie Enzymatique
CNRS-Universite Lyon 1
Batiment 308
Villeurbanne, France

Dileep K. Dandge
ST & E, Inc.
Pleasanton, California

DeLyle Eastwood
Lockheed Enginering & Sciences Co.
Las Vegas, Nevada

Lawrence A. Eccles
United States Environmental
 Protection Agency
Environmental Monitoring Systems
 Laboratory
Las Vegas, Nevada

Kisholoy Goswami
ST & E, Inc.
Pleasanton, California

Kenneth T. V. Grattan
Measurement and Instrumentation Centre
School of Electrical Engineering
The City University
London, England

G. D. Griffin
Health and Safety Research Division
Oak Ridge National Laboratory
Oak Ridge, Tennessee

Nelson R. Herron
Lockheed Engineering & Sciences Co.
Las Vegas, Nevada

Stanley M. Klainer
FiberChem, Inc.
Las Vegas, Nevada

Ernst Koller
Lambda Fluoreszenzechnologie GmbH
Grottenhof-Str.3
Graz, Austria

Ulrich J. Krull
Department of Chemistry
Erindate College
University of Toronto
3359 Mississauga Road N.
Mississauga, Ontario, Canada

Mark J. P. Leiner
AVL-List GmbH
Kleist-Str. 48
Graz, Austria

Robert A. Lieberman
AT & T Bell Laboratory
Murray Hill, New Jersey

Patrick Mauchien, CEA
Centre d'Entudes Nucleaires
Boite Postale 6
Fontenay, France

Fred P. Milanovich
Optical Sensor Consultants
Livermore, California

Douglas Modlin
Optical Sensor Consultants
Livermore, California

Olivier Parriaux
Centre Suisse d'Electroonique et de
 Microtechnologie S.A.
Maladiere 71
CH-2007 Neuchatel, Switzerland

Jean-Jacques Perez, CEA
Centre d'Etudes Muchleaires
Boite Postale 6
Fontenay, France

John I. Peterson
National Institutes of Health
Biomedical Engineering and
Instrumentation Branch
Bethesda, Maryland

W. Rudolf Seitz
University of New Hampshire
Parsons Hall
Department of Chemistry
Durham, New Hampshire

M. J. Sepaniak
University of Tennessee
Department of Chemistry
Knoxville, Tennessee

Stephen J. Simon
Lockheed Engineering & Sciences
 Company
Las Vegas, Nevada

Einar Stefansson
Duke University Eye Center
Durham, North Carolina
and
University of Iceland
Reykuavik, Iceland

Elaine T. Vandenberg
Department of Chemistry
Erindale College
University of Toronto
Mississauga, Ontario, Canada

Tuan Vo-Dinh
Bldg. 4005 S
MS 5258
Oak Ridge National Laboratory
Oak Ridge, Tennessee

Julie Wangsa
Department of Chemistry
The University of Iowa
Iowa City, Iowa

Otto S. Wolfbeis
Analytical Division
Institute of Organic Chemistry
Karl-Franzens University
Graz, Austria

TABLE OF CONTENTS

Volume I

Volume II

Chapter 9

OPTICAL ION SENSING

W. Rudolf Seitz

TABLE OF CONTENTS

I. INTRODUCTION

In principle, the methods used to sense hydrogen ion activity can also be applied to the measurement of metal ion activities. This involves immobilizing an indicator, In, that changes optical properties when it combines with a metal ion, M, to form a complex ion, MIn:

$$M + In \rightleftharpoons MIn \qquad (1)$$

The response range is defined by the equilibrium constant, K_f for the metal-indicator reaction,

$$K_f = [MIn]/[M][In] \qquad (2)$$

and corresponds roughly to $pM = \log K_f + 1$ (analogous to the response range of $pH = pKa \pm 1$ observed for immobilized pH indicators). When $pM < \log K_f - 1$, the indicator is essentially saturated with metal ion, and further increases in metal ion concentration produce little if any change in observed signal. When $pM > \log K_f + 1$, most of the indicator is in the uncombined state, and further decreases in metal ion concentration produce little change in observed signal. These principles have been discussed in more detail in an earlier review of optical metal ion sensing.[1]

In practice several difficulties arise in attempting to implement the above approach. The first is lack of selectivity. Reagents that serve as optical indicators for metal ions generally react with several metal ions and will not provide selective response to a single metal ion unless all interfering ions are absent from the sample. Selectivity for a particular metal ion can often be enhanced by adjusting pH and adding auxiliary reagents such as masking agents. If these measures are implemented by adding the appropriate reagents before exposing the sample to the sensor, then one loses the principal advantage of using a sensor, the ability to make measurements *in situ* in the unmodified sample. Indeed, one can question whether it is worthwhile to use a sensor at all if it is necessary to add auxiliary reagents to the sample for the sensor to function properly.

A second problem is that many reagents that can serve as indicators for metal ions also combine with hydrogen ion. As a result metal ion binding involves displacement of proton, which may be represented:

$$M + HIn \rightleftharpoons MIn + H^+ \qquad (3)$$

As a result the MIn formation constant is conditional, decreasing as the hydrogen ion concentration increases. This means that metal ion sensing is inherently pH dependent. The pH has to be either controlled or be measured so that an appropriate correction factor can be applied.

A third problem is that the formation constant for a particular indicator may not be in the right range for a particular application. A good example is the use of chromogenic ionophores to sense physiological levels of potassium.[2] The equilibrium constant for potassium binding is too small to get significant potassium binding unless special structural features are introduced into the ionophore to enhance binding.

A fourth problem is that the stoichiometry of reactions used to detect metal ions is often not 1:1. Instead, they may involve the formation of 2:1 or 3:1 ligand-to-metal complex. Immobilization of the ligand is likely to interfere with the formation of higher complexes, thus affecting response in a way that is not easily predicted from solution behavior.

Anion detection is even more challenging than cation sensing. Ligand exchange reactions provide an avenue for selectively detecting those anions which act as ligands. However, selectivity and finding a system with the appropriate equilibrium constant for a particular application remain serious problems.

To date the development of cation sensors has attracted considerably more attention than anion sensor development. Sensing based on several conventional metal ion indicators has been evaluated. The limitations of these systems have inspired some creative new approaches to metal ion sensing. Particular attention has been paid to generic approaches that will enable neutral ionophores to be used for optical ion sensing. The hope is to develop a general approach to optical ion sensing in which selectivity can be varied simply by choosing a different ionophore.

This chapter will consider cations first, then anions. The section on cations will be organized according to the strategy used to detect the cation.

The subject of optical sensors for ions has been treated previously in a short review article.[3]

II. CATION SENSING

A. INTRINSIC CATION SENSING

Many transition metal ions have absorption bands due to d-d transitions in the uv/visible/near IR. Because molar absorptivities for these transitions are low, typically around 10 to 100 l/mole-cm, these transitions are useful for *in situ* sensing only if the metal ion concentration is high. Coordination to various ligands causes the wavelength of these absorption bands to shift, a factor that must be accounted for in designing an *in situ* measurement system.

Cu(II) in plating baths has been determined *in situ* based on the Cu(II) absorption band at 820 nm.[4] Separate fibers conduct light to and from the sample. A reflector serves to redirect incident light into the fiber that leads to the detector. Because the Cu(II) absorption band is at long wavelengths, the source for this measurement can be an LED with a silicon photodiode as the detector resulting in an inexpensive monitoring system. The response depends on sulfuric acid concentration presumably because of changes in Cu(II) coordination.

A few ions luminesce. One important example is the uranyl ion, $UO_2(II)$. In order to minimize the hazards associated with exposure to radioactivity, a system has been developed for remote *in situ* detection of uranyl ion in radioactive samples.[5,6] Laser light at 416 nm is used to excite uranyl ion luminescence through an optical fiber. Luminescence is observed at 513 nm. Both luminescence intensity and lifetime are measured. Reduction in the luminescence lifetime serves as a means of identifying and correcting for the effect of dynamic quenching.[7] Because the lifetime of uranyl luminescence is over 100 μsec, the use of lifetime measurements to correct for quenching is readily implemented instrumentally.

Uranyl luminescence is greatly enhanced by adding a small amount of phosphoric or nitric acid.[5,6] To take advantage of this for sensing, a so-called "reservoir optrode" has been employed. The idea is illustrated schematically in Figure 1. A reservoir of dilute acid is separated from the sample by a membrane. The acid diffuses through the membrane changing the composition of the medium of the sample in the immediate vicinity of the membrane surface, the area where luminescence is measured through the optical fiber. Unfortunately, a critical evaluation of this arrangement has not appeared in the literature, although a related approach to chemiluminescence measurements has been described.[8] It is important that there be a slow net flow of reagent through the membrane, rather than having sample diffuse into the reservoir.

Some lanthanide ions, notably Tb(III) and Eu(III), luminesce. The efficiency is high when these ions complex with ligands that absorb light efficiently and transfer it to the lanthanide ion which then emits. Because the lifetime of lanthanide luminescence is much longer than the lifetime of most sources of background signal, time resolved measurements can be used to increase the ratio of signal to background, thereby lowering detection limits. The Eu(III)-2-naphthoyltrifluoroacetonate complex has successfully been used as a label for highly sensitive time-resolved immunochemical measurements through fiber optics.[9]

B. ION EXCHANGE MEMBRANE-BASED SENSING

An interesting approach to cation sensing has been developed using Nafion, a perfluorinated cation exchange membrane with sulfonate sites.[10] Rhodamine 6G, a cationic fluorophor, is immobilized on the membrane. Because perfluorinated ion exchange membranes have the property of strongly binding hydrophobic counterions through a combination of electrostatic and hydrophobic attraction, the rhodamine 6G adheres strongly to the membrane without being covalently bonded to it. The level of rhodamine 6G is kept low, so that the majority of the sites on the Nafion are available to interact with other cations. Many cations, including Co(II), Cr(III), Fe(III), Cu(II), Fe(II), Ni(II), and $NH_4(I)$, quench fluorescence upon combining with the membrane. Others, including alkali metal ion, alkaline earth ions, Zn(II), and Mn(II), have no effect. Response is reversible. Ions that do not quench luminescence can be determined by preloading the membrane with quencher and measuring the rate at which fluorescence intensity increases when exposed to a nonquenching ion. Detection limits are on the order of 1 μmol for both quenching and nonquenching ions.

Although this approach lacks selectivity, it has a number of interesting and attractive features. In particular, the dye involved in the optical readout does not have to interact directly

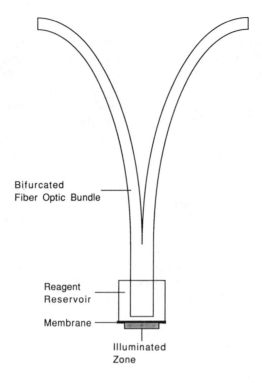

FIGURE 1. Reservoir Optrode. Reagent diffuses through a membrane to modify the sample environment in the illuminated zone.

with the ion. This means that the dye can be chosen for other properties, such as stability, high luminescence efficiency, and favorable wavelengths for excitation and emission. In principle, at least, the selectivity issue can be addressed by preparing membranes with selective ion binding sites. Note, however, that the ion binding site and the luminescent dye will have to be close together so that the ion will quench luminescence.

C. CONVENTIONAL REAGENTS

One approach to the development of metal ion sensors is to adapt existing spectrophoto-metric and fluorimetric reagents to sensing. This involves immobilizing the reagent and measuring the resulting optical changes through fiber optics. Particular attention has been paid to fluorimetric reagents because fluorescence is more conveniently measured through fiber optics and is usually more sensitive than absorption. This subject has been covered in an earlier review.[1]

1. Fluorigenic Ligands

Table 1 lists fluorigenic metal ion reagents that have been evaluated for optical sensing along with the method of immobilization. Most of these are well established solution reagents. While immobilization of known reagents is a natural first step in sensor development, it should be recognized that the requirements for sensing differ in important respects from solution measurements. The reagents listed below are all subject to the general limitations described in the introduction. Furthermore, they are designed such that only one form of the reagent fluoresces, either the ligand or the complex, but not both. While this is preferred for sensitive single wavelength solution measurements, it is not as satisfactory for continuous sensing applications where it is desirable to be able to measure an intensity ratio which is less subject to drift than a single intensity measurement. Ultimately, it may prove to be worthwhile to synthesize new reagents specifically tailored to sensing applications.

TABLE 1
Fluorigenic Reagents for Metal Ions

Str. A		Covalently immobilized to cellulose using cyanuric chloride
Str. B		Electrostatically immobilized to anion exchange resin
Str. C		Covalently immobilized to cellulose using cyanuric chloride
Str. D		Covalently immobilized to silica using cyanuric chloride
Str. E		Electrostatically immobilized on an anion exchange membrane

a. Morin

The first system to be investigated[11] was the use of immobilized morin (3,5,7,2',4'-pentahydroxyflavone) to sense Al(III). The morin was covalently bonded to a cellulose substrate using cyanuric chloride as a coupling reagent. This particular system serves to illustrate the problems associated with metal ion sensors based on conventional reagents. The first is that metal binding properties of the reagent change as a consequence of immobilization. Cyanuric chloride binds through hydroxy groups. Since morin has five hydroxy groups, the immobilized material is almost certainly a complex mixture of several structures. Experimentally, the Al(III) binding capacity was shown to be significantly less than the total amount of immobilized morin, indicating that a significant fraction of the morin was reacting at the hydroxy group involved in complexation and thus losing its ability to complex with Al(III).

Although flavones can form 2:1 complexes in solution,[12] it was shown that the experimental response of the Al(III) sensor fitted a model that assumed a 1:1 complex. The steric constraints imposed by immobilization impede the formation of higher complexes. The conditional equilibrium constant for the formation of the 1:1 complex was shown to vary as a function of pH, as shown in Table 2. The conditional equilibrium constant decreases at pHs below 5.1 because complexation involves displacement of a proton. At pH above 5.1, the constant decreases due to competitive complexation of Al(III) by hydroxide. The consequence of this is that response is inherently pH dependent. Selectivity also depends on pH, favoring Al(III) at pHs in the 4 to 5 range. At higher pHs, metals which also complex with morin and are less subject to complexation by hydroxide are more likely to interfere. In fact, at higher pH immobilized morin may be used to sensing beryllium(II).[13]

b. 8-Hydroxyquinoline Sulfonate

A sensor has been prepared by electrostatically immobilizing 8-hydroxyquinoline sulfonate (8HQS) on an anion exchange resin.[14] The behavior of this reagent is similar to morin. Response can be described in terms of the formation of 1:1 metal:8HQS complexes. Equilibrium constants for the formation of immobilized complexes are similar to those reported for 1:1 metal:8HQS complexes in solution. These values can then serve to predict how the sensor will respond in a given sample.

The sensor based on 8HQS was shown to reversibly respond to Al(III), Mg(II), Zn(II), and

TABLE 2
Experimental Values of the Conditional
Formation Constant for Al (III) Binding
to Immobilized Morin

pH	K_r
3.8	4.0×10^3
4.1	1.2×10^4
4.6	1.4×10^4
4.8	1.7×10^4
5.1	3.3×10^4
5.5	2.8×10^4
5.9	3.0×10^4
6.5	1.8×10^4

Cd(II). The pH range for most sensitive response depended on the metal ion, varying from pH 5 for Al(III) to pH 10 for Mg(II). This is as expected from the relative affinities of these metal ions for 8HQS and for hydroxide.

The excitation wavelength for 8HQS is below 400 nm and is only marginally compatible with glass optical fibers and incandescent sources. This is a disadvantage compared to the other conventional fluorigenic reagents in Table 1, which are all excited in the visible.

c. Calcein

Calcein (also known as fluorexon) has been covalently immobilized to cellulose via cyanuric chloride.[15] In solution calcein is most widely used for the determination of Ca(II). This application requires high pH where calcein is nonfluorescent but forms fluorescent complexes with calcium. The cyanuric chloride linkage hydrolyzes at this pH, precluding calcium sensing with immobilized calcein. Instead, the response of immobilized calcein was evaluated at lower pH levels where calcein fluoresces by itself but forms nonfluorescent complexes with several metal ions. Conditional equilibrium constants at pH 5.15 and 6.85 for formation of calcein complexes with Ni(II), Cu(II), and Co(II) were measured by a competitive technique. Values of the log conditional formation constant ranged from 5.9 for Co(II) at pH 5.15 to 12.4 for Cu(II) at pH 6.95. Because they are so large, calcein can only be used for reversible sensing at very low concentrations. The problem at these concentrations is that the number of immobilized calcein molecules on the sensor may exceed the number of metal ions in solution. As a result, a significant fraction of the analyte has to be removed from the sample to reach equilibrium. The sample is necessarily perturbed and response times are extremely slow. At lower pH values, the conditional formation constants will be lower and continuous reversible sensing may be more practical.

d. 2,2′-Dihydroxyazobenzenes

An important class of ligands for analytical applications is the 2,2′-dihydroxyazobenzenes. These nonfluorescent ligands change color upon complexing with a metal ion. Some ligands also become fluorescent upon complexation. One of these ligands has been immobilized and evaluated[16] as a solid state fluorigenic reagent for Al(III). It was shown to respond to Al(III). In principle this reagent could serve for continuous Al(III) sensing. The equilibrium constant for Al(III) binding would establish the range of analytical response.

e. Chlortetracycline

Chlortetracycline immobilized on an anion exchange resin has been successfully used as a reagent for sensing Ca(II) at pH 7.5.[17] Response is based on the formation of a fluorescent Ca(II) complex with the indicator. The system also responds to Mg(II), Zn(II), and Sr(II).

2. Chromogenic Reagents

Relatively little has been done with chromogenic reagents. In work yet to be published the author and coworkers have successfully immobilized eriochrome black T and shown that it responds to Mg(II) at pH 9.

Although not used for metal ion sensing, some interesting work has been done using electrostatically immobilized orthophenanthroline derivatives as reagents for Cu(I) and Fe(II).[18] It was found that rate of formation of the 2:1 Cu(I) complex was faster than the rate of formation of the 3:1 Fe(II) complex, presumably because the 3:1 complex requires a greater degree of steric rearrangement. Thus, immobilized indicator phases for metal ions based on complex formation should be designed either to allow only the 1:1 complex to form or to allow a high degree of ligand mobility so that higher order complexes can form quite readily.

D. IONOPHORE-BASED CATION DETECTION

Because of the limitations of conventional spectrophotometric and fluorimetric reagents for cations, several researchers have taken the approach of trying to find ways to couple selective cation binding by ionophores to an optical readout. The term "ionophore" is used to describe ligands that selectively bind ions. Typically, ionophores are macrocyclic molecules with an ion binding cavity of discrete dimensions. Selectivity is based on the "goodness of fit" between the ion and the ion binding site. Crown ethers are the best known class of ionophores. However, many other types of molecules, both synthetic and naturally occurring, act as ionophores. The ion binding properties of ionophores are summarized in a major review.[19]

In analytical chemistry, ionophores have had the greatest impact in the development of ion selective membrane electrodes. A lipophilic uncharged ionophore is incorporated into a membrane. The ionophore establishes the relative rates of transport of various ions across the membrane, which in turn establishes selectivity when the membrane is used in an electrode. In addition to providing selectivity, the neutral ionophores used in ion selective electrodes are not subject to protonation, and thus have equilibrium constants for ion binding which are not affected py pH. It would be highly desirable if the advantages of optical sensing could be combined with the selectivity and insensitivity to pH that can be attained with neutral ionophores.

The goal of combining ionophore selectivity with an optical readout has been approached in several ways. The most direct approach is to prepare chromogenic ionophores in which ion binding is accompanied by a change in the optical properties of the chromophore. While this approach may result in reagents suitable for specific applications, it is subject to important limitations and is not generic, i.e., a new chromogenic ionophore will have to be developed for each ion and each range of concentrations. Therefore, other approaches have been evaluated, including the development of indicators based on ion pairing reactions and the use of potential sensitive dyes. These various approaches are considered separately below.

1. Cation Sensing Based on Chromogenic Ionophores

The development of ionophores in which ion binding is accompanied by a change in color or fluorescence has been the object of considerable research. Most of this work has been directed at methods for analysis of metal ions on a one-time basis rather than for continuous optical sensing. The possibility of using these reagents for continuous optical sensing has been investigated in depth by Al-Amir et al.[2] Table 3 lists ligands evaluated for optical sensing along with pK_as and selected binding constants for potassium ions. Because reagents I through III bind cations very weakly they are restricted to sensing at very high concentrations. In reagents IV through VII, metal ion binding is accompanied by loss of a proton from the reagent. In reagent IV, the proton is lost from the nitrogen, while in the other reagents the phenolic proton is displaced by metal ion binding. The tendency of these reagents to form anions enhances their affinity for cations. The effect is greatest for reagents VI and VII which

TABLE 3
Chromogenic Ionophore Indicators

		pKa	Kf(K⁺)
Str. F	I		<0.1
Str. G	II		0.3
Str. H	III		2.6
Str. I	IV	9.0	13.7
Str. J	V	9.65	45
Str. K	VI	9.48	113

Note: Data for IV, VI, and VII are for immobilized reagent. Data for the other compounds are for dissolved reagent.

lose a proton from the ionophore part of the molecule. This maximizes the electrostatic interaction between the metal ion and the ionophore. Unfortunately, it also causes the formation constant for the metal ion-ionophore complex to be inherently pH dependent.

Reagent VI has been evaluated for potassium sensing.[20] The sensor consists of reagent VI adsorbed on Amberlite XAD-2 resin held on the end of an optical fiber by a porous Teflon membrane. The measured parameter is reflectance. At pH 8.0, this sensor responds reversibly and continuously to potassium concentrations from a detection limit of 0.5 up to 100 mmol/l. The K(I)/Na(I) selectivity ratio is 6.4:1.

The color change in reagent VI is due to the loss of the proton to form the phenolate ion. In essence, reagent VI is a pH indicator in which deprotonation is promoted by ion binding. If the pH is too high, the reagent will deprotonate without ion binding, leading to a background signal. When the pH is high enough to cause complete deprotonation, sensing is no longer possible. The pK_a of reagent VI was measured as 7.22 in solution and 7.42 after immobilization.

In addition to complexing with K(I), reagent VI has a moderately high affinity for Ca(II) and can be used to sense Ca(II) in the 5 to 50 mM range at neutral pH.[21]

A chromogenic crown ether selective for Ca(II) and Ba(II) has been described as a potential reagent for optical sensing based on surface plasmon resonance.[22] However, the binding properties of the reagent were investigated in methanol rather than water.

2. Indicator Systems Based on Ion Pair Formation

Complexation of a metal ion by a neutral ionophore results in the formation of a hydrophobic cation. This cation can interact with a hydrophobic anion to form an ion pair which can be extracted into a nonpolar solvent. This has been exploited to develop selective methods for alkali metal ions based on ion pair extraction. A colored or fluorescent anion is used. The amount of metal ion is equal to the amount of the anion which is extracted. The reactions are summarized below:

$$M^+ + I \rightleftharpoons MI^+ \tag{4}$$

$$MI^+ \text{ (aq)} + F^- \text{ (aq)} \rightleftharpoons MI^+ F^- \text{ (org)} \tag{5}$$

where M^+ is the metal ion, I is the ionophore, and F^- is an anionic chromophore or fluorophor.

Ion pair formation has several attractive features for ion sensing. Although the equilibrium constant for complex formation (Reaction 4) is small, the equilibrium constant for ion pair formation (Reaction 5) is large. The overall effect is to strongly favor the solvent extraction process. The thermodynamic driving forces for Reaction 5 include both the electrostatic attraction between the positive charge on the complex and negative charge on the chromophore plus "hydrophobic" interactions between the uncharged parts of the chromophore and ionophore. The more lipophilic the ionophore and/or the chromophore, the stronger the hydrophobic interaction and the more strongly Reaction 5 tends to the right.

Because ion pair formation is thermodynamically more favored than metal ion-ionophore complex formation, it overcomes the equilibrium constant limitations encountered with chromogenic ionophores. Furthermore, the reagent used to achieve selectivity, i.e., the ionophore, is separate from the reagent used to provide an optically detectable signal, i.e., the chromophore. This means that ion pair formation can be applied with any neutral ionophore, greatly enhancing the possibilities for selectivity. The effective equilibrium constant can be varied simply by changing the chromophore concentration.

Implementing ion pairing reactions reversibly in a sensor that responds on a continuous basis presents several challenges. Both the chromophore and the ionophore have to be immobilized in a manner that does not interfere with ion pair formation. The system that has been developed is based on fluorescence and is shown schematically in Figure 2.[22] It involves the following components: (1) neutral ionophore adsorbed on silica, (2) the ammonium salt of 8-anilino-1-naphthalenesulfonic acid (ANS), and (3) the Cu(II) complex of poly(ethylenimine) (Cu(II)-PEI). The Cu(II)–PEI is a cationic polyelectrolyte. In the absence of added alkali metal cation, the ANS binds to Cu(II)–PEI by a combination of electrostatic and hydrophobic attractions. This increases the effective size of the ANS so that it can be immobilized by entrapment behind a dialysis membrane. Because Cu(II) is paramagnetic, it completely quenches the fluorescence from bound fluorophors. Addition of an alkali metal ion leads to the formation of a cationic metal ion-ionophore complex, which forms an ion pair with some of the ANS, drawing it away from the Cu(II)–PEI. Because this takes ANS away from the quenching effect of Cu(II), it is accompanied by an increase in fluorescence. Figure 3 shows intensity as a function of sodium ion concentration for a system with a sodium selective ionophore.[23] The response function depends on the concentration of Cu(II)–PEI. As Cu(II)–PEI increases, ANS is less available for ion pairing and the slope of the response curve decreases. This effectively gives the analyst some control over the range of metal ion concentrations that can be detected. The ion pairing-based sensor for sodium is only slightly dependent on pH between 3 and 11. It is believed that this is associated with protonation of the poly(ethylenimine) and can be eliminated.

While the ion pairing approach is considerably more versatile than preparing chromogenic ionophores, it has a different set of problems. In the reported sodium ion sensor, neither the ANS nor ionophore is covalently attached to a substrate. They are, therefore, subject to leaching which will limit sensor lifetime. Also, it was found that the kinetics of response depend on the structures of both the ionophore and the chromophore. Many otherwise attractive systems have response times of several hours. In particular, attempts to develop a potassium ion sensor using valinomycin as the ionophore were thwarted by kinetic problems. The pros and cons of the ion pairing approach along with alternative implementation schemes have been discussed.[24]

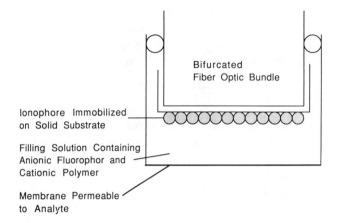

FIGURE 2. Schematic of sodium sensor based on ion pairing.

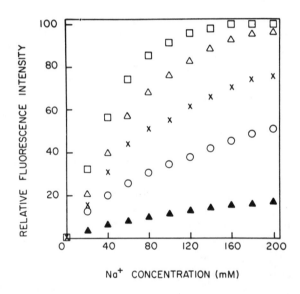

FIGURE 3. Fluorescence intensity as a function of Na⁺ concentration for a sodium selective sensor based on ion pairing. The difference curves correspond to different amounts of cationic polyelectrolyte in the filling solution. As the concentration of cationic polyelectrolyte is reduced, the anionic fluorophor is more available to form ion pairs with the sodium-ionophore complex and the slope of the response curve increases.

3. Potential Sensitive Dyes

Potential sensitive dyes provide an alternative approach to generic ionophore-based ion sensing. These dyes are widely used to measure potential changes across biological membranes.[25,26] They are typically highly delocalized aromatic ions that fluoresce with high efficiency. Although ions, they are sufficiently hydrophobic to permeate through biological membranes and have a strong tendency to adsorb onto lipophilic surfaces. In the absence of a potential across a biological membrane, there will be equal surface coverages on the inside and outside surfaces of the membrane. When a potential develops across the membrane, the relative coverages on either side of the membrane change. If, for example, the dye is cationic, it will be preferentially drawn to the negative side of the membrane. Figure 4 shows the structure of a typical potential sensitive dye and illustrates the mechanism responsible for changes in fluorescence intensity with potential.

Because potential sensitive dyes tend to form weakly fluorescent aggregates and are subject to concentration quenching, their fluorescence is highly concentration-dependent. Therefore, changes in cell potential are frequently accompanied by a change in fluorescence intensity associated with the changes in the surface coverage of dye on either side of the membrane. The direction and magnitude of the change depend on the ratio of dye to membrane as well as on the specific fluorescence characteristics of the dye. An important feature of the response mechanism is that there is no direct association between the potential sensitive dye and the ion. Instead, the ion concentrations on either side of the membrane and the permeability of the membrane to various ions establish a potential by the same mechanism that operates in an ion selective membrane electrode. The potential gradient is responsible for the dye distribution on either side of the membrane which in turn affects the observed fluorescence intensity.

Because of the complexity of the processes causing a change in fluorescence, the variation in fluorescence intensity with potential is not readily calculated. Instead, it is typically determined by a calibration procedure. This involves the addition of valinomycin, a natural ionophore with high potassium selectivity. The valinomycin partitions into the membrane, causing it to become selectively permeable to potassium. After the valinomycin is added, the potential across the membrane is assumed to be due only to potassium and, therefore, can be calculated from known potassium gradients.

The calibration procedure involves the use of an ionophore to impart selectivity to the fluorescence response. It is generic for any neutral ionophore that partitions into the lipid membrane. Thus, this is a generic approach for optical ion detection with ionophore selectivity. The mechanism by which selectivity is achieved is identical to that involved in ion-selective electrodes. The difference is that the readout is optical rather than electrical. The advantage of using an optical readout is that no reference electrode is required. In addition, it may be possible to miniaturize the sensor and achieve calibration stability by designing the system so that the measured parameter is an intensity ratio.

The possibility of ion-selective sensing based on potential sensitive dyes was first demonstrated by Wolfbeis and Schaffar.[27] The approach is shown schematically in Figure 5. The system includes the octadecyl ester of rhodamine B as the potential sensitive dye, valinomycin as the ionophore, and a Langmuir-Blodgett film supported on a glass slide. An important advantage of this approach is that the reagents are quite stable. However, as shown in Figure 6, the total change in intensity is quite small, approximately 12% for a four order of magnitude change in potassium concentration.

The effect of varying the composition of the Langmuir-Blodgett film has since been investigated in more depth.[28] The use of a reference sensor consisting of the film without added valinomycin improves selectivity by compensating for nonspecific effects on intensity accompanying changes in ion concentrations in solution. Intensity decreases linearly with increases in the log of potassium ion concentrations from 0.01 to 100 mM. There is a 17% change in total intensity over this range, a slight improvement over earlier results. A selectivity factor for potassium over sodium of 10^4 is claimed. However, this implies that the authors could actually measure a 0.0017% change in relative intensity, something this reviewer strongly doubts.

Similar results have been achieved applying the same approach to calcium[29] and sodium sensing.[30] Selectivity is excellent, but the overall decrease in intensity is small, severely limiting precision.

Because the lipid bilayer is supported on a solid surface, it is unlikely that the potential sensitive dye actually permeates across the membrane. Instead it has been postulated that the changes in intensity are due to the Stark effect, i.e., the change in optical properties of a fluorophor when an external electric field is applied induced by the ion carrier.[29]

Larger changes in intensity as a function of concentration have been achieved using a liposome-based sensor.[31] Liposomes are vesicles in which a spherical lipid bilayer surrounds

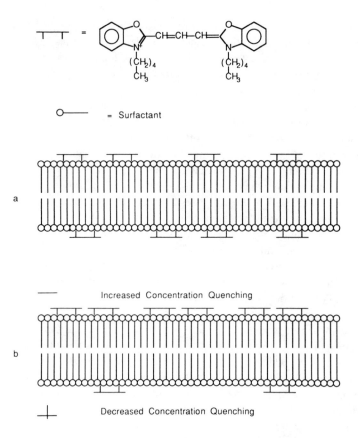

FIGURE 4. Structure of a typical potential dye and response mechanism. The dye is represented by T T. The two vertical lines represent the aliphatic hydrocarbon chains which associate with the lipid layer helping to bind the dye to the lipid bilayer surface. In a, there is no potential across the lipid bilayer and the surface coverage of dye is the same on both sides. In b, the top side of the bilayer is negative relative to bottom side, preferentially drawing the cationic dye to the top surface. The changes in surface coverage are accompanied by changes in the degree of

an interior aqueous compartment. Figure 7 shows fluorescence intensity as a function of potassium concentration for a system consisting of liposomes, valinomycin as the ionophore, and merocyanine 540 as the potential sensitive fluorophor. Both the magnitude and direction of the intensity change depend on the ratio of dye to liposomes. This system can be used for sensing by confining the liposomes behind a dialysis membrane. The feasibility of using this approach for reversible sensing has been confirmed.

A second advantage of the liposome approach is that the interior composition of the liposome can be manipulated. Specifically, the liposome/valinomycin/merocyanine 540 system has been modified to include oxazine 1 in the interior of the liposome.[32] In this system there is Förster energy transfer from merocyanine 540 on the inside surface of the liposome lipid bilayer to the oxazine. Merocyanine 540 on the outside surface of the liposome is too far from the oxazine for energy transfer to occur. The measured parameter in this system is the ratio of merocyanine 540 emission intensity to oxazine emission intensity. By designing systems based on intensity ratio measurements, it should be possible to develop sensors with excellent calibration stability.

The challenge with the liposome approach is to achieve stability. Cyanines and oxonols, the most frequently used potential sensitive dyes, are subject to rapid decomposition. The octadecyl ester of rhodamine B was originally chosen to avoid this problem.[27] Another source of instability is ion leakage through the lipid bilayer, resulting in a change in the interior ion

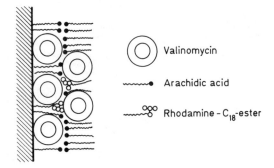

FIGURE 5. Schematic of indicator system for potassium based on potential sensitive dyes. (Reprinted from *Anal. Chim. Acta*, 198, 1, 1987.)

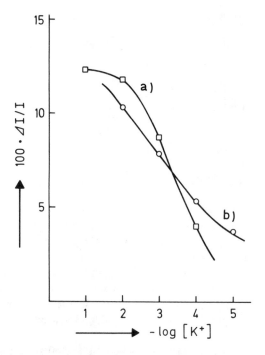

FIGURE 6. Response to potassium for sensor based on potential sensitive dye. The response is plotted in terms of I, the change in intensity, divided by the intensity in the absence of potassium times 100. Curves a) and b) are for sensors formulated with and without arachidic acid. (Reprinted from *Anal. Chim. Acta*, 198, 1, 1987.)

concentration. Both ionophore and dye may also leach out of the system. The full capabilities and limitations of this approach remain to be established.

A third approach is to use an electrochromic dye to sense membrane potential. These dyes incorporate into lipid bilayers and undergo a wavelength shift as a function of the electrical potential applied across the membrane. The possibility of using this type of potential sensitive dye to sense ions has been suggested.[33] However, the wavelength shifts are small, making it unlikely that high sensitivity can be achieved.

III. ANION SENSING

The development of anion selective indicators is a considerable challenge. One approach

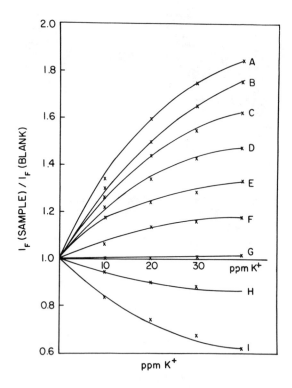

FIGURE 7. Relative fluorescence intensity vs. potassium concentration for a liposome suspension in the presence of merocyanine 540, a potential sensitive dye. Curves A through I correspond to successively greater amounts of merocyanine 540 in the reagent formulation.

to anion selectivity is to take advantage of differences in the ability of various anions to quench fluorescence. In particular, paramagnetic anions and anions with high atomic number atoms will tend to be good quenchers. Anions that act as oxidizing or reducing agents can be sensed via their effect on colored dyes. A more general approach is to base detection on reagents that selectively bind anions. Ionophores that selectively bind anions are being developed, but this area lags well behind the development of cation selective ionophores. Instead, the anion selective reagents that have been developed to date for optical sensing have all been based on the relative ability of different anions to act as ligands to bind to a metal center.

A. ANION SENSING BASED ON FLUORESCENCE QUENCHING

Halide ions have been reversibly sensed using a fluorescence quenching approach.[34] The indicator is a cationic fluorophor covalently immobilized on a glass substrate. The cationic site attracts anions close to the fluorophor, facilitating quenching. Fluorescence intensity decreased with increasing halide concentration. Because the heavy atom effect is responsible for quenching, sensitivity follows the order I > Br > Cl. Sulfite and pseudohalides interfere but common anions such as sulfate, phosphate, nitrate, and perchlorate do not affect response.

B. REDOX-BASED ANION SENSING

Sulfide has been determined based on its ability to reduce 2,6-dichlorophenolindophenol from the blue azoquinoid form into the colorless leuco form.[35] Response of the probe does not involve an equilibrium. Instead, the rate of the reaction depends on sulfide concentration. The reflectance measured 2 min after initiating the reaction between sulfide and the indicator varies with sulfide concentration. The indicator can be regenerated by oxidation.

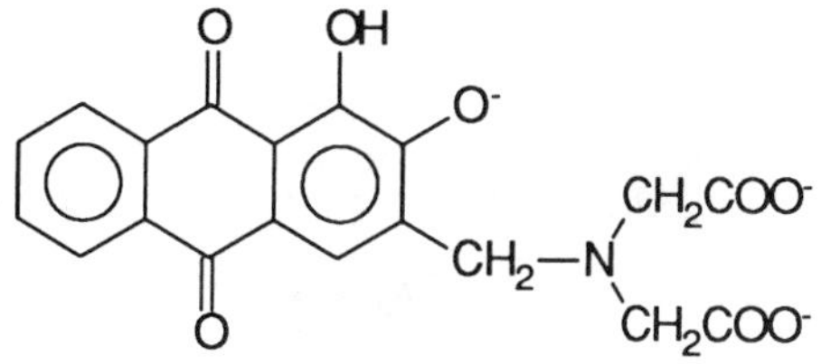

Alizarin Complexone

FIGURE 8. Structure of alizarin complexone.

C. ANION SENSING BASED ON LIGAND EXCHANGE

Sulfide has also been determined based on its ability to release dithiofluorescein from complexes with silver nitrate and with o-hydroxymercuribenzoic acid.[35] The complex is immobilized by adsorption on a solid surface. Added sulfide combines with silver or mercury displacing dithiofluorescein into solution. The result is a reduction in reflectance of the immobilized reagent. The reaction involved in detection is irreversible and cannot be used for continuous sensing. However, the indicator can be regenerated by contact with fresh reagent.

In a related approach, halides have been probed based on their ability to displace fluorescein from silver fluoresceinate,[36] thus overcoming the quenching effect of silver on fluorescein fluorescence. Sensitivity depends on the affinity of the halide for Ag(I) and follows the order I > Br > Cl. This system is inherently nonreversible.

Fluoride has been sensed using lanthanide complexes of alizarin complexone.[37] The structure of the ligand is shown in Figure 8. When it combines with a lanthanide ion, the resulting complex is uncharged and remains on a solid surface rather than dissolving in water to any significant extent. Fluoride displaces water coordinated to the lanthanide ion, causing a shift in the reflectance spectrum of the immobilized complex. Selectivity is based on the tendency of lanthanide ions to coordinate with fluoride.

The ion pairing approach implemented with cations has also been applied to the detection of fluoride.[38] The ionophore is a pentacoordinate Al(III) complex. Reversible, selective response to fluoride was achieved, but response times were greater than an hour and reagent leaching was a serious problem.

REFERENCES

1. **Seitz, W. R., Saari, L. A., Zhujun, Z., Pokornicki, S., Hudson, R. D., Sieber, S. C., and Ditzler, M. A.,** in *Advances in Luminescence Spectroscopy*, Cline Love, L. J., and Eastwood, D., Eds., ASTM, Philadelphia, 1985, 63.
2. **Al-Amir, S. M. S., Ashworth, D. C., and Narayanaswamy, R.,** Synthesis and characterisation of some chromogenic crown ethers as optical potassium ion sensors, *Talanta*, 36, 645, 1989.
3. **Narayanaswamy, R. and Russell, D. A.,** Optical sensors for dissolved chemical species, *Sensors Actuators*, 13, 293, 1988.
4. **Freeman, J. E., Childers, A. G., Steele, A. W., and Hieftje, G. M.,** A fiber-optic absorption cell for remote determination of copper in industrial electroplatng baths, *Anal. Chim. Acta*, 177, 121, 1985.
5. **Malstrom, R. A. and Hirschfeld, T.,** On-line uranium determination using remote fiber fluorimetry, in *Analytical Spectroscopy*, Lyon, W. S., Ed., Elsevier, Amsterdam, 1983, 25.
6. **Malstrom, R. A.,** Uranium analysis by remote fiber fluorimetry, Rep. DP-1737, prepared for the U.S. Depatment of Energy under contract DE-AC09-76SR00001, 1988.
7. **Hieftje, G. M. and Haugen, G. R.,** Correction of quenching errors in analytical fluorimetry through use of time resolution, *Anal. Chim. Acta*, 123, 255, 1981.
8. **Nau, V. and Nieman T. A.,** Application of microporous membranes to chemiluminescence analysis, *Anal. Chem.*, 51, 424, 1979.
9. **Petrea, R. D., Sepaniak, M. J., and Vo-Dinh, T.,** Fiber-optic time-resolved fluorimetry for immunoassays, *Talanta*, 35, 139, 1988.
10. **Bright, F. V., Poirier, G. E., and Hieftje, G. M.,** A new ion sensor based on fiber optics, *Talanta*, 35, 113, 1988.
11. **Saari, L. A. and Seitz, W. R.,** Immobilized morin as fluorescence sensor for determination of Al(III), *Anal. Chem.*, 55, 667, 1983.
12. **Katyal, M. and Prakash, S.,** Analytical reactions of hydroxyflavones, *Talanta*, 24, 367, 1977.
13. **Saari, L. A. and Seitz, W. R.,** Optical sensor for beryllium based on immobilised morin fluorescence, *Analyst*, 109, 655, 1984.
14. **Zhujun, Z. and Seitz, W. R.,** A fluorescent sensor for aluminum(III), magnesium(II), zinc(II), and cadmium(II) based on electrostatically immobilized quinolin-8-01 sulfonate, *Anal. Chim. Acta*, 171, 251, 1985.
15. **Saari, L. A. and Seitz, W. R.,** Immobilized calcein for metal ion preconcentration, *Anal. Chem.*, 56, 810, 1984.
16. **Ditzler, M. A., Doherty, G., Sieber, S., and Allston, R.,** Fluorescence study of an immobilized ligand-metal ion complex, *Anal. Chim. Acta*, 142, 305, 1982.
17. **Kawabata, Y., Tahara, R., Imasaka, T., and Ishibashi, N.,** Fiber-optic calcium(II) sensor with reversible response, *Anal. Chim. Acta*, 212, 267, 1988.
18. **Ditzler, M. A., Pierri-Jacques, H., and Harrington, S. A.,** Immobilization as a mechanism for improving the inherent selectivity of photometric reagents, *Anal. Chem.*, 58, 195, 1986.
19. **Izatt, R. M., Bradshaw, J. S., Nielsen, S. A., Lamb, J. D., and Christensen, J. J.,** Thermodynamic and kinetic data for cation-macrocycle interaction, *Chem. Rev.*, 85, 271, 1985.
20. **Alder, J. F., Ashworth, D. C., Narayanaswamy, R., Moss, R. E., and Sutherland, I. O.,** An optical potassium ion sensor, *Analyst*, 112, 1191, 1987.
21. **Ashworth, D. C., Huang, H. P., and Narayanaswamy, R.,** An optical calcium ion sensor, *Anal. Chim. Acta*, 213, 251, 1988.
22. **van Gent, J., Sudholter, E. J. R., Lambeck, P. V., Popma, T. J. A., Gerritsma, G. J., and Reinhoudt, D. N.,** A chromogenic crown ether as a sensing molecule in optical sensors for the detection of hard metal ions, *J. Chem. Soc. Chem. Commun.*, 893, 1988.
23. **Zhujun, Z., Mullin, J. L., and Seitz, W. R.,** Optical sensor for sodium based on ion-pair extraction and fluorescence, *Anal. Chim. Acta*, 184, 251, 1986.
24. **Seitz, W. R., Zhujun, Z., and Mullin, J. L.,** Reversible indicators for alkali metal ion optical sensors, *SPIE Proc.*, 713, 126, 1986.
25. **Sims, P. J., Waggoner, A. S., Wang, C. H., and Hoffman, J. F.,** Studies on the mechanism by which cyanine dyes measure membrane potential in red blood cells and phosphatidylcholine vesicles, *Biochemistry*, 13, 3315, 1974.
26. **Waggoner, A. S.,** Dye indicators of membrane potential, *Annu. Rev. Biophys. Bioeng.*, 8, 47, 1979.
27. **Wolfbeis, O. S. and Schaffar, B. P. H.,** Optical sensors: an ion-selective optrode for potassium, *Anal. Chim. Acta*, 198, 1, 1987.
28. **Schaffar, B. P. H., Wolfbeis, O. S., and Leitner, A.,** Optical sensors. 23. Effect of Langmuir-Blodgett layer composition on the response of ion-selective optrodes for potassium based on the fluorimetric measurement of membrane potential, *Analyst*, 113, 693, 1988.

29. **Schaffar, B. P. H. and Wolfbeis, O. S.,** A calcium selective optrode, *Anal. Chim. Acta*, 217, 1, 1989.
30. **Schaffar, B. P. H. and Wolfbeis, O. S.,** A sodium-selective optrode, *Mikrochim. Acta*, III, 109, 1989.
31. **Zhujun, Z. and Seitz, W. R.,** Ion-selective sensing based on potential sensitive dyes, *SPIE Proc.*, 906, 74, 1988.
32. **Zhujun, Z. and Seitz, W. R.,** Unpublished work.
33. **Opitz, N. and Lübbers, D. W.,** Electrochromic dyes, enzyme reactions and hormone-protein interactions in fluorescence optic sensor (optrode) technology, *Talanta*, 35, 123, 1988.
34. **Urbano, E., Offenbacher, H., and Wolfbeis, O. S.,** Optical sensor for continuous determination of halides, *Anal. Chem.*, 56, 427, 1984.
35. **Narayanaswamy, R. and Sevilla, F., III,** Flow cell studies with immobilised reagents for the development of an optical fibre sulphide sensor, *Analyst*, 111, 1085, 1986.
36. **Hirschfeld, T., Deaton, T., Milanovich, F., Klainer, S. M., and Fitzsimmons, C.,** The feasibility of using fiber optics for monitoring groundwater contaminants, EPA Report AD-89-F-2A074, 1984.
37. **Narayanaswamy, R., Russell, D. A., and Sevilla, F., III,** Optical-fibre sensing of fluoride ions in a flow-stream, *Talanta*, 35, 83, 1988.
38. **Zhujun, Z., Mullin, J. L., Tang, Y., and Seitz, W. R.,** Optical ion detection via ion pairing, in *Biosensors International Workshop 1987*, Schmid, R. D., Ed., VCH, Weinheim, 1987, 229.

Chapter 10

OXYGEN SENSORS

Otto S. Wolfbeis

TABLE OF CONTENTS

I. FUNDAMENTALS AND GENERAL ASPECTS

The major advantages of oxygen optrodes over amperometric electrodes such as the Clark electrode[1] are small size, lack of oxygen consumption and reference cells, and inertness against sample flow rates and stirring, as well as high external pressure (as they occur in seawater studies). The following chapter discusses optical oxygen sensors first from a more general point of view. Later, the variety of existing types of optrodes will be described in some detail.

A. SENSING PRINCIPLES

There are numerous reports on the "interference" of oxygen in the precise determination of luminescence quantum yields and lifetimes, because oxygen is a voracious dynamic quencher of fluorescence and phosphorescence. However, there are only a limited number of dyes known to be subject to quenching at a rate and efficiency large enough to make them useful as indicators. Theory predicts the quenching process to obey the Stern-Volmer equation

$$I_o/I = 1 + K_{sv} \cdot pO_2 \qquad (1)$$

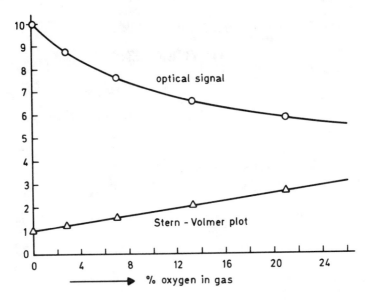

FIGURE 1. Plot of fluorescence intensity vs. quencher concentration, and corresponding Stern-Volmer plot (I_o/I vs. quencher concentration) according to Equation 1.

where I_o and I are the fluorescence intensities in the absence and presence, respectively, of oxygen. K_{sv} is the overall dynamic quenching constant which is equal to $k_d\tau_o$, with k_o being the diffusional bimolecular rate constant and τ_o the natural lifetime of the indicator. For oxygen in liquid solvents, pO_2 is sometimes replaced by its concentration $[O_2]$, which is related to pO_2 by the Henry-Dalton solubility coefficient α, so that Equation 1 reads

$$I_o/I = 1 + K_{sv}\cdot\alpha[O_2] \tag{2}$$

Because this type of quenching is a dynamic event involving the collision of dye and oxygen, the process is sometimes called collisional quenching (in contrast to static quenching which is caused by the formation of a stable ground state complex). Typical values are 10^8 to $10^{10} M^{-1}s^{-1}$ for k_d and $5\cdot10^{-7}$ to $1\cdot10^{-9}$s for τ_o. Equation 1 may also be written in the lifetime form

$$\tau_o/\tau = 1 + K_{sv}\cdot pO_2 \tag{3}$$

with τ_o and τ being the decay times of the fluorophore in, respectively, the absence and presence of oxygen. All three equations hold for phosphorescence as well.

Assuming the average lifetime of a quenchable fluorophore to be 1 to $20\cdot10^{-8}$ s, one sees that K_{sv} can assume values between 2000 and 1 M^{-1} in case of exclusive dynamic quenching. Larger values for K_{sv} are mostly indicative of mixed dynamic and static, or even exclusively static, quenching. Smaller values are useless from an analytical point of view in most cases because of the resulting lack of sensitivity. A typical plot of luminescence intensity vs. quencher concentration as obtained in a fluorescence quenching process, and the respective Stern-Volmer plot are shown in Figure 1.

K_d (and therefore K_{sv}) depend on the viscosity of the solvent, its temperature, and, of course, the nature of the fluorophore. Although Equation 1 predicts a linear relation between I_o/I and pO_2, deviations of linearity are mostly observed at oxygen partial pressures above 200 torr (26.6 kPa). An important consequence results from the fact that Equations 1 to 3 are linear and have a fixed intercept: in the complete absence of straylight, a one-point calibration of the sensor becomes possible.

The process of dynamic fluorescence quenching is usually interpreted in terms of formation of a charge-transfer complex ("oxciplex") upon collision of oxygen and the excited fluorophore (indicator). As a result, nonradiative relaxation of the oxciplex gives the respective ground state species which is thermodynamically instable and rapidly dissociates to give the fluorophore (in the S_o state) and oxygen (in either the S_o or T_o state). Obviously, no oxygen is consumed in this process. This is a distinct feature of this kind of oxygen sensor and in contrast to the amperometric electrode where the analyte is slowly consumed by chemical reduction.[1]

Aside from fluorescence quenching,[2] oxygen has been detected optically by phosphorescence quenching,[3] chemiluminescence,[4] reflectometry of immobilized hemoglobin,[5] and thermally stimulated luminescence of poly(ethylene-2,6-naphthalenedicarboxylate).[6] However, the majority of oxygen optrodes is based on fluorescence quenching, and they will be treated first.

Two types of oxygen optrodes have been described in the literature that differ entirely in the way they are fabricated. In the first type, the oxygen-sensitive material is placed on a rigid and optically transparent support, and its fluorescence is monitored with the help of fiber optics, or even without these. In the second type, the sensor chemistry is directly attached to the fiber tip, either at the distal end or on the core (after removing coating and clad). In both types of sensors, the sensing layer changes its fluorescence in accord with the oxygen tension, and this is monitored instrumentally. Depending on their field of application, each type has its merits. In both approaches an equilibrium has to be established between sample and reagent phase which frequently is a hydrophobic material. This has two beneficial consequences: no reagent is consumed once an equilibrium is established, and potentially interfering substances usually cannot interfere because they cannot enter the membrane, the anesthetic halothane being a notable exception.

B. INDICATORS

A variety of chemically quite different indicators has been applied for probing oxygen. Polycyclic aromatic hydrocarbons such as pyrene, pyrenebutyric acid, fluoranthene, decacyclene, diphenylanthracene, and benzo(g,h,i)perylene are viable indicators because they are efficiently quenched and fairly soluble in silicone (the most frequently used polymer). Alternatively, a variety of longwave absorbing dyes such as perylene dibutyrate, and heterocycles including fluorescent yellow, trypaflavin, and porphyrins (e.g., chlorophyll) are known to be quenchable by oxygen when adsorbed on polar supports (silicagel or the like). They have favorable absorptions and emissions, but their photostability is generally poor (except perhaps for the first). In addition, fluorescence decay is very fast, and quenching graphs are distinctly nonlinear. A third group of indicators comprises metal-organic complexes of ruthenium, osmium, iridium, and platinum. These have strong metal-to-ligand energy transitions and long-lived excited states (up to 5 µs) which makes them useful for lifetime-based oxygen sensors.

Most of these indicators are highly specific. No interferences are found for PAH-based oxygen optrodes with water vapor, nitrogen, noble gases, carbon monoxide, carbon dioxide, methane, and higher alkanes at realistic pressures, although these usually can pass the polymer membranes. Some olefins quench, however. Despite the fact that heavy metal atoms such as Pb(II), Fe(III), and the like also quench the fluorescence of PAHs, they usually cannot enter the polymer membrane and therefore remain inert. Major interferents are sulfur dioxide (a very efficient quencher), halothane (an inhalation narcotic), chlorine, and some nitrogen oxides. Oxygen sensors based on ruthenium(II)tris(bipyridyl) complexes are even less prone to interferents because halothane does not quench their luminescence.

C. DYE IMMOBILIZATION AND MEMBRANE MATERIALS

In order to obtain an oxygen sensor it is necessary to immobilize the oxygen-sensitive dye on a rigid support that can be fixed at the end of a fiber, or on the fiber itself. This, however, is not an easy task: most indicators do not have appropriate functional groups suitable for covalent immobilization, so that physical immobilization or immobilization on ion-binding membranes is frequently employed. In addition, it has been found that the quenching constants (K_{sv} in Equation 1) are reduced by 30 to 50% upon covalently binding the dye onto a rigid surface. Finally, all kinds of indicators not covalently immobilized are slowly washed out from the sensing membrane, particularly when the sample is blood with its binding sites for lipophilic molecules.

A simple way of immobilizing a lipophilic indicator is to dissolve it in a hydrophobic polymer such as poly(vinyl chloride) (PVC) or silicone. Silicone, in particular, has excellent oxygen permeability and solubility but is a poor solvent for most dyes. PVC, on the other hand is a good solvent for most polycyclic aromatic hydrocarbons (PAHs), but has slow oxygen diffusion. In order to make PAHs better lipid-soluble and less water-soluble, they may be fitted with tertiary butyl groups, which results in a 5- to 20-fold improved solubility in silicones and other materials.[7] Cross-linked poly(hydroxyethyl methacrylate) is another solvent which retains polycyclic aromatics such as diphenylanthracene.[8]

Silicones have excellent optical properties, can be handled easily, and allow the fabrication of extremely thin films. One disadvantage in case of minute sample volumes is its solubilizing power for oxygen. This may result in a depletion of oxygen when small sample volumes are brought into contact with silicone membranes. On the other hand, the observed quenching constants for a given indicator are highest in silicone, when compared with any other polymer materials including PVC, poly(methyl methacrylate), polystyrene, polybutadiene, or polyethylene.

Other methods of dye immobilization include adsorption of dye on porous surfaces[9,10] and soaking polystyrene particles[11] or foils[12] with dye solutions. The covalent immobilization of pyrene on glass supports[13] and in silicone matrix[14] has also been described. In the latter case, a bifunctional silicone bearing a triethoxy silyl group and an isocyanato group is used to link a hydroxyalkyl PAH to a silicone polymer. A quite different way of immobilization is to encapsulate a dye solution in gas-permeable polyurethane nano-capsules[15] (o.d. in the order of a few nanometers).

The widely used oxygen indicator ruthenium-tris(bipyridyl)(II) may be immobilized electrostatically on a cation exchange membrane such as Nafion, or physically entrapped in silicone.[16] The incorporation of decacyclene into a Langmuir-Blodgett quadruple layer has been reported.[17] The dye is not washed out by aqueous samples, and an excellent sensitivity and response time towards oxygen was observed.

D. VISCOSITY EFFECTS, DIFFUSION, AND OXYGEN SOLUBILITY

Theory predicts quenching efficiency to be a function of viscosity η, since, according to the Stokes-Einstein equation, the diffusion coefficients depend on η. Diffusion coefficients (D) may be obtained from the Stokes-Einstein equation

$$D = kT/6\pi \cdot \eta \cdot R \tag{4}$$

where k is the Boltzmann constant and R is the collision radius (usually taken as the sum of the radii of fluorophore and quencher). This equation frequently underestimates the diffusion

TABLE 1
Effect of Solvent Viscosity on Biomolecular Rate
Constants of Dynamic Quenching (K_d) and Relative
Stern-Volmer Quenching Constants
(Water = 1)

Solvent	Viscosity (cP)	K_d (M⁻¹s⁻¹)	rel. K_{sv}
Gas phase	0.02	1.3×10^{12}	67
Water	1.0	2.5×10^{10}	1
Glycerol	1500	1.7×10^7	0.00

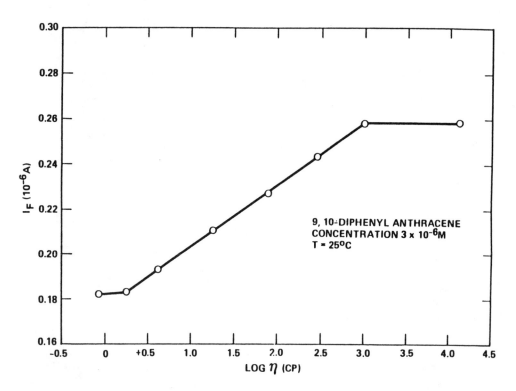

FIGURE 2. Viscosity dependence of the fluorescence of 9,10-diphenyl anthracene in silicone rubber at 25°C. (From Reference 18.)

coefficients in case of oxygen. Since D directly influences the dynamic quenching constant k_d via the number of collisions that occur per time unit, viscosity has a strong effect upon the quenching constant as can be seen from Table 1.

A careful study on the effects of viscosity on the quenching of the fluorescence of diphenylanthracene by oxygen in poly(dimethylsiloxane) solvent revealed[18] a distinct viscosity dependence of fluorescence intensity (Figure 2), but the slope of the Stern-Volmer plot is not a function of viscosity, as it is constant for viscosities of 10^3, 10^4, 10^5 cS fluids and fully polymerized silicone. Another remarkable finding is the concentration dependence of the slope of the response curve, with k_d increasing with higher fluorophore concentrations. This change in slope cannot be due to concentration effects, since adding another bimolecular process to the rate equations predicts an increase in the slope. One possible interpretation is the assumption of an active sphere surrounding the oxygen molecule. The more fluorescent molecules are within this active sphere, the more will be quenched.

TABLE 2
Permeability Coefficients, Diffusion Constants, and Solubility Coefficients of Oxygen in Various Polymers

Polymer	$P \times 10^{6}$ [a]	$D \times 10^{4}$ [b]	$s \cdot 10^{-3}$ [c]	Temp. (°C)
Polybutadiene	143	150	9.6	25
Polyisoprene	175	173	9.9	25
Polyethylene, d = 0.914	21	46	4.7	25
Polyethylene, d = 0.964	3	17	1.8	25
Poly(ethyl methacrylate)	9	10	8.5	25
Poly(acrylonitrile)	0.0015	—	—	25
Poly(tetrafluoro ethylene)	32	15	0.27	23
Poly(vinyl acetate)	4	5.5	6.2	30
Poly(vinyl alcohol)	0.07	—	—	30
Poly(vinyl chloride)	0.34	1.2	2.9	25
Nylon 6	0.29	—	—	30
Poly(oxy dimethyl silylene[s])	4537	1200[f]	30.6[f]	25
Poly(ethylene terephthalate[e])	0.44	—	—	25
Cellulose acetate	5.85	—	—	30

Note: From Reference 23.

[a] cm^3 (STP) • cm • cm^{-2} • s^{-1} • Pa^{-1}.
[b] cm^2 • s^{-1}.
[c] cm^3 (STP) cm^{-3} • Pa^{-1}.
[d]   With 10% filler.
[e]   Amorphous.
[f]   At 0°C.

When oxygen permeates through a polymer membrane, the rate of permeation is given by the permeability coefficient P, defined as (amount of permeant)/(area)(time)(driving force across the membrane). The dimension of P is usually cm^3(STP)/cm^2·s·(cmHg). Since the permeation of small molecules through flawless polymer films occurs by consecutive steps of solution of a permeant in the polymer and diffusion of the dissolved permeant, the permeability coefficient can be given by P = D·S, where D is the diffusion constant and S is the solubility coefficient. The usual dimensions for D and S are given in Table 2. Both P and S show a temperature dependence which can be described by an Arrhenius-type equation.

Similarly, the solubility of oxygen in all polymers decreases with increasing temperature above the glass temperature and obeys an Arrhenius relationship

$$S(T) = S_o \exp(- \delta H/RT) \qquad (5)$$

where δH is the enthalpy of solution and S_o the pre-exponential term. The enthalpy can be calculated from the temperature dependence of the relative solubility and was found[19] to be about −3 kcal/mol in both fluid and fully polymerized silicone. Figure 3 shows the temperature dependence of the solubility of oxygen in silicone rubber. Typical data for P, S, and D for oxygen in various polymers are compiled in the Table 2. The unique properties of silicone are obvious.

The diffusion constant of oxygen in silicone also obeys an Arrhenius relationship with

$$D = D_o \exp(- E_D/RT) \qquad (6)$$

Experimentally[19] it was found for a series of poly(dimethyl siloxanes) of various viscosity that oxygen has a large diffusion constant D_o (0.115 cm^2 s^{-1}) and a low activation energy E_D

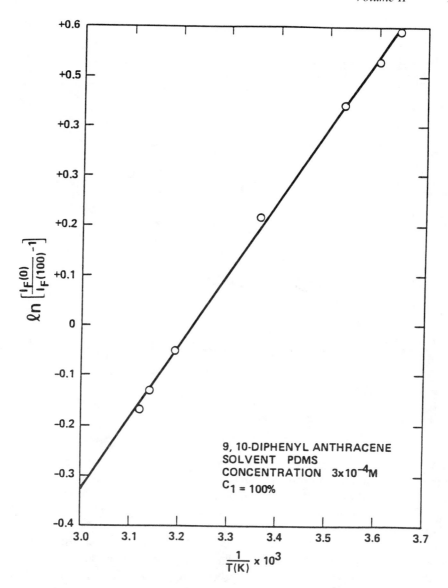

FIGURE 3. Solubility of oxygen in silicone vs. $1/\tau$ as measured by fluorescence. (From Reference 18.)

(4.77 kcal/mol) which is not temperature-dependent between 5 and 45°C. For comparison, the diffusion coefficient for oxygen in water is 2.5×10^{-5} cm^2 s^{-1} at 25°C. The diffusion coefficient is independent of oxygen concentration and fluorophore concentration in silicone over the usual pressure and temperature range. In the presence of small weight fractions of fumed silica, the diffusion of oxygen is reduced, but the activation energy is not affected at all.[20] Other materials with excellent oxygen permeability include silicone rubber, poly(tetrafluoroethylene), and polyacetylene.

Experiments on the diffusion of oxygen in poly(methyl methacrylate) (PMMA) have previously been performed by Shaw,[21] using the room temperature phosphorescence technique. Oxygen diffusion in poly(hydroxyethyl methacrylate) (PHEMA) has been studied more recently by fluorescence quenching techniques.[8,22] The diffusion of oxygen through PMMA as studied by phosphorimetry is very slow, with coefficients varying from 2.7 to 5.5×10^{-9} cm^{2}s^{-1} at 20°C. It is constant within the used dye concentration and the nature of the phosphor.

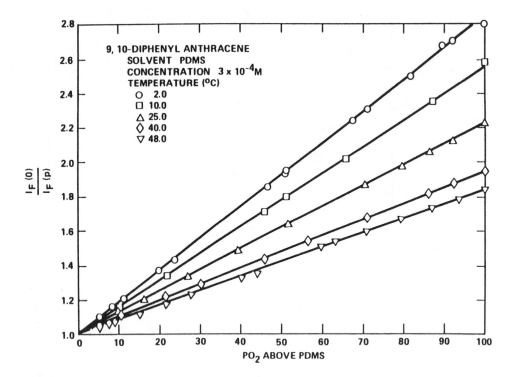

FIGURE 4. Temperature dependence of the Stern-Volmer plots of the quenching of diphenylanthracene by oxygen in poly(dimethylsiloxane). (From Reference 18.)

The activation energy ($D°$ in Equation 6) is 250 cal mol^{-1}. In (PHEMA), the diffusion, as measured fluorimetrically,[22] is 1.36×10^{-7} cm^{2}s^{-1} at 20°C. Figures 4 and 5 show the Stern-Volmer plots for the quenching of diphenylanthracene in silicone rubber and PHEMA. Evidently, the process is distinctly less efficient in PHEMA than in silicone.

It was calculated that the maximum possible oxygen concentration in the aqueous PHEMA matrix is 4.07×10^{-4} M under pure oxygen at atmospheric pressure. The value of this study lies in the fact that a material was studied that closely resembles the composition of a sensing membrane composed mainly of hydrogel and water. This type of material can be used in biosensors using immobilized enzymes. Quenching of fluorescence obeys Stern-Volmer kinetics with a K_{sv} of 0.016 torr^{-1}. Diffusion data for other hydrogels have been reported by Kubin and Spacek.[24]

An interesting observation is the finding that mixtures of two polymers do not necessarily display properties that would be expected by averaging the properties of the pure components. As an example, the quenching of pyrenebutyric acid by oxygen in mixtures of silicone rubber and PVC decreases almost linearly as the percentage of PVC increases.[25] If, however, the percentage of PVC exceeds 25%, quenching efficiency becomes very high again, being almost the same at 35% PVC as at 5% PVC.

It is probably useful at this stage to compare polymer solubilities for oxygen with the solubility of oxygen in water and blood. Only 0.25 mM oxygen (8 ppm) dissolve in water at 22°C under air, and 1.2 mM (40 ppm) under pure oxygen. In arterial blood with 90 torr oxygen partial pressure the concentration is similarly low (0.143 mM). Blood solubility may also be expressed in terms of gas volumes: 100 ml blood dissolve 3.1 µl oxygen at 37°C per torr applied oxygen pressure. The largest fraction of oxygen in blood, however, is bound to hemoglobin. A typical value for bound oxygen is 20 to 21 ml oxygen per 100 ml blood.

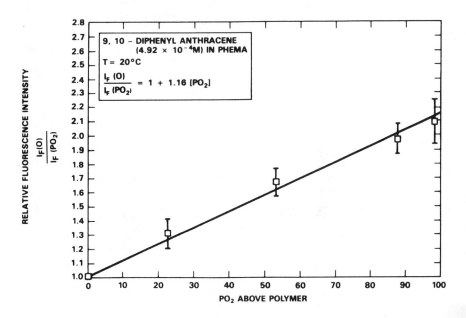

FIGURE 5. Quenching of the fluorescence of diphenylanthracene in poly(hydroxyethyl methacrylate) by oxygen. Note the distinctly smaller efficiency when compared with Figure 4. (From Reference 22.)

E. TEMPERATURE EFFECTS

There are a few reports on the temperature dependence of oxygen optrodes. For 9,10-diphenylanthracene in poly(dimethyl siloxane), a decrease of about -37% in the quenching constant is reported in going from 2 to 48°C (see Figure 4). At constant oxygen level, a temperature change of 1°C, therefore introduces an error of 4% in oxygen determination when assuming a temperature-independent fluorescence quantum yield. Since, however, in most other cases fluorescence intensity increases with temperature, this effect may partially be blurred.

This example demonstrates two of the problems associated with the determination of temperature coefficients ("tempcos") of sensors. Temperature is known to affect (1) the luminescence quantum yield of the dye (usually a negative tempco), (2) the quenching constant (positive tempco), (3) the solubility of oxygen in the membrane (negative tempco), (4) the diffusion of oxygen into the membrane (positive tempco), and (5) in case of blood measurement, the oxygen/hemoglobin binding curve (negative tempco). Furthermore, temperature affects fluorescence lifetime (τ_0). All these factors contribute to the "apparent tempco" of an oxygen optrode. Most of the reports on the tempco of sensors refer to one of these effects only. Throughout, temperature improves the response time of oxygen probes. Typically, the 37°C response time is around two thirds of the 22°C value.

Kroneis[26] has measured the tempco of the quenching constant of benzo(g,h,i)perylene in silicone rubber over the 25 to 60°C range. K_{sv} increases from 0.0112 to 0.0140 torr^{-1}, and the slope of a plot of temperature vs. K_{sv} is almost the same as the slope of a plot of oxygen permeability vs. temperature (Figure 6). Peterson et al.,[11] on the other hand, observed a 0.6% decrease in pO_2 indication per degree Celsius increase with their fiber probe based on a dye adsorbed on resin beads. Under nitrogen, the signal was found to be temperature-independent. Bacon and Demas[12] report the lifetime of their fluorophore to decrease from 5.8 μs at 0°C to 3.3 μs at 60°C. The temperature dependence of the overall quenching constant was not given. The fluorescence of surface-adsorbed ruthenium(II)tris(dipyridyl) drops by almost -30% in going from 25 to 37°C, but the quenching constant remains practically unchanged.

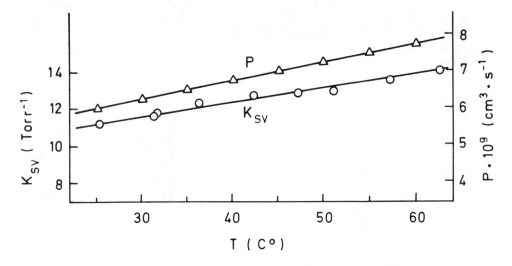

FIGURE 6. Temperature dependence of the quenching constant of benzo(g,h,i)perylene in silicone rubber, and plot of oxygen permeability vs. temperature. (From Reference 26.)

A most useful study on the temperature dependence of the fluorescence of silicone membranes with dissolved indicator (pyrenebutyric acid) was performed by Opitz et al.[27] The results show an unexpectedly small temperature effect (–20% signal loss in changing from 20 to 200°C). The overall quenching constant K_{sv}, in contrast, varies strongly with temperature. It was concluded that it is the temperature variation of k_d which accounts for this effect because of an enhanced oxygen diffusion.

The fiber optic oxygen probe described by Miller et al.[28] displays an error of +7 torr at 100 torr oxygen, when the temperature used in calculations is 36 instead of 37°C (i.e., a temperature measurement error of –1°C). This seems to be the only report on a true temperature coefficient of a complete sensor system.

F. SENSOR STABILITY

As mentioned above, practically all indicators are washed out by the sample (in particular blood) unless covalently immobilized. The result is a slow signal drift even when the arrangement is carefully thermostatted and the opto-electronic system has no drift. Thus, it was stated[29] that approximately 10^{-10} mol of a perylene dye leached out of a round sensing membrane of 22 mm diameter over a period of 2 d at 37°C. Similarly, the ruthenium dipyridyl complex used in some sensors has been found to leach out from silicone with organic solvents such as acetone.[12] As an alternative, it has been suggested to first adsorb the dye onto hydrophilic solid material such as silica gel which then is entangled into a hydrophobic polymer.[16]

No leaching (but some bleaching) has been observed when the oxygen-sensitive dye was covalently attached to the surface of the rigid support.[13] Bleaching can indeed be an unpleasant source of error. Practically all organic and metal-organic fluorophores are subject to some photochemistry, i.e., light-induced chemical transformation. In this context it is interesting to note that optrodes have sometimes been referred to as *photochemical* sensors, but this seems to be the least expression to be used since photochemistry is very much undesired in all kinds of on-line sensors. Photophysical (or optochemical) sensor is probably more justified in this context, although there seems to be no need for a new word.

Photodecomposition is particularly efficient under strong UV radiation and when powerful lasers are used as light sources. It can be reduced by applying pulsed light excitation as well as low radiant powers. One major species responsible for photodecomposition is singlet oxygen ($^1\Delta$) which can be formed from triplet ($^3\Sigma$) oxygen as a result of the dissociation of ground state oxciplexes (see above). We found that addition of β-carotene (which quenches singlet oxygen) or allylurea can considerable reduce the extent of photochemistry. Tertiary amines may serve the same purpose.[30]

Photodecomposition does not severely affect the performance of oxygen optrodes based on measurement of lifetime,[12,31] because lifetime is independent of dye concentration (within reasonable ranges). Thus, leaching and bleaching have no severe effects, as have lamp intensity and detector sensitivity fluctuations. Finally, the effects of fiber bending remain small. It is obvious that this kind of sensor has considerable stability advantages over optrodes based on intensity measurement.

Several attempts have been made to compensate for lamp fluctuations as well as leaching and bleaching effects: Peterson et al.[11] relate green fluorescence to the intensity of blue scattered light. This can compensate for lamp fluctuations and detector sensitivity drifts. Zhujun and Seitz[4] measure the ratio of reflected light (at 405 and 435 nm) of an oxygen-sensitive layer composed of immobilized hemoglobin plus deoxyhemoglobin. Lee et al.[6] developed indicators with two luminescence bands, one of which is quenched by oxygen while the other remains relatively unaffected. A useful system was found in some 4-bromo-1-naphthoyl derivatives extracted into γ-cyclodextrin and bonded to cellulose which showed oxygen-dependent phosphorescence and oxygen-independent fluorescence. The ratio of luminescence intensities as measured at two wavelengths is a parameter for oxygen pressure which is independent of dye concentration. It was also suggested to use the ratio of the intensities of monomer or excimer bands of pyrenes (which are differently affected by oxygen). Another possibility is to include an oxygen-insensitive fluorophore (with spectral data different from the indicator) in the sensing membrane. It is desirable (though not imperative) that the reference fluorophore bleaches at a rate similar to that of the indicator.

G. ANALYTICAL RANGES, ACCURACY, AND DETECTION LIMITS

The analytical range of an oxygen optrode is governed by the respective quenching curve and the Stern-Volmer constant, respectively (Figure 7). For fluorescence-based sensors a typical range is from 0 to 200 torr oxygen. Phosphorescence-based sensors (Section 3 herein) are much more sensitive, with typical ranges from 0.0005 to 0.004 torr.

Figure 7 shows a plot of fluorescence intensity vs. oxygen pressure for two quenching constants (A, 0.05 torr^{-1}, and B, 6.005 torr^{-1}). It is obvious from such a plot that (1) higher quenching constants result in better accuracy at low levels (because of a larger relative signal change per torr oxygen), and (2) the dynamic range is smaller because at levels above 200 torr the signal change becomes vanishingly small. Thus, the relative intensity change in going from 150 to 500 torr in curve B is from 57 to 28, i.e., –29 intensity or lifetime units, whereas in curve A it is –7.9 units only. In other words, a signal obtained with sensor A at 200 torr is obtained with sensor B at 2400 torr only. In view of this situation it can be said that variation of quenching constants always is a compromise between analytical range and accuracy.

It is also obvious from Figure 7 that the sensitivity of oxygen optrodes based on luminescence quenching is highest at the initial part of the curve, i.e., at low oxygen tensions where $\Delta I/pO_2$ is maximal and almost linear. At high pO_2, on the other hand, $\Delta I/pO_2$ is small and sensitivity poor. One might therefore assume that optical fluorosensors are most precise in this range. This is largely true. However, at low pO_2, only minor changes in the pO_2 during the calibration process result in large errors in the determination of I or even I_o. This introduces a second type of error that makes measurements in the steep part of the calibration curve less precise.

The precision and accuracy of an optrode are governed by the uncertainities in the determination of I_o, K_{sv}, and I. As a rule of thumb, the first two govern the precision, and I the accuracy. A precise determination of I_o is obviously essential for obtaining sufficiently precise

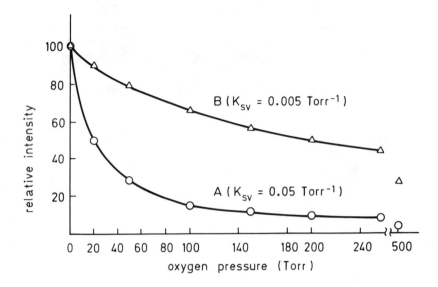

FIGURE 7. Plots of fluorescence intensity vs. oxygen partial pressure for two different quenching constants. (A) K_{sv} = 0.05 torr^{-1}; (B) K_{sv} = 0.005 torr^{-1}.

calibration curves and K_{sv} values. I_o is obtained with oxygen-free solutions or gases. In practice, it is found not to be easy to keep a calibrant gas free of oxygen. Aqueous solutions of sodium dithionite have therefore been used. In view of this situation it has become practice to establish the quenching curve not with help of I_o, but of two values I_1 and I_2, at two different oxygen pressures, since the quenching curve can be reconstituted from two or more experimental values for I.

A final source of error results from contributions of straylight to the fluorescence signal which gives an apparent signal I' composed of I (the true intensity at given pO_2) and straylight I_x. A fairly simple method has been worked out to precisely determine K_{sv} and I_o from three sets of intensity data.[33]

The detection limits are governed by both the initial slope of the quenching curve and instrumental resolution. Assuming a $\pm 0.1\%$ uncertainty in light intensity measurement (which is realistic in a well-thermostatted device), the detection limit is $0.003/K_{sv}$, when K_{sv} is expressed as torr^{-1} and the signal-to-noise ratio is 3. K_{sv} values ranging from 0.001 to 0.1 torr^{-1} have been reported, which results in detection limits from 3 to 0.03 torr for fluorosensors. Phosphorescence based sensors, in contrast, are much more sensitive (see Section III).

II. FLUOROSENSORS

A. PYRENE-BASED SENSORS

Obviously, the first oxygen fluorosensor is the one described by Bergman[9] in 1968. Fluoranthene, a strongly fluorescent polycyclic aromatic hydrocarbon (PAH) was adsorbed on a porous glass support (Vycor glass) and excited with an UV light source. The resulting fluorescence is strongly quenched by oxygen and was measured with a photocell. Upon changing from nitrogen to oxygen, 63.5% of the fluorescence was quenched. Solutions of fluoranthene in polyethylene and silicone rubber were also studied, but quenching in polyethylene amounts to −10% only and the porous glass type sensor was therefore preferred.

The fluorescence of a variety of other PAHs is known to be quenched by molecular oxygen.[34] Among these, pyrene and, less so, pyrenebutyric acid (PBA) are probably most efficiently quenched by virtue of their long radiative lifetimes of >100 ns. PBA in dimethylformamide[35] or silicone polymer[13,29] has been applied in a device called "optode" or "optrode",

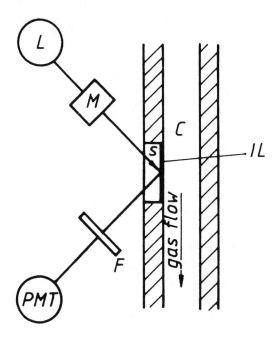

FIGURE 8. Schematic of a flow-through cell for continuous sensing of oxygen. L, light source; M, monochromator; S, solid support (glass); IL, indicator layer; C, flow-through cell; F, secondary filter; PMT, photomultiplier tube. (From Reference 13.)

for example the one schematically shown in Figure 8. Note that no fibers are used in this case. Rather, the sensing layer forms part of the wall of the flow-through cell. When PBA is dissolved in a silicone membrane, the Nernst distribution strongly favors its presence in the polymer rather than in the aqueous sample phase, but it was noted that the dye is slowly washed out, resulting in a small signal drift.[36]

The problem was overcome by immobilizing PBA on controlled porous glass that was previously sintered onto a glass support.[13] This gives a mechanically stable oxygen optrode with a very fast response (less than 20 ms for 90% of the total signal change). The sensor is suited for breath gas analysis, although there is some cross sensitivity to humidity. This effect is generally observed in case of oxygen sensors where the dye is bound to surfaces with varying degrees of hydration. The optrode is highly specific for oxygen: no interferences were observed with nitrogen, carbon monoxide, dinitrogen oxide, methane, CO_2, and noble gases. Main interferents are halothane, sulfur dioxide, chlorine, and N_2O_3. Detection limits are in the order of 1 torr (as with most types of fluorescence-based oxygen optrodes).

An interesting approach is to entrap the indicator in small particles of sub-micrometer thickness. Two experimental techniques were used: in the first,[15] pyrenebutyric acid was dissolved in benzene/poly(dimethylsiloxane) and entrapped in polyacrylate capsules (average diameter 150 to 250 nm). In the second,[37] the dye was dissolved in dioctyl phthalate, encapsulated in polyurethane, and suspended or embedded in silicone rubber or water. The quenching efficiency was found to be slightly smaller than in homogeneous solution, an effect that was attributed to boundary layer phenomena at the interface between the different phases.

B. OTHER PAH-BASED SENSORS

Notwithstanding the stability and long decay time of PBA and pyrene, it suffers from shortwave excitation (340 nm) and emission (390 to 440 nm). Kroneis[26] has tested a variety of other PAHs and found decacyclene and benzo(g,h,i)perylene to be more suitable indicators. In lieu, a steam-sterilizable fiber optic oxygen sensor was constructed for use in bioreactors,[29]

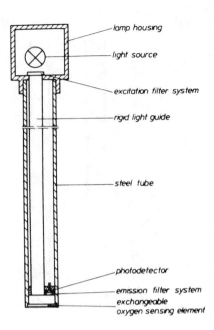

lamp housing

light source

excitation filter system

rigid light guide

steel tube

photodetector

emission filter system

exchangeable
oxygen sensing element

FIGURE 9.   Fiber optic oxygen probe for use in bioreactors. (From Reference 29.)

consisting of a rigid light guide for insertion into a bioreactor (Figure 9). The oxygen-sensing layer, a filter, and a photodetector were placed at the fiber tip, and excitation was through the fiber rod. The sensor layer was covered with a black layer that acts as an optical isolation to avoid interferences by the fluorescence of biomatter. When immersed into a bioreactor, response times were 9 to 65 s, and the drift was −0.01 to −0.09% signal loss per hour. No effects of stirring were observed.

Peterson et al.,[11] in a search for longwave absorbing and emitting indicators, found perylene dibutyrate adsorbed on polystyrene beads to be a viable sensing material. It has excitation and emission maxima of, respectively, 468 and 514 nm, is stable, and is efficiently quenched by oxygen, thereby allowing a resolution of ±1 torr up to 150 torr (20 kPa). A double-fiber optrode was constructed which is schematically shown in Figure 10. It consists of two thin fibers, one guiding blue excitation light to the dyed particles in a tubing at the common end of the two fibers, the other guiding scattered blue light and green fluorescence to a photodetector. The dyed polystyrene beads are contained in a 25 μm polypropylene tubing. In a modified version,[38] a reference dye with spectral properties different from the indicator was used and the single fiber technique applied. The end of the tubing was sealed with epoxy.

The sensor measures the ratio of scattered blue light (I_o) and green fluorescence (I). An electronic circuit processes the blue and green signal intensities in accord with the following relation:

$$pO_2 = (gain)(I_o/I - 1)^m \qquad (7)$$

which is the Stern-Volmer equation rearranged with an exponent m added for curvature since plots are usually not linear. This ratio method can eliminate temperature effects and drifts resulting from photobleaching.

While ionic quenchers do not interfere owing to the polyethylene envelope, the sensor has some sensitivity towards nitrous oxide (which has 2% of the effect of an equal oxygen partial pressure) and halothane. The interference by halothane is severe, cumulative, related to both concentration and time of exposure, and changes both the sensitivity and zero adjustment of

Illumination

Fluorescent Dye Sensitive to Oxygen

Optical Fibers

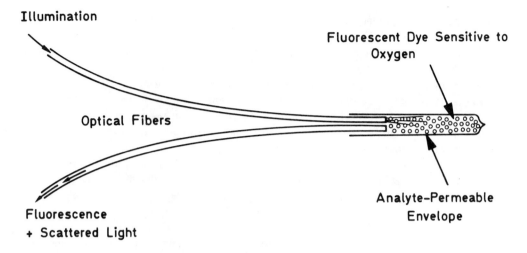

Fluorescence + Scattered Light

Analyte-Permeable Envelope

FIGURE 10. First fber optic oxygen catheter for use *in vivo*. (From Reference 11.)

the optrode. If exposure is not too severe, the effect is reversible. The performance of this sensor in *in vivo* studies is described in detail in Chapter 16.

Cox and Dunn[18] investigated the oxygen quenching of five PAHs, including 9,10-diphenylanthracene, decacyclene, and rubrene. The first was considered the most useful indicator because it is highly fluorescent in viscous solvents, stable, displays a good Stokes' shift, presents no health hazard, and has a temperature-independent quantum yield. Therefore, when 9,10-diphenylanthracene is dissolved in polysiloxanes, its fluorescence intensity at a given oxygen concentration is due to changes in oxygen solubility only.

The effects of temperature and solvent viscosity on the quenching by oxygen of this system have been discussed in Section I.D. Most noteworthy, the quenching efficiency increases strongly with decreasing temperature (Figure 4), which is in contrast to the behavior of PAHs in fluid solution or in the gas phase. Possibly, this is due to the strong increase in oxygen solubility in silicone when the temperature is lowered to 2°C (Figure 3). The almost linear relations between I_o/I and oxygen pressure indicate that when the fluorophore is entangled or entrapped in poly(dimethylsiloxane), the only bimolecular deactivation process is quenching by oxygen.

In a study[8,39] on oxygen diffusion through poly(dimethylsiloxane) (PDMS) and poly(hydroxyethyl methacrylate) (PHEMA) it was found that oxygen has a diffusion constant of 3.56×10^{-5} cm²s⁻¹ in PDMS (at 25°C), but 1.2×10^{-6} cm²s⁻¹ only in PHEMA (at 20°C). While this finding makes PDMS a superior material, its disadvantage is that it is not suitable for incorporation of enzymes and is not penetrated by hydrophilic enzyme substrates such as glucose. PHEMA, on the other hand, contains 37% water, can be heavily loaded with glucose oxidase (GOx), and is readily penetrated by glucose. However, the steeper slope of the oxygen response of PDMS-based sensor materials indicate that a device based on PDMS may be more sensitive. A theoretical model was established that accounts for the mass balance and diffusion kinetics of the GOx-catalzed oxidation of glucose by oxygen in a planar layer.

A fiber optic oxygen sensor was described[8] that uses PHEMA plus 9,10-diphenylanthracene as the oxygen-sensitive material which was covalently grafted via silane coupling techniques onto fused silica fibers. A special base-catalyzed condensation process made the sensor particularly resistant to hydrolysis. Fluorescence was stimulated by the evanescent wave of a UV dye laser or a xenon lamp and quenched by –20% when the sensor went from an oxygen-free to an oxygen-saturated (40 ppm) aqueous environment. When GOx was copolymerized with HEMA, a hydrophilic material was obtained that eventually may be used as a sensing material for glucose by coupling it to an oxygen probe. This may result in a single-layer hydrophilic sensor material that contrasts another type of oxygen-based enzyme sensor which consists of two layers, namely a hydrophobic oxygen optrode and a hydrophilic reaction layer.[33,40,41]

Oxygen optrodes with fluorophores dissolved in silicone rubber have found practical application in extracorporeal blood oxygen measurement,[42] in invasive catheters,[28] in breath gas analysis, and as transducers for biochemical reactions, e.g., in biosensors for glucose,[35,40,41] lactate,[43] or ethanol.[44] Respective biosensors and biomedical applications will be described in more details in the respective chapters.

An interesting application[45] of this sensor type is the simultaneous determination of oxygen and an interfering quencher such as halothane, both of which act as dynamic quenchers of the fluorescence of PAHs. In fact, halothane has always been considered as an interferent in blood oxygen assay, both by electrochemical and optical methods. A double sensor method is applied, with two fibers having two differently responding sensors at their ends. The one comprises a highly halothane-sensitive indicator layer, attached to the fiber end, the other has an identical chemistry but is covered with poly(tetrafluoro ethylene) which is impermeable to halothane. Its concentration ([H]) can be calculated with the help of an extended Stern-Volmer equation

$$[H] = (\alpha - \beta)/{}^{H}K_a \qquad (8)$$

where α and β, respectively, are the optical signals from the two sensors, and ${}^{H}K_a$ is the dynamic quenching constant (K_{sv} in Equation 1). Oxygen, of course, is assayed by the PTFE-covered sensor in the usual way. This two-sensor technique allows determination of halothane, or oxygen, or both with a precision of $\pm 5\%$ relatively for halothane, and $\pm 3.5\%$ relatively for oxygen. Detection limits are 0.1% halothane and 0.4% oxygen. Fluorans and N_2O do not interfere. The method is an application of a more general theory[46] on the assay of a quenching analyte in the prescence of interfering quenchers, using either two fluorophores or perm-selective membranes (as in the halothane case).

The feasibility of sensing oxygen in the 300 to 500 K temperature range has been investigated.[27] A silicone membrane doped with pyrenebutyric acid and attached to the end of a quartz light guide was exposed to varying temperatures and the optical signal measured. The signal showed an almost linear decrease in going from 20 to 200°C under nitrogen, 10% oxygen, or 100% oxygen (total signal change –18%). However, after being corrected for straylight, it showed a much more complex behavior. The temperature dependence of the quenching constant K was found to be

$$K(T) = K_o(T_{crit} + T)/(T_{crit} - T) \qquad (9)$$

with values of 1.16×10^{-2} torr^{-1} for K_o and 600 K for T_{crit}.

Semiconductor lasers have unique advantages over other types of conventional and laser light sources: they have a fairly large output (sufficient for most fiber applications), narrow band width, and low price. Unfortunately, their wavelengths do not match the absorption spectra of PAHs. Okazaki et al.[47] have therefore frequency-doubled the 780 nm emission of a semiconductor laser to obtain a 390 nm line with 50 nW intensity which is suitable to excite benzo(g,h,i)perylene. The beam was launched into a fiber which guides light to the sensing material (the indicator dissolved in silicone grease) at its end. A second fiber was used to collect fluorescence at 430 nm. Oxygen is determinable in the 0 to 30% range at atmospheric pressure.

A useful application of glass-immobilized PAHs was found in oxygen detection in seawa-ter.[48,78] An instrument has been built consisting of a xenon light source, bandpass filters, a dichroic beam splitter, collimating and focussing lenses, a plastic clad silica rod with an unspecified fluorophore immobilized at the tip, an emission bandpass filter, a photomultiplier tube, a reference PIN photodiode, and the electronics for signal amplification and processing. The response time (defined as the 2/e fraction of the time required for the total signal change to occur, i.e., 63%) is less than 1 s for gases and less than 10 s for water, with a background noise of 1% only. The performance of the embodiment was tested in marine environment and compared with that of a Clark electrode. The sensor compares favorable with the electrode both in the lab and in the ocean (Figure 11). Using these data, it was possible to measure the oxygen depth profile in the ocean near Victoria, B.C., Canada. In contrast to the slow-responding electrode, the fluorosensor provided a stable recording after 10 s.

C. SENSORS BASED ON SURFACE ADSORBED DYES

Aside from PAHs which have been applied in covalently immobilized, surface-adsorbed, or polymer-dissolved form, metalorganic complexes of the ruthenium(II)tris(dipyridyl) type have been used as oxygen indicators for several reasons: they have long lifetimes (0.2 to 5.0 μs) and are therefore efficiently quenched by oxygen. Second, they have favorable excitation wavelengths (about 450 to 490 nm) which makes them excitable by the argon ion laser 488 nm line, with blue light-emitting diodes (LEDs), or even tungsten halogen lamps. Finally, they display extraordinarily large Stokes' shifts, with emission maxima centered above 600 nm.

In view of these advantages, an oxygen-sensitive material was prepared by entrapping an aqueous solution of the ruthenium dye inside kieselgel particles and entrapping the beads in a silicone polymer.[16,49] The fluorescence of such an emulsion is related to the oxygen partial pressure, but a Stern-Volmer plot is not linear over the whole pressure range (Figure 12). Fluorescence is quenched by around 50% in going from nitrogen to air.

Figure 13 shows the response time (12 s for 63% signal change), reversibility, and relative signal change. No interference was observed with halothane. The sensor may be sterilized at 130°C with almost no signal loss and change in quenching constants. Particular features of this sensor type are (1) complete lack of leaching, (2) high K_{sv} values, (3) favorable analytical wavelengths, and (4) the possibility of incorporation of enzymes into the aqueous phase. As an example, alcohol oxidase may be incorporated onto the water-filled beads. Since ethanol penetrates silicone, it is oxidized inside the membrane and thereby causes the oxygen partial pressure to decrease. This is the basis for a fiber optic ethanol biosensor.[50]

A variety of optical oxygen sensors have been described where the dye is adsorbed on rigid surfaces, particles, and the like. As a matter of fact, some dyes are quenched by oxygen when surface-adsorbed only, but not in fluid solution. Thus, the Peterson method[51] for gas flow visualization has been applied[52] for testing the oxygen-sensitivity of fluorescent yellow (excitation/emission at 440/510 nm) on silica gel. The particles were spread as a monolayer onto a gluing tape, which then served as the sensing membrane. Quenching is very efficient, but reproducible results can only be obtained with dry gases. The method is therefore not suitable for oxygen determination in exhalation gases. Considerably improved results were obtained when porous polystyrene particles (Porapak Q) was used as a support.[53] Both water vapor and N_2O do not interfere. Some interference was observed with anesthetics such as ethrane and halothane, but this could be eliminated in a calibration process. The response time is less than 5 ms for 90% of the total signal change, which is distinctly quicker than any other oxygen sensing device.

Another dye whose quenching by oxygen was studied during this work is tetraphenylporphyrin (TPP). It has the advantage of being excitable at 520 to 550 nm, e.g., with LEDs, when adsorbed on Porapak Q. The two fluorescence bands maximize at 650 and 720 nm, and probably originate from the monomer and dimer species. Fluorescence is quenched by about –60 and –55% at 650 and 720 nm, respectively, in going from pure nitrogen to pure oxygen. TPP on Porapak Q is again moisture-insensitive, but the strong temperature dependence of

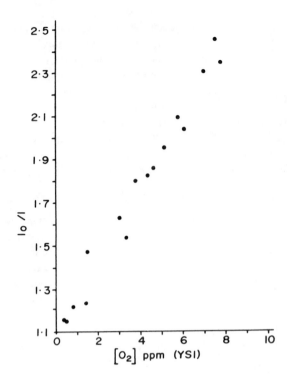

FIGURE 11. Comparison of the data of an oxygen optrode with those of an Yellow Springs Instrument oxygen electrode. ($I_o/I - 1$) is linearly related to pO_2 via Equation 1. Note that the intercept is >1. (From McFarlane, R. and Hamilton, M. C., Proc. SPIE, 798, 324, 1988.

TPP emission may be considered as a problem. The devices have been used for breath gas analysis (see Chapter 19).

Aside from the longwave absorbing dyes mentioned above, pyrene, coronene, and other shortwave absorbing dyes were also adsorbed and tested.[51] Pyrene was found to be very suitable. Rather surprisingly it was found[53] that quenching of surface-adsorbed pyrene by oxygen is the same for gaseous oxygen and water-dissolved oxygen, although the theory of diffusion-controlled reactions predicts an influence of viscosity. One interpretation of this finding would be that the dye is dissolved in the polymer particles rather than surface-adsorbed, but this possibility was ruled out because of the extremely short response time (5 ms) of the sensing material.

In the author's laboratory, a variety of dyes was tested for use as oxygen indicators.[54] Excellent quenching by oxygen was observed with the following fluorophores (excitation/ emission maxima given in parenthesis) when adsorbed on kieselgel chromatographic plates: acridine yellow (458/530 nm), perylene-tetracarboxylic acid N-alkylimide (PTAI) (495/555 nm), Greengold (460/520 nm), and anthranilic acid (330/410 nm). In a typical experiment for quenching studies, PTAI acid imide was adsorbed or covalently immobilized via established silyl reagent methods to give a sensing material that can be excited at 480 to 500 nm with either a blue or green LED. Figure 14 shows a recording of fluorescence vs. varying oxygen tension. Response times are in the order of 1 s.

The immobilized dye exhibits two emissions centered at 540 and 575 nm, whose respective quenching constants are different (0.022 and 0.028%$^{-1}$ oxygen at 760 torr). Similar effects were found with the other dyes. However, in all cases a marked influence of humidity was

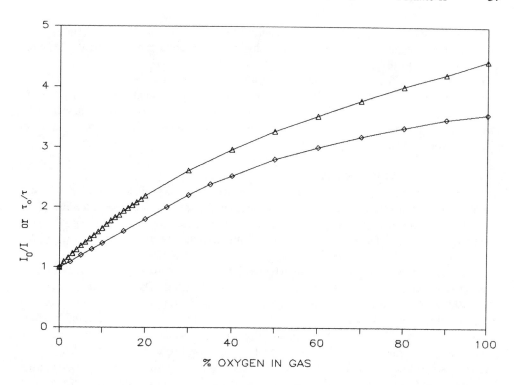

FIGURE 12. Stern-Volmer plot of the fluorescence quenching by oxygen of kieselgel-adsorbed ruthenium(II)tris(dipyridyl) entrapped in silicone rubber. (A) intensity measurements; (B) lifetime measurements. (From Reference 16.)

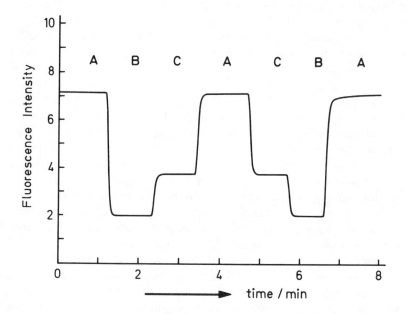

FIGURE 13. Response time, relative signal change, and reversibility of the sensor described in Figure 12.

observed (Figure 15) which makes the material useful only at constant humidity, for instance in underwater experiments. Possibly, the use of hydrophobic supports may eliminate the cross-sensitivity towards moisture.

Eventually, the effect was exploited[55] to devise a fiber optrode for on-line measurement of humidity in air with its constant oxygen content (although not constant pO_2). In an extension of the applicability of LED-excitable oxygen sensors it seems feasible that they may be utilized for monitoring very rapid changes in atmospheric pressures which, for instance, in airplanes would cause the release of oxygen masks.

D. DECAY TIME SENSORS

Equation 3 predicts that oxygen not only affects the intensity of the luminescence of a luminophore, but also its lifetime (defined as the time required until the initial luminescence intensity has dropped to 1/e of its initial value). For practical considerations regarding the complexity of an instrument, but also for quenching efficiency reasons, it is advisable to use an indicator with a lifetime exceeding 100 ns.

Bacon and Demas[12] found ruthenium(II)tris(4,7-diphenyl-1,10-phenanthroline) to be a suitable and long-lived fluorophore soluble in silicone polymer. The resulting membranes, about 100 μm in diameter, were found to be a useful material for optical oxygen sensing and were investigated with respect to the effect of oxygen on fluorescence intensity and lifetime. The respective Stern-Volmer plots are practically superimposable, but not linear (Figure 16). A quenching constant K_{sv}, as high as 0.057 torr^{-1} can be extracted from the data. Major interferents are chlorine and SO_2, and solvents acetone and dichloromethane destroy the film. No interferences were found with aqueous surfactants and ethanol. The sensor has a rapid response (t_{95} = 3 s) and has therefore been used for breathing studies (see Chapter 19).

The long lifetimes of metal-organic oxygen indicators render them particularly suitable for lifetime studies. Beside the above studies on the lifetime reduction of ruthenium complexes in silicone foils, there is another report on the determination of oxygen by lifetime measurement.[31] An opto-electronic device was constructed for determination of lifetimes by the phase fluorimetric method and is schematically shown in Figure 17. A frequency-modulated (455 kHz) blue LED served as a light source, and the time delay of the red fluorescence of the sensor material at the tip of the fiber is compared with the pulses of the LED. Ruthenium(II)tris(bipyridyl), adsorbed on kieselgel particles and embedded in silicone rubber, was used as the oxygen-sensitive material. The signal response of the sensor is shown as a Stern-Volmer plot in Figure 12. It is obvious that the lifetime plot has a more linear response up to 35% oxygen than the intensity plot, but that quenching is less efficient in the former case.

More recently it was found[56] that ruthenium(II)-tris(bipyrazine) is even more oxygen-sensitive than the respective dipyridyl complex in having a lifetime of 1.04 μs in argon-purged aqueous solution (vs. 0.69 μs of the dipyridyl complex). An aqueous solution of the dye was placed at a fiber end and covered with teflon. Its fluorescence served as a parameter for determination of oxygen in seawater. Probably because of a large sample volume and the slow diffusion of oxygen in water, the response time is rather long (about 35 min). The use of other ruthenium dyes has been described as well.[57]

Stern-Volmer plots of most sensors based on fluorescence quenching and intensity measurement are nonlinear at oxygen pressures above 300 torr, mainly because both static and dynamic quenching occur simultaneously, and because the indicators are embedded in an inhomogeneous environment. At least the first reason for nonlinearity is eliminated by the lifetime method because lifetime is governed only by the dynamic (collisional) quenching process. Despite a smaller K_{sv}, oxygen can be determined in the 0 to 150 torr range with ±2.8 torr resolution. The response time to achieve 99% of its final value is 30 to 40 s.

Compared to former sensor types, the lifetime-based sensors display a few decisive advantages: they have negligible signal drift arising from leaching and bleaching because the

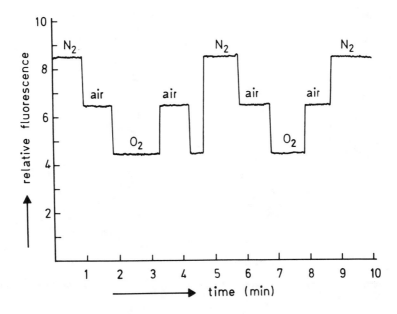

FIGURE 14. Fluorescence quenching of glass-immobilized perylenedicarboxylic acid imide. (From Reference 54.)

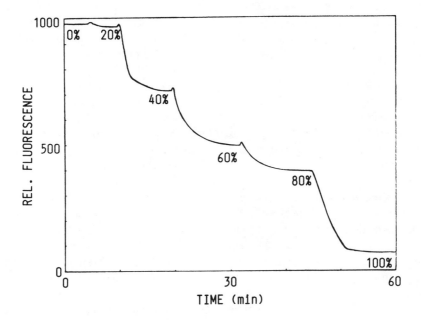

FIGURE 15. Effect of humidity on the fluorescence of kieselgel-adsorbed perylenedicarboxylic acid imide. (From Posch, H. E. and Wolfbeis, O. S., *Sensors Actuators,* 15, 77, 1988.)

decay time is independent of fluorophore concentration (in a first approximation). Second, they display excellent long-term stability (no change in response after 3 weeks of operation)[31] because of the internal reference system, in which the fluorescence phase is internally compared several 100,000 times a second to the excitation (LED) phase. Finally, no drift arising from light source intensity and photodetector sensitivity fluctuations are to be expected because it is not the absolute intensity that is measured, but rather the time lag between excitation and fluorescence.

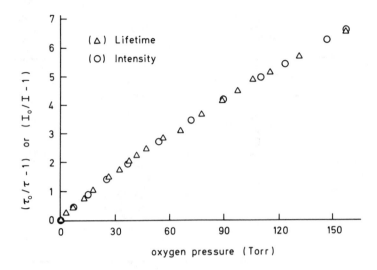

FIGURE 16.   Stern-Volmer plots of the quenching of ruthenium(II)tris(4,7-diphenyl-1,10-phenanthroline) dissolved in a silicone film. (From Reference 12.)

E. ENERGY TRANSFER SENSORS

A new type of oxygen sensor has been described more recently[57] that is based on electronic energy transfer from a donor (whose fluorescence is efficiently quenched by molecular oxygen) to an acceptor (which is less affected by oxygen). Pyrene is used as the donor and perylene as the acceptor. The fluorescence emission band of the donor shows good overlap with the absorption band of the acceptor (Figure 18). When excited at 320 nm, the two-fluorophore system shows strong fluorescence at 476 nm, where pyrene itself is nonfluorescent. Although perylene is not efficiently quenched by oxygen, the system strongly responds to oxygen because fluorescence is quenched with an efficiency that by far exceeds the quenching efficiency for pyrene or perylene alone. There is an almost fourfold increase in the quenching constants of the energy transfer system (as compared to the conventional system). Typical values for K_{sv} in silicone are $0.0065\%^{-1}$ for perylene, $0.043\%^{-1}$ for pyrene, but for pyrene/perylene it is $0.16\%^{-1}$ at 760 torr.

The principle has been applied to devise a fiber optic oxygen sensor by incorporating the two dyes in a silicone polymer matrix that has been attached to the end of an optical fiber. Oxygen can be detected in the 0 to 150 torr range with 0.5 to 3 torr precision. The detection limit is 0.5 torr oxygen. Figure 19 shows the Stern-Volmer plot of the quenching by oxygen of pyrene and the pyrene/perylene couple in silicone, and Figure 20 response time, reproducibility, and the very large signal change in going from nitrogen to oxygen. Also noteworthy is the short response time.

The dramatic increase in the quenching constant has caused some speculations as to the origin of the effect and it was concluded[59] that it is the donor-acceptor excited state complex ("exciplex") that is deactivated very efficiently by oxygen. Because the Stern-Volmer constant directly influences the resolution and detection limits, a sensitivity of 0.5 torr oxygen was found, which is distinctly less than in case of most other fluorosensors.

For an unusual type of oxygen sensor, in which a fluorescent and oxygen-sensitive cladding of a fiber transfers its energy to a fluorescent fiber core (whose absorption overlaps the emission of the clad dye), see Chapter 5.

F. SENSING TWO SPECIES SIMULTANEOUSLY

Attempts have been made to determine, with one sensor, two analytes simultaneously. Two approaches are known: the first is based on the capability of fibers to guide a multitude of

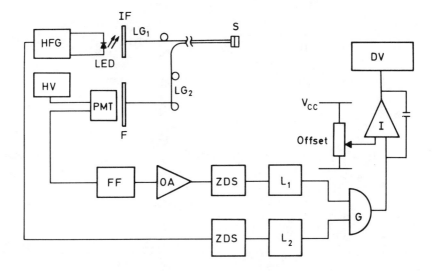

FIGURE 17. Schematic of the opto-electronic arrangement for sensing oxygen by the lifetime method. (From Lippitsch, M. E., Pusterhofer, J., Leiner, M. J. P., and Wolfbeis, O. S., *Anal. Chim. Acta*, 205, 1 1988.)

spectral information simultaneously. This concept is realized in an optrode for oxygen and CO_2, with two analyte-sensitive layers and chemistries having quite different emission wavelengths.[60] The other is based on multiwavelengths or multiparametric analysis of a single indicator. These two types of oxygen sensors shall be discussed in the following.

In an approach based on spectral separation, the oxygen sensitive material (kieselgel-adsorbed ruthenium(II)-tris-dipyridyl) and the CO_2-sensitive material (an immobilized pH indicator in bicarbonate buffer) are immobilized at the end of a fiber bundle using silicone rubber polymer. Both indicators have the same excitation wavelength (in order to avoid energy transfer), but quite different emission maxima (Figure 21). Figure 22 shows a cross-section through the sensing membrane at the fiber end. The two emission bands can easily be separated optically and give independent signals (Figures 23 and 24). Oxygen can continuously be determined in the 0 to 200 torr range with ±1 torr accuracy, and CO_2 in the 0 to 150 torr range with ±1 torr. The accuracy is higher at low partial pressure, so that the detection limits are about 0.5 to 1.0 torr in both cases.

The response times strongly differ from each other, as can be seen from Figures 23 and 24. It is well known that oxygen sensors are faster than CO_2 sensors because of both diffusional and chemical processes involved in the second case. Moreover, CO_2 sensors tend to become destabilized once they are exposed to low CO_2 levels for some while and that they need hours for regeneration (until the drift has settled). Typical response times until the total signal change has occurred are 40 s for the oxygen sensor, but 3.8 to 5.2 min for the CO_2 sensor, both in the presence of a 20 μm optical isolation. Note in Figure 24 that the response time is faster in going from lower to higher levels of CO_2 than reverse. The longer response of the CO_2 sensor is, of course, partially also due to the fact that the CO_2-sensing layer is covered by the oxygen-sensitive layer which prolongs the time necessary for equilibration.

Sensing two analytes with one sensor requires two indicators having no cross-sensitivity. At least three kinds of cross-sensitivity in fluorescence-based multiple sensors have to be discerned: (1) a strong overlap of the fluorescence emission bands of the indicators, so that there is no analyte-specific wavelength with sufficient signal intensity; (2) the non-specificity of the indicator for a given analyte (oxygen, for instance, is a quencher of the luminescence of some pH indicators); (3) the overlap in the emission band of one dye with the absorption band of the other, giving rise to energy transfer between analyte-sensitive and analyte-insensitive dye. The second type of energy transfer (the dipole-dipole interaction) is easily

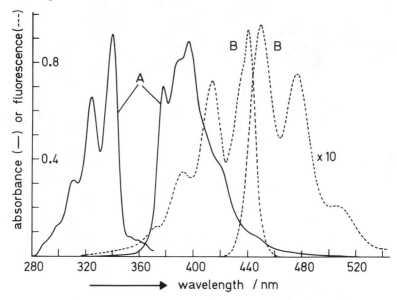

FIGURE 18.   Absorption and fluorescence of pyrene (————) and perylene(- - - -) in methanol, demonstrating the overlap of the pyrene (donor) emission with the perylene (acceptor) absorption. The arrows indicate the analytical wavelengths for excitation and emission. (From Sharma, A. and Wolfbeis, O. S., *Appl. Spectrosc.*, 42, 1009, 1988.)

avoided in this case by spatial separation of oxygen- and CO_2-sensitive layers, since resonance energy transfer inversely depends on the sixth power of distance.

Another method for measuring two species with one indicator is based on the analysis of the emission spectra of an indicator which suffers both pH induced spectral shifts and dynamic luminescence quenching.[61] Figure 25 shows the spectral shifts of a dye at varying pH and oxygen pressure. A_s is the acid form, and A_b the base form of the emission or excitation spectrum of a pH indicator. A_x is the spectrum at a pH in the range of the pK_a value. I_x is the pH-independent emission intensity at the isosbestic wavelength. The ratio of the intensities measured at λ_m and λ_{is} is a parameter for the actual pH (or CO_2 pressure when used in a CO_2 sensor). Since the luminescence of the dye is quenched by oxygen, its excitation or emission spectrum will be lowered in intensity with increasing partial pressure. The intensities of the spectra of acid form, base form, and unknown sample will become reduced due to dynamic quenching (A'_s, A'_b, A'_x). Measurement of intensity at λ_{is} (which is independent of pH or CO_2) is a parameter of oxygen. A typical example of such a dye is uranine which has a strongly pH-dependent phosphorescence which is strongly quenched by traces of oxygen when adsorbed on kieselgel.

In another type of sensor, the ratio of emission intensities of a luminophore as measured at two wavelengths is used as the first information, and the change in luminescence lifetime (as affected by another parameter) is the second information. Typical couples of parameters that may be determined by measurement of both intensity (I) and lifetime (τ) of one indicator are the following: pH values (via I) and oxygen (via τ), CO_2 (I) and oxygen (τ), ammonia (I) and halides (τ), and pH (I) and temperature (τ).

III. PHOSPHORESCENCE-BASED SENSORS

Every treatise on phosphorescence-based oxygen sensors has to start with the early work of Kautsky in the 1930s. He observed that a variety of dyes adsorbed on solid supports displays room temperature phosphorescence (RTP) which is observable under pure nitrogen. Traces of

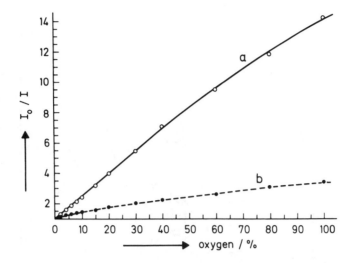

FIGURE 19. Stern-Volmer plot of the quenching of the fluorescence of pyrene (a) and the pyrene/perylene energy transfer system (b) by oxygen. (From Sharma, A. and Wolfbeis, O. S., *Anal. Chim. Acta*, 212, 261, 1988.)

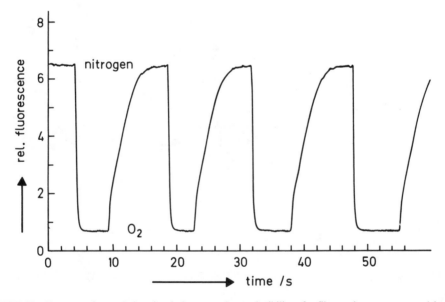

FIGURE 20. Response time, relative signal change, and reproducibility of a fiber optic oxygen sensor based on energy transfer. (From Sharma, A. and Wolfbeis, O. S., *Anal. Chim. Acta*, 212, 261, 1988.)

triplet oxygen ($^3\Sigma$) quench the emission and thereby are converted into singlet oxygen ($^1\Delta$). Typical dyes that show this effect[62,63] are trypaflavine, benzoflavine, euchrysine, rheonine 3A, rhoduline yellow, safranine, chlorophyll, and hematoporphyrin. Silica gel, aluminium oxide were found to be the most suitable adsorbents with respect to oxygen determination via RTP because of the extreme efficiency with which the dyes are quenched on these supports. Cellulose, silk, or synthetic fibers, in contrast, showed no sensitivity towards oxygen[64] when completely dry, but after exposure to air humidity. However, increasing absorption of water also leads to an extinction of phosphorescence which is complete in aqueous surrounding. Consequently, the adsorbates on silica gel or aluminium oxide were considered more viable.

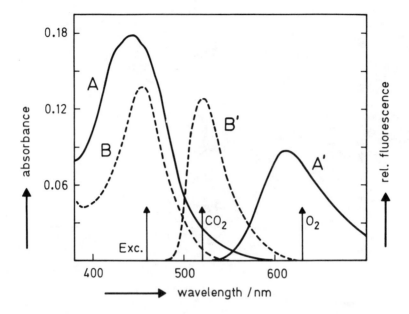

FIGURE 21.   Absorption and emission spectra of the sensing material shown in Figure 22. (From Reference 60.)

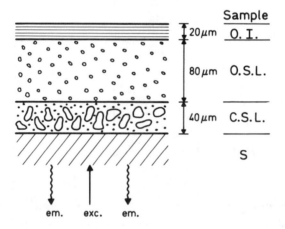

FIGURE 22.   Cross-section through a sensing material responsive to both oxygen and carbon dioxide. (From Reference 60.)

The findings are useful from a present-day point of view since all dyes used by Kautsky can be excited with LEDs and phosphorescence has an extremely large Stokes' shift and long lifetime.

The preparation of such adsorbates is experimentally simple in that the particles are placed in aqueous or methanolic solutions of about 0.1 mM of the dyes and left there for a defined period of time. After washing with water and heat-drying they are ready for use. A typical experiment with oxygen is the following: when trypaflavin is adsorbed on silica gel ("silicic acid") and the adsorbate is placed in a carefully evacuated flask, it shows a green *fluorescence* when illuminated with a UV lamp. When the lamp is switched off, a yellow-green *phosphorescence* is observed for several seconds. As soon as oxygen is admitted to the flask,

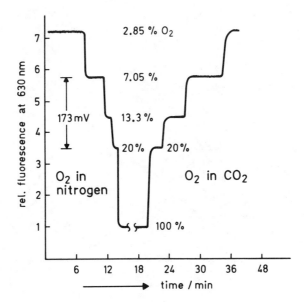

FIGURE 23. Response of the sensor for oxygen and carbon dioxide towards oxygen. (From Wolfbeis, O. S., Weis, L. J., Leiner, M. J. P., and Ziegler, W., *Anal Chem.,.*)

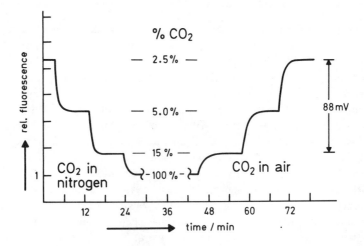

FIGURE 24. Response of the sensor for oxygen and carbon dioxide towards carbon dioxide. (From Wolfbeis, O. S., Weis, L. J., Leiner, M. J. P., and Ziegler, W., *Anal Chem.,.*)

fluorescence is partially quenched, but phosphorescence completely extinguished. These effects are fully reversible. Fluorescence is much less susceptible to quenching than phosphorescence, and all luminescence is quenched when the temperature is lowered to −190°C, probably because the oxygen concentration on the surface becomes extremely enhanced. Table 3 lists the effects observed with various dyes.

While fluorescence quenching by oxygen can vary considerably (see the difference between safranine and uranine), phosphorescence is totally quenched by oxygen at levels of around 0.001 to 0.01 torr. The effects were used to detect traces of oxygen in the range between 0.004 and 0.0005 torr.[10]

It was also observed that the dyes were bleached by oxygen (which implies that oxygen is

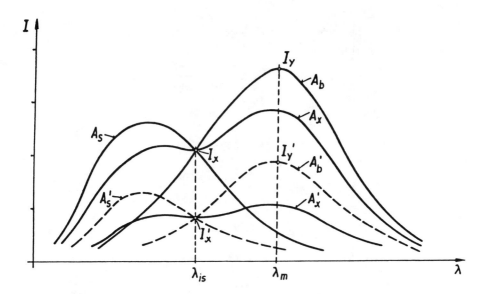

FIGURE 25. Excitation spectra of an indicator suffering both pH-induced spectral shifts and luminescence quenching. A_s, excitation spectrum of the acid form; A_b, excitation spectrum of the base form; A_x, spectrum of unknown sample; A_s', A_b', and A_x' respectively, spectra in the presence of a dynamic quencher of fluorescence; I_x and I_x', intensities at the isosbestic (pH-invariable) wavelength λ_{is}; I_y and I_y', intensities at wavelength of maximum (λ_{max}).

consumed), unless singlet oxygen is transferred to another species. Such an acceptor was found in isoamylamine, which therefore may be called an inhibitor of photosensitized oxydation. Other inhibitors (which in optical sensors can act as stabilizers) are the leuko forms of triphenylmethane dyes such as leukomalachit green and allylurea.[65] A brief account on Kautsky's work is found in Reference 3.

An even more sensitive oxygen sensor is based on the finding[66] that the slow-decaying phosphorescence at temperatures below −70°C is converted into a rapidly decaying fluorescence when oxygen is admitted. In a typical experiment, trypaflavin is adsorbed on kieselgel and placed in an evacuated flask in liquid air or solid CO_2. When excited with visible light, the material has an orange phosphorescence with a decay time as long as 40 to 50 s. If, after switching off the excitation light source, traces of oxygen are allowed to enter the flask, phosphorescence is instantaneously quenched, but a flash of green fluorescence is observed. The effect is observed with traces of oxygen and has been exploited to detect it at pressures around 5×10^{-6} torr (i.e., 1×10^{-11} mol) in a volume of 50 ml. The effect is the same under vacuum, hydrogen, or nitrogen, and the working range is 10^{-6} to 10^{-4} torr under visual absorption. Other dyes that showed the effect at −180°C are rhodamines B and 6 G, uranine, phosphine, benzoflavine, 1-hydroxypyrene-3,6,8-trisulfonate, and 7-hydroxycoumarins.

Pollack et al.[67] have modified the method. Since Kautsky's procedure can be used to follow a varying rate of oxygen production (as, for instance, in photosynthesis) over a short time only because of the sensitivity of the method, the authors have continuously transported the oxygen from the site of generation to the detection site (the phosphorescent gel) with a carrier stream of pure nitrogen. They were able to detect the production of 4×10^{12} oxygen molecules when a leaf is exposed to a flash of light. Another field of application was found in the detection of the onset of oxygen production in water electrolysis when the voltage is raised from 0 to 2.6 V. As in Kautsky's work, it was noted that it takes 5 to 15 min for the phosphorescent gel to regain its full intensity after an oxygen production lasting minutes. This is not due to the inertia of the material but to the time needed to flush all the oxygen out of the system.

The quenching of the phosphorescence (excited by steady-state and flash methods) in poly(methyl methacrylate) has been studied in order to determine the diffusion coefficients of

TABLE 3
Luminescence and Fluorescence Quenching by Oxygen of Silicagel-Adsorbed Dyes[63]

Dye	Phosphorescence intensity	Fluorescence intensity	Fluorescence quenching by oxygen
Trypaflavin	strong, 5-10 s	strong	strong
Benzoflavin	strong, 5-10 s	strong	strong
Euchrysin 3R	strong, 5-10 s	strong	strong
Rheonine A	strong, 5-10 s	strong	strong
Rhodulin yellow	strong, 5-10 s	strong	strong
Isoquinoline red	strong, < 1 s	strong	weak
Safranine	not observable	weak	very strong
Rhodamine B	not observable	strong	very weak
Rhodamine G	not observable	strong	very weak
Uranine	strong, 5-10 s	strong	not observable
Eosin	not observable	medium	very weak
Erythrosine	not observable	weak	very weak

oxygen.[21] Triphenylene and phenanthrene were used as phosphorescent probes. It was found that reduction of the phosphorescence lifetime of the probes in the presence of oxygen is much smaller (by –10%) than the reduction of phosphorescence intensity (–80%). This is explained in terms of both static and dynamic quenching, and a theory was established for the relationship between measured phosphorescence intensity and the diameter of the phosphorescent sample.

Low oxygen concentrations are frequently encountered in waste water. Zakharov and Grishaeva[68,69] have utilized the phosphorescence quenching effect of oxygen to devise an optosensor for low oxygen levels in water. It was found that the phosphorescence was retained even when the material was immersed in water or various organic solvents,[68] with lifetimes ranging from 50 to 100 ms. Thus, activated samples of silica gels of various structures and celluloses of varying viscosity phosphoresced in water, ethanol, isoamyl alcohol, heptane, and chloroform. Deoxygenation led, in most cases, to significant increase in intensity. In addition, there is a dependence of the emission intensity upon the structure of the silica gels, in that those having fine pores display the strongest intensity. The effect of pH was also studied along with some long-time experiments. The authors give contradictary results as to the linearity of Stern-Volmer plots, but it may be expected from experience that these surface-adsorbed dyes have the same complex decay and quenching behavior as have others of this type because the surface strongly affects the microenvironment and accessibility.

The findings have been applied for sensing oxygen in waters. The afterglow of acridine orange adsorbed on Silochrome S-120 was most sensitive to quenching by oxygen and used for continuous assay in the range of 0.06 to 1.00 μg oxygen per liter.[69,70] With acriflavine, the sensitivity is 0.35 μg l^{-1} when the dye is adsorbed on silica gel and 10 to 100 μg l^{-1} when adsorbed on cellulose. Other kinds of supports have also been studied and dynamic ranges from 0.4 to 400 μg l^{-1} were reported.[71] Sensitivity can considerably be improved by using hydrophobic supports such as silanized silica gel.[72] Detection limits are 5×10^{-4} μg l^{-1}, but it was observed that the dye undergoes "photoabsorption of oxygen" which resulted in a distortion of the Stern-Volmer plots at high illumination intensity.[73] This effect is negligible when the exciting light is sufficiently attenuated. The method has not been coupled to fiber optics but easily could.

More recently, the quenching of zinc(II)tetraphenylporphin (ZnTPP) phosphorescence by molecular oxygen was studied in detail and found to be markedly different for dye supported on silica gel and on NaCl crystals.[74] On NaCl, quenching shortens the phosphorescence lifetime and obeys Stern-Volmer kinetics. The quenching rate constant is approximately 1100 $torr^{-1}s^{-1}$ at room temperature. On silica gel, quenching of ZnTPP phosphorescence takes place at lower oxygen pressure and is not accompanied by a change in the luminescence rate constant. The data agree with a modified Langmuir model which assumes that only oxygen adsorbed on the silica gel surface next to a ZnTPP molecule will quench luminescence, and that oxygen competes with water molecules for adsorption sites on the substrate. If ZnTPP on silica gel is to be used as oxygen-sensitive material in optical sensors, the data suggest a dynamic range from 5×10^{-5} to 5×10^{-2} torr.

The phosphorescence of a palladium metalloporphyrin, when decomposed along with arachidic acid and a diyne acid in organized monolayer assemblies using the Langmuir-Blodgett technique, varies with oxygen tension.[75] The visible absorption of the assemblies suggests that the dye has become agglomerated during the decomposition process which results in excited states which have different oxygen susceptibilities and produce strongly nonlinear Stern-Volmer plots.

IV. OTHER SENSOR TYPES

An interesting, though not fully understood, observation is the thermally stimulated luminescence of poly(ethylene-2,6-naphthalene-dicarboxylate) (PEN) which occurs when the material is first exposed to UV radiation and oxygen.[6] After switching off the light source and heating to above 71°C, PEN begins to emit luminescence that increases with temperature up to a maximum output at 130°C. The samples would not luminesce if kept in darkness or in a vacuum. The prime wavelengths causing the reaction were found to be 330 to 390 nm. The emitted light had a maximum at 465 nm. The luminescence observed during this study was attributed to a photochemical reaction involving UV light, oxygen, and PEN, but no detailed mechanistic investigations were reported. From a present day view it cannot be excluded that the actual species that causes luminescence to occur is ozone rather than oxygen, since ozone is known to react with various polymers by chemical addition. Subsequent thermal decomposition of endoperoxides can result in luminescence. Ozone may be present in ambient air but also may be produced by the UV light of the light source.

An oxygen sensor for exhaust gases[76] comprises a supported film of heat-responsive sensing material having a light reflective surface which reversibly reacts with oxygen to form an oxide which causes a predictable change in reflectivity of the surface. The sensor also includes a temperature sensing means and a three-way optical system.

Tetraamino-ethylenes without aromatic functions react with oxygen to produce bright chemiluminescence (CL). Since CL intensity is directly proportional to the concentration (partial pressure) of oxygen and because low levels of CL can easily be measured, a device has been constructed[4] in which the oxygen diffuses through a Teflon membrane before reacting with the CL reagent (a 10% hexane solution of 1,1′,3,3′-tetraethyl-bi-imidazoline). Steady state CL intensity is proportional to pO_2, and response to pure oxygen decays gradually over a 12-h period because of reagent consumption. The detection limit of oxygen in the gas phase is estimated to be as low as 1 ppm (v/v). The effect of temperature is an increase in signal with increasing temperature, a fact that is primarily attributed to the temperature dependence of the diffusion coefficient. Response times are in the order of 10 to 20 s.

Zhujun and Seitz[5] have made an interesting approach to determine oxygen via the well known differences in the absorption spectra of hemoglobin (Hb) and oxyhemoglobin. Hb was immobilized electrostatically on cation exchange resin and positioned at the common end of a bifurcated fiber bundle. A Teflon membrane separated the immobilized reagent phase from the sample. Upon association of oxygen with Hb, the Soret band is shifted from 405 to 435 nm. The ratio of the intensities of reflected light at these two wavelengths is the analytical information that can be used to measure oxygen at levels between 20 and 100 torr. A steady state is reached within 3 min after introduction of an oxygen sample. The sensor must be stored in a reducing medium to prevent oxidation of Hb to met-Hb. Unfortunately it is stable for only 2 d when stored at room temperature and for a week when stored at 4°C. An attractive feature of this system is the possibility of measuring at two wavelengths which provides an internal reference system that can compensate for various potential drift sources.

V. OXYGEN OPTRODES AS TRANSDUCERS

The importance of all kinds of oxygen sensors not only lies in the determination of oxygen, but also in being capable of acting as transducers for various chemical and biochemical reactions during which oxygen is consumed or produced. Typical examples are enzymatic reactions involving oxidases, oxygenases, and, indirectly, dehydrogenases. The oxidases are particularly popular because they have, as a second substrate, oxygen which therefore makes the reaction fully reversible without the need for additional enzymes. Typical examples for biosensors based on oxygen transducers have been mentioned[33,40,41,43,44,49,50] and are described in more detail in Chapters 3 and 17, together with other types of biosensors.

Oxidases, during their enzymatic action, shuttle electrons from the coenzyme to molecular oxygen and thereby produce hydrogen peroxide (HP). As a result, HP sensors may serve as transducers for oxidase-catalyzed reactions. It has been found[77] that some of the oxygen optrodes described above may also serve as transducers in the assay of HP. When HP is decomposed by a catalyst to give water and molecular oxygen, an increase in pO_2 will be observed which may be monitored via an oxygen optrode.

Three types of HP sensors were described[77] based on this principle. In the first type, the enzyme catalase (which very efficiently decomposes HP) was co-adsorbed with an oxygen-sensitive indicator on kieselgel particles and then embedded into silicone in a fashion similar to the one described before.[16] Since HP penetrates silicone, it can be decomposed directly at the site where the indicator is placed. In another type, catalase and indicator were immobilized on different beads which were mixed in the membrane. In the third one, a conventional oxygen sensor was covered with a layer of silicone containing finely dispersed silver particles (which is a very effective catalyst). Figure 26 shows a cross-section through the different types of membranes.

Interestingly, HP itself acts as a quencher of the fluorescence of ruthenium(II)tris(bipyridyl), but the quenching constant is too small to be of analytical utility (Figure 27). With the silver sensor type, HP can continuously be recorded in the 0.1 to 10.0 mM concentration range, with a precision of ±0.1 mM at 1 mM HP. The response time (Figure 28) varies from 2.5 to 5 min. One problem, however, is the interference by varying levels of molecular oxygen, and it was suggested[77] to make use of a two-sensor technique, with one sensor containing the catalyst, and the other not. By setting the signals ($I_o/I - 1$) of sensors A and B to α and β, respectively, it was derived that the concentration of HP can be calculated according to

$$[HP] = 2(\beta - \alpha)/f \cdot K_{sv} \tag{10}$$

where factor 2 accounts for the fact that two molecules of HP convert to one O_2 and f is a factor that considers the unknown distribution of oxygen between sites of production and detection and the unknown rate of conversion of HP into water and oxygen. K_{sv} is the oxygen quenching constant.

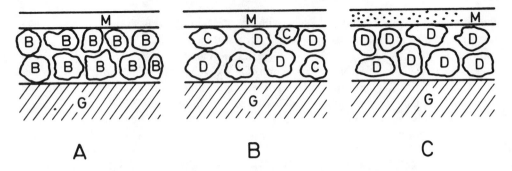

FIGURE 26. Cross-section through the three types of hydrogen peroxide-sensitive membranes. In type (A), the silica gel beads (B) contain both the surface-adsorbed indicator and the enzyme catalase; the sensing layer is further covered with a thin silicone membrane (M). In type B, the catalase is contained in one half of the beads (C), and the indicator in the other (D). In type C, the dyed silica gel particles (D) are in the silicone rubber layer that is covered by a membrane (M) that contains finely dispersed silver powder. (From Posch, H. E. and Wolfbeis, O. S., *Mikrochim Acta*, I, 41, 1989.)

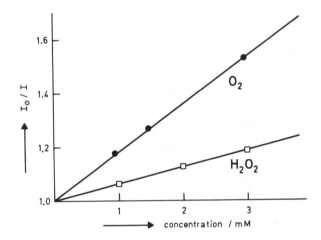

FIGURE 27. Quenching of ruthenium(II)tris(dipyridyl) by hydrogen peroxide and oxygen. (From Posch, H. E. and Wolfbeis, O. S., *Mikrochim Acta*, I, 41, 1989.)

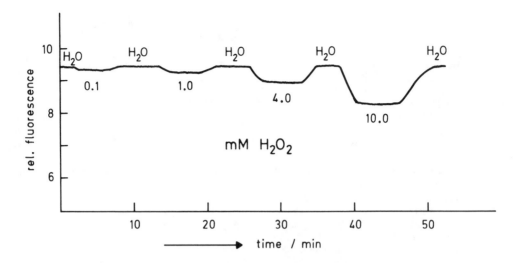

FIGURE 28. Response of a sensor for hydrogen peroxide based on catalytical decomposition. (From Posch, H. E. and Wolfbeis, O. S., *Mikrochim Acta*, I, 41, 1989.)

REFERENCES

1. **Hitchman, M. L.,** *Measurement of Dissolved Oxygen*, John Wiley & Sons, New York, 1978.
2. **Wolfbeis, O. S.,** Optical and fiber optic fluorosensors in analytical and clinical chemistry, in *Molecular Luminescence Spectroscopy; Methods and Applications*, Vol. 2, Schulman, S. G., Ed., John Wiley & Sons, New York, 1988, chap. 3.
3. **Kautsky, H.,** Quenching of luminescence by oxygen, *Trans. Faraday Soc.*, 35, 216, 1939.
4. **Freemann, T. F. and Seitz, W. R.,** Oxygen probe based on chemiluminescence, *Anal. Chem.*, 53, 98, 1981.
5. **Zhujun, Z. and Seitz, W. R.,** Optical sensor for oxygen based on immobilized hemoglobin, *Anal. Chem.*, 58, 220, 1986.
6. **Hendricks, H. D.,** Thermally stimulated luminescence of poly(ethylene 2,6-naphthalenedicarboxylate), *Mol. Phys.*, 20, 189, 1971.
7. **Marsoner, H., Kroneis, H., and Wolfbeis, O. S.,** U.S. Patent 4,657.736, 1987.
8. **Shah, R., Margerum, S. C., and Gold, M.,** Grafted hydrophilic polymers as optical sensor substrates, *Proc. SPIE*, 906, 65, 1988.
9. **Bergman, I.,** Rapid response atmospheric oxygen monitor based on fluorescence quenching, *Nature*, 218, 396, 1968.
10. **Kautsky, H. and Hirsch, H.,** Detection of minutest amounts of oxygen by extinction of phosphorescence, *Z. Anorg. Allg. Chem.*, 222, 126, 1935; (in German).
11. **Peterson, J. I., Fitzgerald, R. V., and Buckhold, D. K.,** Fiber-optic probe for *in vivo* measurement of oxygen partial pressure, *Anal. Chem.*, 56, 62, 1984.
12. **Bacon, J. R. and Demas, J. N.,** Determination of oxygen concentrations by luminescence quenching of a polymer immobilized transition metal complex, *Anal. Chem.*, 59, 2780, 1987.
13. **Wolfbeis, O. S., Offenbacher, H., Kroneis, H., and Marsoner, H.,** A fast responding fluorescence sensor for oxygen, *Mikrochim. Acta*, I, 153, 1984.
14. **Hsu, L. and Heitzmann, H.,** U.S. Patent 4,712,865.
15. **Lübbers, D. W., Opitz, N., Speiser, P. P., and Bisson, H. J.,** Nano-encapsulated fluorescence indicator molecules measuring pH and pO_2 down to submicroscopical regions on the basis of the optode principle, *Z. Naturforsch.*, 32C, 133, 1977.
16. **Wolfbeis, O. S., Leiner, M. J. P., and Posch, H. E.,** A new sensing material for optical oxygen measurement, with the indicator embedded in an aqueous phase, *Mikrochim. Acta*, III, 359, 1986.
17. **Schaffar, B. P. H. and Wolfbeis, O. S.,** The development of optical sensors using the Langmuir-Blodgett technique, *Proc. SPIE*, 990, 122, 1989.

18. **Cox, M. E. and Dunn, B.,** Detection of oxygen by fluorescence quenching, *Appl. Optics*, 24, 2114, 1985.

19. **Cox, M. E. and Dunn, B.,** Oxygen diffusion in poly(dimethylsiloxane) using fluorescence quenching, I. Measurement technique and analysis, *J. Polymer Sci.*, 24A, 621, 1986.

20. **Cox, M. E. and Dunn, B.,** Oxygen diffusion in poly(dimethylsiloxane) using fluorescence quenching, II. Filled samples, *J. Polymer Sci.*, 24A, 2395, 1986.

21. **Shaw, G.,** Quenching by oxygen diffusion of phosphorescence emission of aromatic molecules in poly(methyl methacrylate), *Trans. Faraday Soc.*, 63, 2181, 1967.

22. **Parker, J. W. and Cox, M. E.,** Mass transfer of oxygen in poly(2-hydroxyethyl methacrylate), *J. Polymer Sci.*, 26A, 1179, 1988.

23. **Brandrup, J. and Immergut, E. H. (Eds.),** The Polymer Handbook, 2nd Ed., John Wiley & Sons, New York, 1975, p. III-229.

24. **Kubin, M. and Spacek,** Structure and properties of hydrophilic polymers and their gels. Diffusion in gels, *Coll. Czech. Chem. Comm.*, 30, 3294, 1965.

25. **Opitz, N. and Lübbers, D. W.,** Increased resolution power in pO_2 analysis at low pO_2 levels via sensitivity-enhanced pO_2 sensors (optodes) using fluorescence dyes, *Adv. Exp. Med. Biol.*, 180, 261, 1984.

26. **Kroneis, H. W,** unpublished results, 1982.

27. **Opitz, N., Graf, H. J., and Lübbers, D. W.,** Oxygen sensor for the temperature range 300 to 500 K based on fluorescence quenching of indicator-treated silicone rubber membranes, *Sensors Actuators*, 13, 159, 1988.

28. **Miller, W. W., Yafuso, M., Yan, Ch. F., Hui, H. K., and Arick, S.,** Performance of an in-vivo continuous blood gas monitor with disposable probe, *Clin. Chem.*, 33, 1538, 1987.

29. **Kroneis, H. W. and Marsoner, H. J.,** A fluorescence-based sterilizable oxygen probe for use in bioreactors, *Sensors Actuators*, 4, 587, 1983.

30. **Atkinson, R. S., Brimage, D. R. G., and Davidson, R. S.,** Use of tertiary amino groups as substituents to stabilise compounds towards attack by singlet oxygen, *J. Chem. Soc.*, Perkin I, 960, 1973.

31. **Lippitsch, M. E., Pusterhofer, J., Leiner, M. J. P., and Wolfbeis, O. S.,** Fibre optic oxygen with the fluorescence decay time as the information carrier, *Anal. Chim. Acta*, 205, 1, 1988.

32. **Lee, E. D., Werner, T. C., and Seitz, W. R.,** Luminescence ratio indicators for oxygen, *Anal. Chem.*, 59, 279, 1987.

33. **Trettnak, W., Leiner, M. J. P., and Wolfbeis, O. S.,** Fiber optic glucose sensor with an oxygen optrode as the transducer, *Analyst*, 114, 1519, 1988.

34. **Miller, J. A. and Baumann, C. A.,** The effect of oxygen on the fluorescence of certain hydrocarbons, *J. Am. Chem. Soc.*, 65, 1540, 1943.

35. **Lübbers, D. W. and Opitz, N.,** The pCO_2/pO_2 optode: a new probe for measurement of pCO_2 and pO_2 in gases or liquids, *Z. Naturforsch.*, 30C, 532, 1975.

36. **Lübbers, D. W. and Opitz, N.,** Quantitative fluorescence photometry with biological fluids and gases, *Adv. Exp. Med. Biol.*, 75, 65, 1976.

37. **Opitz, N. and Lübbers, D. W.,** Evidence for boundary layer effects influencing the sensitivity of micro-encapsulated oxygen fluorescence indicator molecules, *Adv. Exp. Med. Biol.*, 169, 899, 1984.

38. **Peterson, J. I. and Stefanson, E.,** Fiber optic oxygen sensor, GBF Monograph Vol. 20 (Biosensor Intl. Workshop), Schmid, R. D., Ed., Verlag Chemie, Weinheim, 1988, 235.

39. **Parker, J. W., Cox, M. E., and Dunn, B.,** Chemical sensors based on oxygen detection by optical methods, *Proc. SPIE*, 586, 156, 1985.

40. **Völkl, K. P., Grossmann, U., Opitz, N., and Lübbers, D. W.,** The use of the oxygen optode for measuring glucose using oxidative enzymes, *Adv. Physiol. Sci.*, 25, 99, 1981.

41. **Lübbers, D. W. and Opitz, N.,** Optical fluorescence sensors for continuous measurement of chemical concentrations in biological systems, *Sensors Actuators*, 3, 641, 1983.

42. **Lübbers, D. W., Gehrich, J., and Opitz, N.,** Fiber optics coupled fluorescence sensors for continuous monitoring of blood gases in the extracorporeal circuit, *Life Supports Syst.*, 4, 94, 1986.

43. **Lübbers, D. W., Völkl, K. P., Grossmann, U., and Opitz, N.,** Lactate measurement with an enzyme optode that uses two oxygen indicators to measure pO_2 gradient, in *Progress in Enzyme and Ion-Selective Electrodes*, Lübbers, D. W., Acker, H., Buck, R. P., Eisenman, G., Kessler, M., and Simon, W., Eds., Springer Verlag, Berlin, 1981, 67.

44. **Völkl, K. P., Opitz, N., and Lübbers, D. W.,** Continuous measurement of alcohol using a fluorescence enzymatic method, *Fresenius Z. Anal. Chem.*, 301, 162, 1980.

45. **Wolfbeis, O. S., Posch, H. E., and Kroneis, H. W.,** Fiber optical fluorosensor for determination of halothane and/or oxygen, *Anal. Chem.*, 57, 2556, 1985.

46. **Wolfbeis, O. S. and Urbano, E.,** Fluorescence quenching method for determination of two or three components in solution, *Anal. Chem.*, 55, 1904, 1984.

47. **Okazaki, T., Imasaka, T., and Ishibashi, N.,** Optical fiber sensor based on the second harmonic emission of a near infrared semiconductor laser as light source, *Anal. Chim. Acta*, 209, 327, 1988.

48. **McFarlane, R. and Hamilton, M. C.,** A fluorescence based dissolved oxygen sensor, *Proc. SPIE*, 798, 324, 1988; also see Reference 78.

49. **Wolfbeis, O. S. and Leiner, M. J. P.,** Recent progress in optical oxygen sensing, *Proc. SPIE*, 906, 42, 1988.

50. **Wolfbeis, O. S. and Posch, H. E.,** Fiber optic ethanol biosensor, *Fresenius Z. Anal. Chem.*, 332, 255, 1988.

51. **Peterson, J. I. and Fitzgerald, R. V.,** New technique of surface flow visualization based on oxygen quenching of fluorescence, *Rev. Sci. Instrum.*, 51, 670, 1980.

52. **Burkhard, O.,** A Novel Detector for Measurement of Oxygen Partial Pressure in Biological Systems, M.D. dissertation, University Mainz, 1987.

53. **Burkhard, O. and Barnikol, W. K. R.,** Fluorescence quenching of dyes adsorbed on surface may be independent of the viscosity of the medium in front of the sample, *Z. Naturforsch.*, 40B, 1719, 1985.

54. **Wolfbeis, O. S.,** Unpublished results, 1983.

55. **Posch, H. E. and Wolfbeis, O. S.,** Fiber optic humidity sensor based on fluorescence quenching, *Sensors Actuators*, 15, 77, 1988.

56. **Goswami, K., Klainer, S. M., and Tokar, J. M.,** Fiber optic chemical sensor for the measurement of partial pressure of oxygen, *Proc. SPIE*, 990, 111, 1989.

57. **Moreno - Bondi, M. C., Wolfbeis, O. S., Leiner, M. J. P., and Schaffar, B. P. H.,** A highly sensitive oxygen optrode for use in a fiber optic glucose biosensor, *Anal. Chem.*, 62, 2377, 1990.

58. **Sharma, A. and Wolfbeis, O. S.,** Fiberoptic oxygen sensor based on fluorescence quenching and energy transfer, *Appl. Spectrosc.*, 42, 1009, 1988.

59. **Sharma, A. and Wolfbeis, O. S.,** Unusually efficient quenching of the fluorescence of an energy-transfer-based optical sensor for oxygen, *Anal. Chim. Acta*, 212, 261, 1988.

60. **Wolfbeis, O. S., Weis, L. J., Leiner, M. J. P., and Ziegler, W.,** Fiber-optic fluorosensor for oxygen and carbon dioxide, *Anal. Chem.*, 60, ***, 1988; in press.

61. **Wolfbeis, O. S.,** Proc. SPIE, 1368, 218, 1990.

62. **Kautsky, H., Hirsch, H., and Baumeister, W.,** Photoluminescence of fluorescent dyes at interfaces, *Ber. Dtsch. Chem. Ges.*, 64, 2053, 1931 (in German).

63. **Kautsky, H. and Hirsch, H.,** Interaction between excited dye molecules and oxygen, *Ber. Dtsch. Chem. Ges.*, 64, 2677, 1931 (in German).

64. **Kautsky, H. and Hirsch, H.,** Phosphorescence of adsorbed fluorescent dyes and its relation to reversible and irreversible changes of gels (celluloses and proteins), *Ber. Dtsch. Chem. Ges.*, 65, 401, 1932 (in German).

65. **Kautsky, H., de Bruijn, H., Neuwirth, R., and Baumeister, W.,** Photo-sensitized oxidation as an effect of an active, metastable state of the oxygen molecule, *Ber. Dtsch. Chem. Ges.*, 66, 1588, 1933 (in German).

66. **Kautsky, H. and Müller, G. O.,** Transformation of luminescence and detection of minutest amounts of oxygen, *Z. Naturforsch.*, 2A, 167, 1947 (in German).

67. **Pollack, M., Pringsheim, P., and Terwoord, D.,** A method for determining small quantities of oxygen, *J. Chem. Phys.*, 12, 295, 1944.

68. **Zakharov, I. A. and Grishaeva, T. I.,** Long-lived phosphorescence of adsorbate phosphors in liquids and its quenching by oxygen, *Zh. Prikl. Spektrosk.*, 31, 1295, 1979.

69. **Zakharov, I. A. and Grishaeva, T. I.,** Determination of microconcentrations of oxygen in water, *Zh. Anal. Khim.*, 35, 481, 1980.

70. **Zakharov, I. A. and Grishaeva, T. I.,** Increasing the sensitivity of the determination of dissolved oxygen in water by afterglow quenching, *Zh. Anal. Khim.*, 36, 112, 1981.

71. **Zakharov, I. A. and Grishaeva, T. I.,** Solid supports of luminescent indicators for the determiantion of oxygen dissolved in water, *Zh. Prikl. Spektrosk.*, 57, 1240, 1984.

72. **Zakharov, I. A. and Grishaeva, T. I.,** Quenching of luminescence of dye adsorbates on hydrophobic silica gels by oxygen dissolved in water, *Zh. Prikl. Spektrosk.*, 36, 697, 1982.

73. **Zakharov, I. A. and Grishaeva, T. I.,** Determination of trace concentration of dissolved oxygen in water, *Zh. Anal. Khim.*, 37, 1753, 1982.

74. **Twarowski, A. J. and Good, L.,** Phosphorescence quenching by molecular oxygen: zinc tetraphenylporphin on solid supports, *J. Phys. Chem.*, 91, 5252, 1987.

75. **Beswick, R. B. and Pitt, C. W.,** Oxygen detection using phosphorescent Langmuir-Blodgett films of a metalloporphyrin, *Chem. Phys. Lett.*, 143, 589, 1988.

76. **Nyberg, G. A.,** U.S. Patent 4,764,343, 1988.

77. **Posch, H. E. and Wolfbeis, O. S.,** Optical sensor for hydrogen peroxide, *Mikrochim. Acta*, I, 41, 1989.

78. **Thomson, R. E., Curran, T. A., Hamilton, M. C., and McFarlane, R.,** Time series measurements from moored fluorescence-based dissolved oxygen sensor, *Oceanic Atmos. Technol.*, 5, 614, 1988.

Chapter 11

GAS SENSORS

Otto S. Wolfbeis

TABLE OF CONTENTS

I. INTRODUCTION

The majority of methods for continuous optical monitoring of gaseous species can be categorized into one of two subgroups. In the first, the intrinsic optical property of the gas (such as its absorption) is utilized in the sensing procedure. The absorption wavelengths range from the UV (200 nm) into the infrared (4 μm). Fluorescence and Raman detection are also applied, albeit much less frequently. These methods rely on conventional spectroscopic techniques which, however, have been coupled to fiber optic technology with its unique features and possibilities. Direct laser spectroscopy has been widely applied in the past to monitor the concentration of pollutant gases in the atmosphere. The fiber methods have various advantages over the previous laser remote monitoring schemes in the open atmosphere in that low-power laser or even nonlaser sources can be used in conjunction with low-loss low-cost fibers. Therefore, a purely optical, economical, real-time, and nonhazardous, i.e., eye-safe, monitoring technique can be realized.

The subgroup of plain fiber methods contrasts the second subgroup, in which an indicator is used to transduce the gas concentration into a measurable optical parameter. This approach

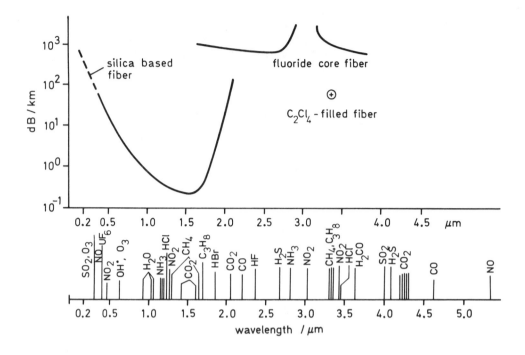

FIGURE 1. Attenuation of a standard silica fiber (top) and typical wavelengths (bottom) at which plain fiber sensing of gaseous species can be performed.

is frequently applied when the gas has no useful intrinsic optical property. The resulting sensors are less rugged and more complicated in fabrication, because they require (1) a specific indicator chemistry reproducibility to be placed at the fiber end and (2) the sensing chemistry to have both a useful shelf and operational lifetime. As will be shown later, some of the chemistries placed on the fibers undergo irreversible reactions, so the probe is useful for a single assay only. These devices are referred to as "probes" in this book, because sensors, by definition, act fully reversibly. Typical examples for indicator-mediated optrodes include those for oxygen, carbon dioxide, and sulfur dioxide. A notable exception is hydrogen which lacks visible or near-infrared (NIR) absorption and for which no reversible indicators are known. However, it can be sensed by other means (Section II herein).

Typical examples for plain fiber sensors include those for chlorine, methane, carbon monoxide, and nitrogen oxides, all of which have an absorption that can can easily be measured via fibers. Figure 1 shows characteristic wavelengths at which direct spectrometry of gaseous chemical species can be performed via fibers, frequently by the evanescent wave technique (see Chapter 3). The 3 to 5 μm wavelength range also holds great promise, but will become practical to work at in fiber sensing only when less expensive IR fibers are available.

A typical experimental arrangement for performing remote fiber absorptiometry with plain fibers is shown in Figure 2; other geometries are found in Chapter 3, Figure 10. In addition to the flow-through-cell method shown there, evanescent wave approaches have also been applied successfully. The concentration of a gas in a mixture is usually measured by comparing the intensities of two light beams having different wavelengths passing through the sample cell, or being absorbed by the surrounding of the fiber. The two wavelengths are chosen in a way that the total absorption of the gas will be of a convenient magnitude and the absorption by other gases will be negligible for the first wavelength, while the other wavelength will be such that absorption by all gases in the mixture is small. Measurement of the ratio of the two intensities accounts for effects of light level changes caused by source ageing, darkening of optical components, and scattering by particles.

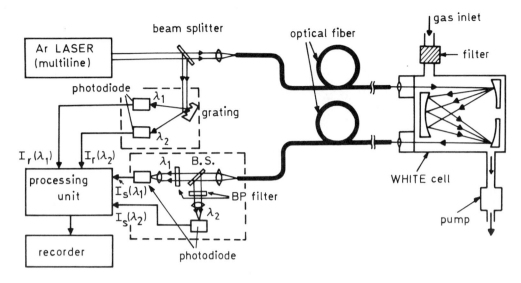

FIGURE 2. Block diagram of an experimental system employing an optical fiber link for differential (ratio) absorption measurement of nitrogen dioxide concentrations in a sample cell (WHITE cell) at a remote location. (From Inaba, H., *Springer Ser. Opt. Sci.,* 39, 288, 1983. With permission.)

Another method for concentration measurement via absorptiometry with plain fibers is gas cell correlation. An attenuator and cell filled with the gas to be measured are alternately moved into the measurement beam. In the correlation method, the wavelengths of the measuring beam and reference beam are identical. The absorption of other gases at the analytical wavelength does not affect measurement. This method has been described in some detail for the case of conventional IR spectrometry.[1]

When an indicator-mediated method is applied, the fiber end or the waveguide surface is equipped with a so-called working chemistry. In this case, it is the immobilized reagent that responds to the analyte by a color change. A typical working chemistry has been shown in Chapter 8, Figure 14. Other sensing schemes such as the interferometric determination of hydrogen (see later) form a minor group in chemical sensing, although they are indispensable in sensing certain rather unreactive analytes.

The following is a representative (though not complete) compilation of existing sensing schemes for the most important gases and vapors, except for oxygen which is treated in Chapter 10. Rather than discussing the existing sensors according to the methods applied for their detection and quantitation, this chapter is organized such that all methods known for a given species are listed under the respective analyte heading. This should facilitate the rapid search for the most appropriate method for a given species.

II. HYDROGEN SENSORS

Hydrogen is an exceptional molecule in that it has no useful absorption in the NIR (where fiber absorptiometry is preferably performed because of the good transmittance of fibers for NIR light). In addition, no indicators are known that would respond to hydrogen by a reversible color change, except for the poorly defined tungsten oxide (WO_3). Therefore, other sensing approaches have to be made. In what seems to have been the first optosensor for hydrogen, an optical fiber coated with palladium was used as the sensing element.[2] The palladium expands on exposure to hydrogen because of the formation of a palladium hydride (PdH_x) having an expanded lattice constant. The expansion depends on the partial pressure of hydrogen and stretches the fiber in both the axial and radial direction. This results in a change in the effective optical path length of the fiber, which can be detected by Mach-Zender

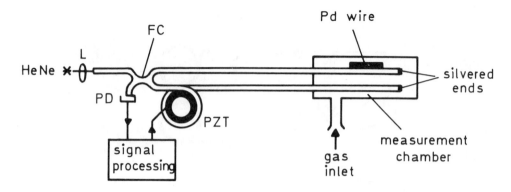

FIGURE 3. Experimental set-up of the interferometric hydrogen sensor based on hydrogen absorption by a palladium metal wire. NeHe, multilongitudinal helium-neon laser; L, lens; FC, fiber directional coupler; PD, photodiode; PZT, piezoelectric transducer. (From Farahi, F., Akhavan-Leilabady, P., Jones, J. D. C., and Jackson, D. A., *J. Phys. E*, 20, 432, 1987. With permission.)

interferometry. A schematic of the experimental arrangement is shown in Chapter 2, Figure 16. A large dynamic range is reported (1 to 30,000 ppm), and the effect is fully reversible. Phase changes of 10^{-6} fringes can be measured on sensitive equipment. Temperature effects are equivalent to 2 ppm H_2 per °C. A simple one-dimensional model was presented[3] that accurately reproduces the fiber response. Using electro-deposited palladium films, a seven-fold enhancement in hydrogen solubility is observed, as well as a deeply bound site which is already saturated by hydrogen in the parts-per-million concentration range. The data suggest that this site may be due to hydrogen binding to a free surface of palladium.

A related type of hydrogen sensor has been developed in which the sensing element is a palladium wire, mechanically attached to a monomode optical fiber[4] (Figure 3). In the presence of hydrogen, the Pd is converted to Pd hydrides, with consequent changes in the physical dimensions of the wire. In turn, these dimensional changes produce a longitudinal strain in the optical fiber. The strain is transduced to a phase retardance in a light beam guided by the fiber and detected interferometrically. In a nitrogen atmosphere, the sensor works over the 0 to 100 Pa pressure range (0 to 0.75 torr) with an accuracy of ±2 Pa.

Another type of hydrogen sensor was obtained[5] by depositing palladium/WO_3 layers on a $LiNbO_3$ single mode waveguide. WO_3, in being an electrochromic material, changes its color when it absorbs hydrogen. The palladium functions as a reducing agent in that it forms a proton and an electron from gaseous hydrogen. The sensor is operated at 1300 nm and is capable of detecting 2000 ppm hydrogen at room temperature. Figure 4 shows the optical arrangement and a cross-section through the transducer element.

The bulk photoluminescence (PL) of certain Schottky-type photodiodes constructed from Pd and CdS is sensitive to molecular hydrogen. Exposure to nitrogen/hydrogen mixtures significantly enhances PL relative to PL in air. In case of cadmium-sulfoselenide diodes, hydrogen also changes the spectral distribution. It was suggested to use the device for optically determining hydrogen,[5] but its selectivity obviously is poor (see Section XII herein).

III. METHANE AND RELATED GASES

Methane is routinely determined in colleries, offshore installations, and various mines. It is highly explosive when present in air at levels above 5%, but otherwise rather unreactive, so that no reversible indicator-mediated detection schemes are known. It can be detected via direct fiber absorptiometry using plain fibers, or via changes in refraction index after surface absorption. Detection limits of 500 ppm are desirable. Methane also is a major constituent of

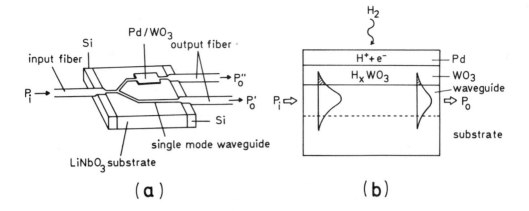

FIGURE 4. Hydrogen sensor based on attenuation of light in a hydrogen-sensitive Pd/WO$_3$-coating. Left: schematic of set-up; right: structure of waveguide and schematic representation of evanescing tail of the electromagnetic wave. (From Nishizawa, K. and Yamazaki, T., Jpn. Kokai, 60, 209, 149. With permission.)

city gas. Fiber optic gas detectors are inherently safe and therefore can serve to detect gas leaks. The aspect of making distributed measurements of flammable or explosive gas concentrations seems to be a highly attractive feature in these considerations.

The formation of alkane hydrates by interaction with water molecules has been exploited to detect alkanes via the technique of multiple total internal optical scattering.[7] The hydrates are formed at the glass surface, and the change in refractive index of the coating causes a change in reflectance characteristics. The instrumental system employed utilizes LED light sources (operated at 560 and 660 nm) and different scattering intensities. For methane, concentrations up to about 1000 ppm may be detected at atmospheric pressure and room temperature.

When platinum-coated fibers are exposed to hydrocarbons in the presence of oxygen, an exothermic reaction of the hydrocarbon with oxygen will occur at the surface. The resultant heat of reaction can be transduced to a phase retardance in a fiber-guided light beam and detected by interferometry. A 100-nm monomode fiber sensing element was used in a typical experiment.[8] Tests have been performed in an oxygen (or air) atmosphere at laboratory temperature. About 300 Pa of hydrocarbon gas appear to be detectable. The experimental arrangement is similar to the one shown in Figure 3, except that palladium has been replaced by platinum.

Infrared-absorption-based detection schemes for alkanes were simultaneously presented at a conference in 1984. Stueflotten et al.[9] developed a detector for remote monitoring of explosive gases, e.g., methane, on offshore installations. The detector comprises three main units: a microcomputer-based signal control and processing unit, a fiber optic sensor, and an optical fiber cable module. The system is based on NIR absorption measurements. At the same time, Inaba et al.[10,11] presented a similar scheme that had been described briefly already earlier. Methane and related gases (such as propane) were detected via plain fiber absorptiometry using fiber links as long as 20 km. Low-loss fibers, a 50-cm long compact absorption cell, and bright LEDs operated at 1.34 and 1.61 μm resulted in detection sensitivities for 25% of the lower explosive limit (LEL), with a detection time of 3 s for the 1.66 μm band. Propane was detected in air at 1.684 μm with a detection limit of 14% of the LEL.

Methane also has been detected via its optical absorption which coincides with the 3.39-μm line of a He–Ne laser. In one case, a conventional gas cell was used.[12] In the other,[13] a single fiber is used as both the sensor and the transmission line. However, the price and attenuation of present-day IR fibers will prevent the use of this wavelength in practice. When a fiber of 1.8 μm diameter and 10 mm length is used, the minimum detectable concentration of methane is less than 5% in air. Possible interferents are ethane, propane, and water. The strong 3.3-μm band, in being a C–H vibration band, not only is given by methane, but by all aliphatic hydrocarbons which hence may interfere in methane detection. Its strength is 261 atm^{-1}cm^{-2} at 297 K in the case of methane, and a minimum measurable concentration of 60 ppm methane can be estimated. Water vapor also absorbs at 3.3 μm and may interfere as well. Also, the band shape of the methane band depends on relative humidity.

The measurement of the absorption by methane and other hydrocarbons in the overtone range (1.3 and 1.65 μm) is definitely more promising at present, mainly because the fiber transmission is much better in this range. Since the 1644.9-nm absorption band of methane coincides with the emission line of the erbium:YAG laser,[14] it does present an excellent light source for remote methane sensing. LEDs operating at this wavelength are also available, along with appropriate low-noise photodetectors. If the sensor is to have the required sensitivity, it will be necessary to carefully compensate for variations in source and detector fluctuations, any interferences by humidity, and possible contamination of the sensor cell or surface.

Via its relatively strong absorption band at around 1.33 μm, methane has been reproducibly determined using a 2-km low-loss optical fiber network and compact absorption cells in conjunction with high radiant InGaAsP LEDs. 1 torr methane was detectable.[15] The system also was operated[16] with fiber cables as long as 10 km. Laser diodes of the InGaAsP type were run below the threshold current level as LEDs, and their emission wavelengths of 1.34 and 1.61 μm neatly matched the NIR absorption bands of methane. An experimental set-up for remote measurement of low-level city gas in a remotely located absorption cell is shown in Figure 5. A similar technique was applied[17] to sense 0.25% methane in air over several kilometers with a response time of 1.5 s.

Samson et al.[18] studied the feasibility of remote Raman spectrometry by optical fiber for the measurement of gas concentrations. It was found that the many advantages of Raman spectroscopy are offset by its inherently weak nature. Experimental results were presented for mixtures of methane, nitrogen, oxygen, and carbon dioxide at atmospheric pressure. Theoretical analysis showed that there is an incident wavelength that maximizes the detection of Raman light when using optical fibers. This optimum incident wavelength increases as the length of the optical fiber increases. Presently, the method is limited to short length fibers.

An optical fiber sensing system to monitor methane at a single point in underground coal mines has been constructed and tested.[19] It uses a white light source, NIR-transmissive fibers, and 50-cm path sensor cells at the fiber ends. The device was regularly calibrated and showed a close correlation with a conventional methanometer in the 0.2 to 1.5% range. To protect it from dust with its detrimental effects on the absorption, it was covered with a gas-permeable membrane. The minimum detectable methane pressure depends, of course, on the fiber length and the operational wavelength (Figure 6).

IV. CARBON MONOXIDE

Carbon monoxide is highly toxic, with a TLV of 50 ppm. It is an important gas to monitor in environmental control systems, in the steel making and coal mining industries, and in parking stations and tunnels. Like the case of methane, CO has a number of useful IR and NIR absorptions. Thus, the band at 4.66 μm with its absorbance of 260 atm^{-1}cm^{-2} (S.T.P.) may be used (by analogy to the tunable diode laser technique employed with methane). A nonfiber approach was shown to exhibit fast response and good sensitivity.[20] Fluoride fibers are required in this wavelength range. However, very long optical path lengths (20 cm) are required in order to obtain necessary detection limits. Assuming the accuracy in absorption measurements to be 1%, the minimum measurable concentration is in the order of 20 ppm. The

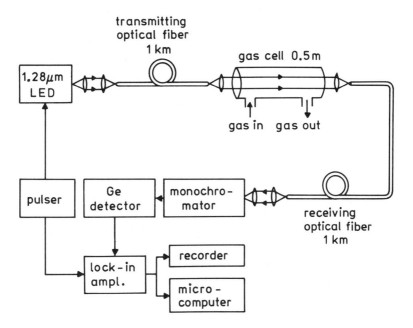

FIGURE 5. Diagram of an experimental set-up for remote absorption measurement of low-level city gas in a remotely located absorption cell using 2-km long, very low-loss silica finers. (From Alarcon, M. C., Ito, H., and Inaba, H., *Appl. Phys.*, B43, 79, 1987. With permission.)

water absorbance slightly overlaps the CO band. A patent[21] describes the fiber optic detection of CO in blood by measuring its absorption at around 5.13 µm, using a second, CO-independent line (e.g., of wavelength 5.05 µm) as a reference signal. In the near-IR region, carbon monoxide has three distinct bands at around 2.2 µm, but this band has not been used so far for sensing purposes.

In a different approach, the optical fiber may be used to guide light to a CO-sensitive reagent immobilized on a porous, translucent glass disc.[22] The absorption of the disc varies in accord with the CO level. A second fiber collects transmitted light. Sensor response was recorded as the ratio of the return signals at 1.4 and 1.7 µm, calculated after correction of the output from the incandescent light source. Equilibrium responses were reversible, and the dynamic range was within the desired 0 to 20 ppm. The technique of optically detecting carbon monoxide using impregnated reagent systems on porous substrates has been described in a patent.[23]

V. CARBON DIOXIDE

Traditionally, gaseous CO_2 has been assayed via infrared absorptiometry, or electrochemically by measuring changes in the pH of a buffer solution as a result of varying CO_2 partial pressure above the solution. IR absorptiometry also has been applied to optical sensors. CO_2 has a strong infrared absorption band extending from 4.2 to 4.4 µm. A patent[21] describes the fiber optic detection of CO_2 in blood by measuring its absorption at around 4.3 µm, using a second, CO_2-independent line (e.g., of wavelength 5.05 µm) as a reference signal.

The IR approach is difficult to adapt to aqueous CO_2 solutions in practice, so that the method with a pH sensor as a transducer (which is the established method in electrochemical CO_2 sensing) has been adapted to optical fiber techniques. The response of a pH sensor to CO_2 occurs as a direct result of the proton concentration in the sensitive layer which is related to the concentration of CO_2 through the following series of chemical equilibria:

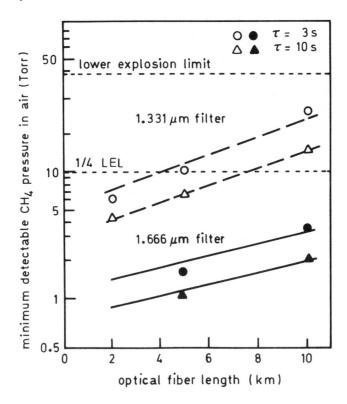

FIGURE 6. Measured minimum detectable methane pressure in air, vs. optical fiber length in the remote measurement employing the 1.34-μm InGaAsP and the 1.64-μm InGaAs light-emitting diodes, respectively, in conjunction with a 50-cm absorption cell. (From Chan, H., Ito, H., Inaba, H., and Furuya, T., *Appl. Phys.*, B38, 11, 1985. With permission.)

(1)	CO_2 (aq) + H_2O	$\rightleftharpoons$	H_2CO_3	(hydration)
(2)	H_2CO_3	$\rightleftharpoons$	$H^+ + HCO_3^-$	(dissociation, step 1)
(3)	HCO_3^-	$\rightleftharpoons$	$H^+ + CO_3^{2-}$	(dissociation, step 2)

These are governed by the following equilibrium constants:

$$K_h = [H_2CO_3]/[CO_2](aqu) \rightleftharpoons 0.0026 \tag{1}$$

$$K_1 = [H^+][HCO_3^-]/[H_2CO_3] \rightleftharpoons 1.72 \cdot 10^{-4} \tag{2}$$

$$K_2 = [H^+][CO_3^{2-}]/[HCO_3^-] \rightleftharpoons 5.59 \cdot 10^{-11} \tag{3}$$

In most studies on the CO_2 sensors, the total analytical concentration of carbon dioxide, i.e., $[CO_2]_{aq} + [H_2CO_3]$, has been related to the response.

In the first optrode for CO_2, Lübbers and Optiz[24] followed the changes in fluorescence of a membrane-covered solution of 4-methylumbelliferone in 1 μM bicarbonate buffer as a function of CO_2 partial pressure in an arrangement shown in Chapter 1, Figure 2(b). This so-called internal buffer (composed of buffer and indicator) was covered with a 6-μm PTFE membrane which is permeable to CO_2, but impermeable to protons and other ionic species. The ratio of fluorescence intensity at 445 nm measured under excitation at 318 nm and 357 nm was related to pressure.

The optrode showed good reproducibility in the 0.1 to 10 pK_a CO_2 range, with a response time of 3 to 4 s for 90% of the final value. It can be accelerated by addition of the enzyme carbonic anhydrase, which catalyzes the establishment of Equilibrium 1. The short response is in striking contrast to the 0.5 to 2.0 min value of the CO_2 electrode and many other CO_2 optrodes. When coupled to a fiber optical system it allows the transcutaneous measurement of CO_2 pressure.[25] The same principle was applied to construct a compact instrument ($5 \times 6 \times 14$ cm in size) for CO_2 consisting of a blue LED as a light source, a longwave absorbing indicator (HPTS) dissolved in bicarbonate and covered with a PTFE layer, two optical filters, and two photodiodes for light detection.[26]

Zhujun and Seitz[27] used HPTS in bicarbonate, covered with a silicone membrane, to sense CO_2. Fluorescence is measured with a bifurcated fiber system. Complete response occurs within a few minutes. Both sulfite and sulfide were found to interfere (probably as their membrane-diffusible forms SO_2 and H_2S). Actually, they could be sensed this way, if CO_2 does not interfere. The equation used to relate the CO_2 partial pressure to hydrogen ion concentration is

$$[H^+]^3 + N[H^+]^2 - (K_1 \cdot C + K_w)\ [H^+] = K_1 \cdot K_2 \cdot C \tag{4}$$

where N is the internal bicarbonate concentration, K_1, K_2, and K_w are the dissociation constants of carbonate acid (steps 1 and 2; see Equations 2 and 3) and water, respectively. C is the analytical CO_2 concentration including both hydrated and unhydrated CO_2. It was shown that, within a limited range, there is linearity between CO_2 pressure and $[H^+]$ according to

$$(H^+) = K_1/CN \tag{5}$$

In practice, the internal HCO_3^- concentration should be such that the CO_2 concentrations of interest yield pH changes in the range of the pK_a of the indicator. In case of the widely used HPTS with its pK_a of 7.3 the external pCO_2 should adjust a pH between 6.5 and 8.0 in the internal buffer.

Heitzmann[28] as well as others[29] have prepared CO_2-sensitive fluorescent membranes by soaking cross-linked polyacrylamide beads with a solution of HPTS in bicarbonate, and embedding them in silicone rubber (Chapter 3, Figure 21). The response to CO_2 was varied by adding different amounts of bicarbonate, carbonate, and HPTS, all of which act as buffers. The poly(acrylamide) beads may be omitted, so that an emulsion of the HPTS/carbonate solution in silicone rubber is obtained. These sensors have excellent long-term performance, but when stored in air or other media with low pCO_2 they tend to become destabilized. It takes several hours to obtain a stable baseline when exposed to higher pCO_2 levels again. This is probably due to some dehydration and also shortage of water molecules available for Reaction 1 which, in turn, results in some contraction and expansion of the droplets and beads because of osmolarity effects.

A typical response curve of a CO_2 sensor is shown in Figure 7. Here, a 20-mM bicarbonate solution acts as the internal buffer. With a given indicator, it is the choice of buffer that primarily determines the slope of the response curve which can be linearized as shown in Chapter 3, Figure 20. The slopes of the graphs also are governed by the pK_a of the indicator (also see Section E) and the total ionic strength of the internal buffer. Unfortunately, high buffer concentrations result in very long response times. These are further prolonged with increasing thickness of the sensor layer (typically 20 μm), additional black poly(tetrafluoroethylene) covers (which serve as optical isolations), slow kinetics of the hydration, and — in particular — slow dehydration of CO_2. For the system described above, the response for a change from 0 to 5% CO_2 was 15 s for 90% of the final value. The reverse change required approximately 34 s. It should be noted that most other optical CO_2 sensors described so far have a much slower response.

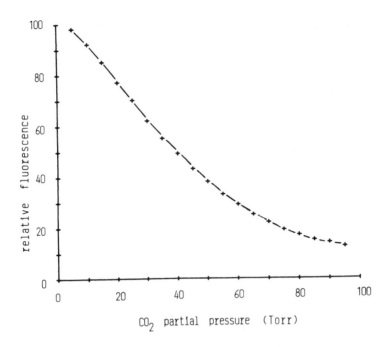

FIGURE 7. Typical plot of signal intensity vs. pCO_2 as obtained with the sensor in Chapter 3, Figure 21, having an internal buffer containing 20 mM bicarbonate and 5 mM 1-hydroxypyrene-trisulfonate. Data extracted from Reference 29. Linearization of the response curve can be achieved by plotting signal intensity in the absence of CO_2 (S_0) over signal intensity at a given pCO_2 (S_x) vs. pCO_2 and is shown on Chapter 3, Figure 20.

A CO_2 sensor with an indicator directly immobilized on the fiber has also been described.[30] A fluorescent dye was covalently bonded onto the glass fiber tip, so the miniature size of the sensor was preserved. The fiber was then coated with a silicone-carbonate copolymer which rejects protons, whereas CO_2 can pass. Direct casting of the polymer onto the sensor chemistry imposed major problems, and finally it was decided to use a preformed cover membrane. When two sensing layers with different spectral properties and being selective for O_2 and CO_2, respectively, are attached to the end of a fiber, a single fluorosensor for both species can be obtained.[31] Blue excitation results in two fluorescences: the green fluorescence reports CO_2, while the red fluorescence reports pO_2. A cross-section through an O_2/CO_2 sensor layer was shown in Chapter 10, Figure 22, and a typical response curve in Chapter 10, Figure 24.

Vurek and colleagues[32] devised an absorbance-based CO_2 sensor that relies on the principle of a previously designed pH sensor. An isotonic solution of salt, bicarbonate, and phenol red was covered with a CO_2-permeable silicone rubber membrane. The device uses two fibers, one carrying the input light, the other reflected light. The sensor responds over the physiological range and its performance was demonstrated *in vivo*. Similarly, a fluorescein-based CO_2-sensitive system was reported by Hirschfeld et al.[33]

Leiner et al.[34] have developed planar CO_2-sensitive chemistries for mass fabrication. Polyester membranes served as planar supports onto which a pH-sensitive material was fixed, using a proton-permeable glue. After soaking the cellulosic material with hydrogel and soaking it with bicarbonate buffer , the membrane is finally covered with a CO_2-permeable silicone-polycarbonate copolymer. CO_2, permeating into the internal buffer, changes the internal pH and hence causes fluorescence intensity to change. The sensing membranes can be prepared in large sheets that later can be punched into small (I.D. 1 to 5 mm) spots and placed at the distal end of a fiber. They also can serve as as sensor sheets in measurement of surface pCO_2 of the skin. Given the low costs for the fabrication of such sensor spots, the main

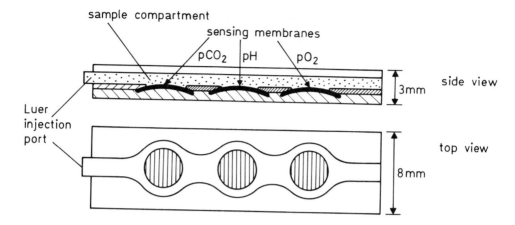

FIGURE 8. Top and side view of a disposable kit for determination of pCO_2, pO_2, and pH in blood. The sample is injected through the Luer port. Three sensing spots (O.D. 4 mm) are mechanically fixed on an injection molded support and have chemistries sensitive to CO_2, O_2, and pH, respectively. The kit is then placed at the end of three fiber bundles which read the fluorescence intensity inside an instrument.

application of these sensors is likely to be in disposable kits for determination of CO_2 along with oxygen and pH in blood. Figure 8 shows the design of such a kit.

Another fiber optic sensor for carbon dioxide was constructed without an inner buffer solution: a dispersion of fluorescein in poly(ethylene glycol) was deposited at the distal end of an optical fiber.[35] Evaporation of the solvent is thus negligible. The response range is from 0 to 28% (v/v) for CO_2, with a detection limit of 0.1%. Full response is achieved within 10 to 20 s. The outer membrane, about 10 μm thick, is composed of poly(ethylene glycol)s with molecular weights of 200 and 1540 Da, respectively, in a 20:80 (w/w) ratio.

A recent patent application[36] describes a method for fluorimetrically sensing CO_2 (and pH) by exciting a pH-insensitive fluorophore whose fluorescence overlaps the absorption of a pH-sensitive fluorophore. The ratio of fluorescences emitted by the two fluorophores is said to be an accurate and stable measure for pCO_2. Two papers related to the performance and selectivity or electrochemical CO_2 sensors are of interest in this context, because they also apply to fiber optic CO_2 sensors based on pH changes in internal buffers.[37,38]

Öhler et al.[39] use the photoacoustic technique to sense carbon dioxide produced in an enzymatic reaction, because CO_2 has an absorption in the IR separated well enough from other bands to be detected with sufficient selectivity. The photo-acoustic cell was separated from the sample liquid by a gas-permeable membrane which provides good acoustical decoupling. The sound produced by the modulated light was detected with a microphone placed in proximity to the sample cell. CO_2, formed from urea by urease immobilized on glass beads, reported the urea concentration range from 10 to 100 mM. Further development of the method is expected to improve the sensitivity by one to two orders of magnitude.

A reflectance based CO_2 sensor has been used in an instrument for detection of microbial activity manufactured by Organon Teknika (Durham, NC). It consist of an LED compatible dye chemistry placed in contact with the blood sample. Upon bacterial contamination and subsequent growth of bacteria, CO_2 is produced and causes the pH of the internal buffer to fall. The resulting change in the color of the indicator is measured by the instrument. It can defect traces of microbial contamination in blood cultures.

VI. AMMONIA

Three major sensing schemes are known for ammonia. In the first, the absorption of light in the NIR by ammonia is exploited in plain fiber sensing. This approach is not very sensitive, and response depends on humidity, but the sensor is simple in design and displays excellent stability. Absorption can be measured in both the transmission mode (in a gas cell) or by the evanescent wave technique. The latter, however, even more strongly depends on the relative humidity of the gaseous samples. In the second approach, ammonia is reacted with a dye such as ninhydrin to yield a purple coloration. This is an irreversible reaction, so that the "sensor" actually is a probe. In the third approach, the basic properties of ammonia are exploited: it is capable of changing the color pH indicators placed on the waveguide. All three sensing schemes have been realized experimentally.

In an NIR absorption spectrometric method,[40] radiation of wavelength 1.18 to 1.22 μm is conducted to a sample cell. Transmitted light is collected by a second fiber and its intensity is determined. A beam having a wavelength outside the above range is used as the reference, and the intensity of the two beams is measured to obtain the ammonia concentration in the gas. Other operational wavelengths[41] include 1.28 to 1.32, 1.46 to 1.56, and 1.61 to 1.67 μm. Ammonia also absorbs the CO_2 laser line which, however, is transmitted by certain fibers only. Silver halide fibers were used to transmit the 10.6-μm radiation from the laser to an absorption cell and farther to an IR detector.[42] The absorption of the line by ammonia gas was measured in the remote cell. For a cell with a pathlength of 19.5 cm, the minimum detectable concentration of NH_3 in air was 35 ppm.

The well-known reaction between ninhydrin and ammonia producing a purple coloration was utilized to develop an ammonia probe. The basis for this analytical technique is the partial attenuation of light signal traversing a cylindrical optical waveguide by successive total internal reflections. A fiber rod was covered with a solution of ninhydrin in poly(vinyl pyrrolidone) and the changes in the absorption of the evanescent wave can be followed.[43] Concentration and relative humidity both determine the slope of the transmission curves. The probe works irreversibly and is able to detect ammonia down to the 50 ppb range.

Another ammonia probe specifically designed for detection of NH_3 in the headspace of bioliquids is also based on the evanescent wave technique and can be applied to vapor phase determination of ammonia in blood and serum.[44] It utilizes the ninhydrin reaction occurring in the polymer coating of the fiber, and the resulting color change is monitored by total internal reflection. The probe is applicable to clinical determinations normally carried out in the vapor phase, but works irreversibly. A linear relationship exists between absorbance and ammonia concentration in the clinically useful range of 0 to 4.0 μg·ml⁻¹. Comparison with the reference method showed the correlation coefficient to be 0.92.

A reversible optical waveguide sensor for ammonia vapors introduced more recently[45] consists of a small capillary glass tube fitted with an LED and a phototransistor detector to form a multiple reflecting optical device. When the capillary was coated with a thin solid film composed of a pH-sensitive oxazine dye,[46] as shown in Chapter 3, Figure 14, a reversible color change occurs upon contact with ammonia. The instrument was capable of reversibly sensing ammonia and other amines. Vapor concentrations from 100 to below 60 ppm ammonia were easily and reproducibly detected. A preliminary qualitative kinetic model was proposed to describe the vapor-film interactions. The method was applied to design a distributed sensor for ammonia.[47]

Ammonia sensors based on the same principle as electrochemical ammonia sensors (that is, the change in the pH of an alkaline buffer solution) have been reported by various groups: Arnold and Ostler[48] followed the changes in the absorption of an internal buffer solution to which p-nitrophenol was added. Ammonia passes by and gives rise to an increase in pH which causes a color change of the indicator to occur. Wolfbeis and Posch[49] entrapped a fine emulsion of an aqueous solution of a fluorescent pH indicator, which simultaneously may act as a buffer, in silicone rubber (Chapter 3, Figure 21). Alternatively, 0.001 M aqueous ammonium chloride may be used as internal buffer. The buffer strength strongly determines both

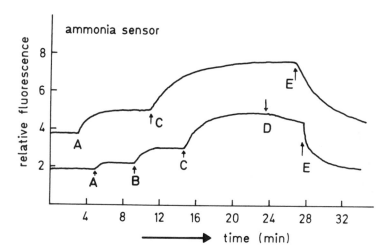

FIGURE 9. Relative signal change and response time in an ammonia sensor based on changes in the pH of an internal buffer. Curve 1: 1-naphthol-4-sulfonate in ammonia buffer (0.01 M NH$_4$Cl in 1 mM NaOH); curve 2: 0.3% 1-naphthol-4-sulfonate in water, adjusted to pH 8.2. Starting with water, the membranes were exposed to the following ammonia solutions: A, 0.1 mM; B, 0.5 mM; 1 mM. Washing was done with pure water (d) and 0.1 M hydrochloric acid (E). (From Wolfbeis, O. S. and Posch, H. E., *Anal. Chim. Acta*, 185, 321, 1986. With permission.)

response time and slope of the response curve. Detection limits are in the order of 5 to 20 μM, and full equilibrium is established very slowly (Figure 9), particularly in the back direction and with aqueous solutions.

In an optimization of these sensors, the various experimental parameters that affect both steady-state and dynamic response characteristics were considered.[50] Figure 10 shows, for instance, the effect of indicator pK_a on the response curve. The use of multiple indicators to enhance steady-state performance was also demonstrated. A sensor for wastewater analysis was developed, and data were compared to those of an ammonia electrode. The use of ammonia-selective optrodes as transducers in enzyme-based biosensors is described in some detail in the chapter on biosensors.

Another type of NH$_3$-sensitive sensor was obtained by immobilization bromothymol blue in a hydrophilic polymer and measuring changes in reflectance induced by ammonia in the gas phase.[51] The working range was from 1.5 to 30 mM and possible interferents were investigated. All kinds of ammonia sensors have a slow response and even slower recovery. The theory of the analyte-to-signal ratio as a function of buffer and analyte concentration is related to the one derived for CO$_2$, except for the fact that pH changes go into other directions, a single protonation step is involved, and pK_a values are different.

Shahriari et al.[52] developed a new porous glass for ammonia detection whose structure imparts a high surface area to the fiber core. Ammonia vapors penetrating into the porous zone pretreated with a reversible pH indicator produce a spectral change in transmission. The resultant pH change is measured by in-line optical absorbance and is said to be more sensitive than sensors based on evanescent wave coupling into a surrounding medium. The signal can be related to the ambient ammonia concentration down to levels of 0.7 ppm (Figure 11). In order to speed up response time, a porous plastic material exhibiting very high gas permeability and liquid impermeability was used in another type of ammonia sensor.[53] The porous plastic fibers were prepared by copolymerization of a mixture of monomers (methyl methacrylate and triethylene glycol dimethyl acrylate) which can be cross-linked in the presence of an inert solvent (such as octane) in a glass capillary. After thermal polymerization, the plastic fibers were pulled out of the capillaries and used in the sensor arrangement shown in Figure 12.

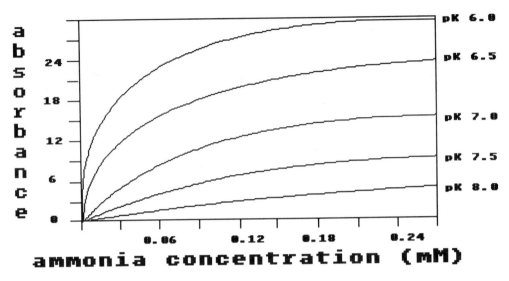

FIGURE 10. Effect of indicator pK_a on the mathematically simulated response of an ammonia sensor based on changes in the absorption of a pH indicator in the internal solution. Obviously, there is a trade-off between dynamic range and resolution within this range. (From Rhines, T. D. and Arnold, M. A., *Anal. Chem.*, 60, 76, 1988. With permission.)

Rather than immobilizing the pH-sensitive dye at the end or on the fiber, it has been suggested[54] that the reagent be delivered under direct control to a sensing volume of approximately 400 nl located at the probe tip. Using the pH-indicator bromothymol blue as the reagent, the sample ammonia concentrations are related to modulations in light intensity with a lower limit of detection of 10 ppb. Compared with previous fiber optic ammonia sensors, the ability to reproducibly renew the reagent is said to have resulted in improvements with respect to response and return times, probe-to-probe reproducibility, lifetime, and flexibility of use. However, the protective and selectivity-imparting properties of the polymer cover are lost in this design.

Rather than entrapping a pH-sensitive dye in silicone, we also have exploited the well-known color reaction that occurs between an alkaline copper(II) solution containing tartrate anion as a complexing agent to prevent copper hydroxide to precipitate. The faintly blue solution of a copper Sulfate Solution entrapped in silicon with an absorption maximum beyond the visible turns deeply blue after exposure to ammonia. The increase in absorption at the emission wavelength of the yellow LED (590 nm) is a direct parameter for ammonia concentration. The probe is not very sensitive (>1 mM) and has a slow response, but is fully reversible.

Generally, all types of ammonia sensors based on pH effects also respond to other uncharged amines such as methylamine, pyridine, or hydrazine because these species have alkaline properties too and can pass almost all polymers used in the respective sensing materials. Second, all acidic gases, including CO_2, SO_2, and HCl, and also vapors of organic acids such as acetic acid will interfere. Therefore, the specificity of such sensors is limited. One way to overcome interferences by acidic species is to make the sample strongly alkaline (if possible) which converts the acids into their nondiffusible salt forms. Alternatively, ammonia may be converted into ammonium ion and assayed as such.[55]

VII. NITROGEN OXIDES

Nitrogen oxides, namely NO and NO_2, are major primary pollutants, so their monitoring is of utmost importance. NO_2 has an absorption maximum at around 405 nm, but the ban is

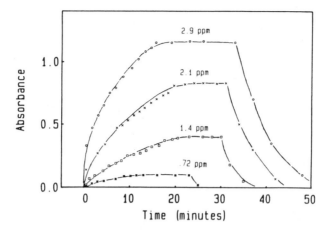

FIGURE 11. Typical porous sensor response curve at different ammonia concentrations. (From Shahriari, M. R., Zhou, Q., anf Sigel, G. H., *Opt. Lett.*, 13, 407, 1988. With permission.)

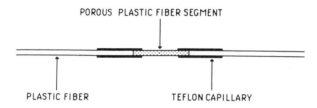

FIGURE 12. Configuration of a porous plastic waveguide used in a sensor for ammonia. The fiber segment is doped with an ammonia-sensitive dye and exposed to the ammonia-containing atmosphere. (From Zhou, Q., Kritz, D., Bonnell, L., and Sigel, G. H., *Appl. Opt.*, 28, 2022, 1989. With permission.)

very broad and extends to about 600 nm. Consequently, it can be determined by absorptiometry using, for example, a blue LED as the light source. From its absorption properties, it can be calculated that the detection limit of NO_2 in a 20-cm cell is in the order of 200 ppm, and the resolution is in the order of ±2 ppm. NO and NO_2 are in thermodynamic equilibrium in the presence of oxygen. Unfortunately, NO_2 is the minor fraction in NO/NO_2 mixtures, the ratio being 90:10 (v/v) in a typical automobile exhaust. NO has no measurable absorption in the visible, but conceivably may first partially or quantitatively be converted into NO_2 by an oxidation reagent, with UV light, by lowering temperatures, or increasing pressure.

The feasibility of an optical fiber system for differential absorption measurement of NO_2 molecules in the visible spectrum using an argon laser has been demonstrated[56] (see Figure 2). The 488-nm laser beam was transmitted to a remotely located multireflection chamber (White cell) of 1-m mirror spacing with optical fiber coupling structure. The output beam was then returned to the transmitter/receiver location by the 20-m or 500-m multimode fibers. After measuring the absorption by NO_2 at 488 nm, the absorption at 514.5 nm was determined as well and used to calculate the differential absorption. The device was used to monitor NO_2 in car exhaust gas, with a detection limit in the order of 20 ppm in this configuration. It was assumed that by using a 1-km White cell, the detection limit could be lowered to 0.03 ppm.

Nitrogen monoxide (NO) has a fundamental IR absorption band at 5.32 μm, and a second harmonic at 2.69 μm. Given the strong absorption of water in the 5.3-μm region and the poor transmissivity of most fibers for IR light, the NIR band is probably the better choice for remote spectrometry using a plain fiber. It can be calculated that roughly 20 ppm NO could be detected in a typical 20-cm cell. Nitrogen dioxide (NO_2) is an asymmetric top molecule with rotation-vibration bands at 4.51, 3.8, 3.44, and 3.09 μm at room temperature. The band width of the 3.44-μm band is 60 cm^{-1} (70 nm). If NO_2 is measured in this spectral region, potential interferences by hydrocarbons and HCl must be taken into account.

NO_2 absorbs the argon ion laser 496.5-nm line. When excitation is chopped, an acoustic field can be excited and thus enables photoacoustic detection of the gas.[57] The periodically varying cell pressure is then coupled, via an aluminum foil, to the fiber wound onto the cell. This results in a periodic variation of the optical pathlength difference between sensor fiber and reference fiber, which is detected in a Mach-Zehnder interferometer. The sensitivity of the device was found to strongly depend on the thickness of the foil.

Nitrogen oxides have also been reported to be dynamic quenchers of the fluorescence of certain polycyclic aromatic hydrocarbons, but a detailed report on both the quenching efficiency and the theoretical limits of detection has not appeared so far. Selectivity, anyway, is likely to be poor because oxygen and sulfur dioxide (which are present in most samples) are known to be potent quenchers, too. An interesting observation[58] is the infrared fluorescence from NO_2 after excitation with a dye laser at 400 to 500 nm. Eventually, this could be exploited in a direct NO_2 sensor.

VIII. HYDROGEN SULFIDE

Hydrogen sulfide is produced in many bacterial processes. Its determination plays a role in fermentation control, bacterial activity proof, bacterial identification via pattern recognition, and food quality assessment. Direct absorptiometry of H_2S (without the use of fibers) has been performed in the infrared at 4.13, 3.83, 3.73, and 2.64 μm. The band strengths are of the same magnitude as water vapor (197 atm^{-1}cm^{-2}), and about 110 ppm seem to be detectable. The NIR absorption range is probably more selective. All published fiber methods for optically sensing H_2S rely on either its weakly acidic properties, or on color reaction given by it, most of which, however, are irreversible, so that probes rather than sensors are obtained.

Thus, when lead hydroxide, dissolved in a solution containing ethylenediamine-tetracetic acid (EDTA) and placed at the fiber end is brought into contact with hydrogen sulfide, the following reaction occurs:

$$Pb^{2+} + 2\ OH^- + EDTA + H_2S \longrightarrow PbS + 2\ H_2O + EDTA$$

Black PbS is formed which precipitates and darkens the solution. The decrease in pH due to consumption of hydroxyl ions is monitored via the decrease in fluorescence intensity of the added pH indicator fluorescein.[59] In such a probe-type optrode, 2.3 ppm H_2S caused a linear 30% signal reduction over a period of 3 h.

When paper is impregnated with lead acetate, it assumes a dark color when in contact with H_2S. This, in turn, changes its reflectance (as measured at 580 nm). Levels as low as 50 ppb H_2S could be assayed[60] with high reproducibility within 10 s, but the response depends on relative humidity. The coloration is irreversible, so the paper has to be slowly moved in case of continuous sensing. Other methods for sensing H_2S are based on the reducing properties of sulfide or hydrogen sulfide.[61,62] These devices are, however, probes rather than sensors because they act irreversibly. The technique of optically detecting hydrogen sulfide using impregnated reagent systems on porous substrates has been described in a patent.[23]

In the only reversible approach,[63] quaternized acridinium was immobilized on cellulose. In slightly alkaline medium, hydrogen sulfide adds to the strongly fluorescent dye and renders it nonfluorescent.[64] Interference by ionic quenchers may be eliminated by covering the sensor with a 4 μm silicone rubber membrane which is permeable to gases but not to ions. The detection limit is in the order of 0.7 mM/l.

Hydrogen sulfide is a weak acid, but the possibility of sensing H_2S via pH changes in an internal buffer solution of slightly alkaline pH (by analogy to CO_2 sensors) has not been realized so far. Such a sensor would, of course, suffer from interferences by many other alkaline or acidic species such as ammonia, sulfur dioxide, or carbon dioxide.

IX. SULFUR DIOXIDE

The strong 101 band of SO_2 at 4.0 µm is used for measuring its concentration in conventional gas cells. There is no obstacle in sight that would prevent the application of the method in combination with IR-transmissive fibers. If the accuracy of the transmission measurement in a 20-cm cell is assumed to be 1%, the smallest measurable SO_2 concentration will be 140 ppm. There are no strong interfering absorption bands in stack gases in the region of the 101 band. The lines of the rotation-vibration bands already overlap strongly in atmospheric pressure. By using a single absorption line in the concentration measurement, it is not possible to improve the accuracy as much as would be possible, for example, as in the case of carbon monoxide.

Continuous determination of SO_2 in air can be performed[65] by fluorimetry owing to its strong intrinsic fluorescence at about 330 nm. This approach has not been adapted to fibers so far. However, in complex samples the method is not very specific. Advantage can be taken of the observation that the fluorescence of benzo(b)fluoranthene and related polycyclic aromatic hydrocarbons (PAHs) is strongly quenched by SO_2. About 70 ppm SO_2 in gases are detectable.[66] The useful range is from 0.01 to 6.0% (v/v) as can be seen in Figure 13. Other gases likely to occur in air were found to be inert in usual concentrations, with the notable exception of oxygen which also acts as a dynamic quencher. Its interference is negligible for SO_2 levels below 6% in air at constant oxygen pressure, because the quenching efficiency of SO_2 is about 26 times higher than that of oxygen. Presumably, nitrogen oxides in higher concentrations also interfere. For varying levels of interferents, a two-sensor technique was suggested.[66] Effects of temperature have not been studied but might impose several problems in practice, too.

An interesting observation[67] is the dramatic increase in fluorescence quenching efficiency when a two-fluorophor system is used in place of a single quenchable fluorophore. Thus, when both pyrene and perylene are dissolved in a silicone rubber matrix, electronic energy is transferred from the first (absorber) to the second (the emitter). The quenching by SO_2 of the energy transfer is more efficient by a factor of three than quenching of either pyrene and perylene alone (Figure 14). The effect has been utilized in another type of SO_2 sensor material which, however, suffers from the same selectivity problems as the above one.

SO_2 is a strongly acidic gas that interferes in all kinds of carbon dioxide and ammonia sensors using a pH optrode as a transducer. However, by proper choice of an internal buffer solution, a sulfur dioxide sensor can be constructed[68] that is highly specific for SO_2. If the internal buffer has a pH lower than 4, CO_2 no longer can interfere. Typical internal buffer compositions are 0.01 M hydrogen sulfite buffer of pH 2 to 3, a pH indicator of pK_a 1.5 to 2.5, and 0.01 M in NaCl. HCl is the only major acidic interferent in this sensor type which, however, lacks the sensitivity of, e.g., a CO_2 sensor and, of course, is affected by alkaline gases.

X. SOLVENT VAPORS

The input grating couplers schematically shown in Chapter 2, Figure 12 were found to be extremely sensitive to humidity and solvent vapors such as acetone and ethanol.[69] This is due to the change in the refractive index of the cover material that acts as a light guide and consequently results in considerable changes in the input efficiencies of the laser. Obviously, this is not an indicator-mediated sensing scheme, but rather relies on changes in the index of refraction of the waveguide.

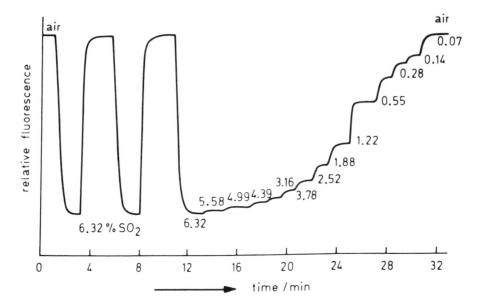

FIGURE 13.   Response time, relative signal change, and reversibility of a fiber optic sensor for sulfur dioxide. (From Wolfbeis, O. S. and Sharma, A., *Anal. Chim. Acta*, 208, 53, 1988. With permission.)

Vapors of polar solvents can be detected on-line by virtue of their decolorizing effect upon certain blue papers as used in thermal printers.[70] Typical vapors that can be detected include ethers, alcohols, esters, and ketones, while hydrocarbons and chlorinated hydrocarbons remain inert (Figure 15). Detection limits (with a device consisting of a yellow LED light source and a phototransistor as the detector) vary from 10 to 1000 ppm, and both conventional flow-through cell sensing (via absorptiometry) and fiber optic sensing (via reflectometry) were found to be applicable. Response times vary from 0.5 to 4 min. The blue material has a red fluorescence whose intensity also varies in accord with the vapor pressure.

Certain triphenylmethane dyes in combination with an acidic component such as bisphenol A have an intense blue color which is decolorized in the presence of vapors of polar solvents.[71] The sensing layer was prepared by dissolving the lactone form of the dye and bisphenol A in PVC and casting it as a 1-μm layer on a solid support. This membrane has been used to optically sense these species by placing it on either the distal end or the core of a waveguide. The decolorization effect was explained in terms of a disturbance of a network of hydrogen bonds in these sensor films. Detection limits as low as a few parts per million were reported, and a small and inexpensive device was constructed. In subsequent work, the type of indicator was varied[72] and the method applied to sense polar solvents dissolved in water.[73] This was achieved by covering the sensing membrane with a gas-permeable but water-impermeable membrane.

Vapors of halothane, a widely used inhalation narcotic, can be detected owing to its quenching effect upon the fluorescence of polycyclic aromatic hydrocarbons.[74] Figure 16 shows the response of such a sensor to halothane in air, nitrogen, and oxygen. Interferences by oxygen can be taken into account by a second sensor sensitive to oxygen only. The two-sensor technique allows the determination of 0 to 4% halothane and/or 10 to 21% oxygen with a relative precision of ±5%. No interferences are reported for CO, CO_2, N_2O, and fluorans. Potential applications include monitoring of (1) halothane in blood (using plastic fiber catheters), (2) oxygen *in vitro* in the presence of halothane, and (3) anesthetic gases in the breathing circuit.

Halothane has also been assayed[75] optically via surface plasmon resonance (SPR). After a theoretical and experimental investigation of the possibilities of using SPR for gas detection,

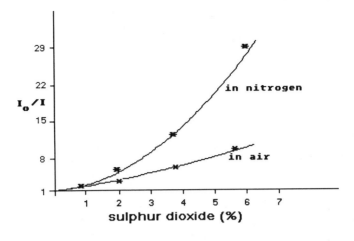

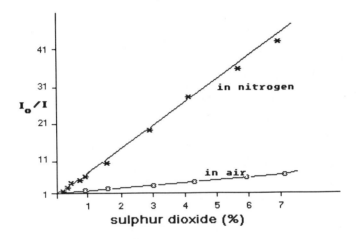

FIGURE 14. Top: Stern-Volmer plot of the quenching of pyrene (dissolved in silicone rubber) by sulfur dioxide in nitrogen and in air. Bottom: quenching of the pyrene/perylene energy transfer system in silicone rubber by the same gases. Note the differences in the curve shapes and in the slope of the graphs. (From Sharma, A. and Wolfbeis, O. S., *Proc. SPIE*, 990, 116, 1989. With permission.)

the authors described an organic layer that reversibly absorbs halothane (see Chapter 2, Figure 13). Specifically, the layer is composed of a 43-nm film of silicone-glycol copolymer on a silver-coated (56 nm) glass substrate. The SPR angle is changed after gas absorption, a fact that can be exploited to sense halothane down to the parts-per-million range. A very good linearity was observed between gas concentration and shift in resonance angle. The signal-to-noise ration in the photodetector was considered the limiting factor in the sensitivity, once the other components are optimized.

XI. HUMIDITY

Surprisingly, no direct IR spectroscopic fiber method for determination of relative humidity (RH) in gases has come to the attention of the author, although water has several absorptions in the IR and NIR, the latter being exploited in sensing, for instance, water in organic solvents. All methods described in the following are based on indicator-mediated sensing schemes.

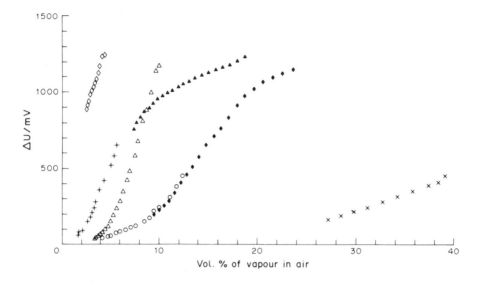

FIGURE 15.   Response of a vapor sensor based on the decolorizing effect of vapors: plot of phototransistor signal on the vapor concentration of several frequently used solvents. (rhomboid), acetone; (x), diethyl ether; (pique), tetrahydrofurane; (triangle), ethyl acetate; (+), ethanol; (o), methanol. (From Possch, H. E., Wolfbeis, O. S., and Pusterhofer, J., *Talanta*, 35, 89, 1988. With permission.)

A simple version for a humidity sensor makes use of the well-known reversible color changes of cobalt dichloride in gelatine (from pink to blue) cast as films on a 12-cm long 600-μm fiber when exposed to humidity.[76] The absorption is measured at 680 nm through the fiber by internal reflection, and RH can be determined at levels between 40 and 80%. Ballantine and Wohltjen[77] describe an optical waveguide humidity sensor that employed the same colorimetric reagent/polymer system on a 9-cm glass capillary. The method is also described in a patent.[78]

The above sensors suffer from poor precision in the <40% RH range. Using the $CoCl_2$ reagent permeated into the porous segments of a porous glass that acts as the fiber core and has a high specific surface enabled[79] the detection of RH down to levels of 1%. The effects of temperature on such a sensor were studied in some detail[80] over the 25 to 90°C temperature range and were found to be substantial. Another sensor type[81] utilizes the effect of water vapor on the fluorescence quantum yield of adsorbed fluorophors. Thus, a silicagel-adsorbed perylene dye, when excited with an LED at 490 nm, shows about 90% reduction in fluorescence in going from 0 to 100% RH (Chapter 10, Figure 15). While this sensor covers the whole humidity range, the response is not linear, and oxygen interferes to some extent. See also Reference 82. In a nonfiber optical sensor for RH, cellulose impregnated with $CoCl_2$ was applied as a sensor material.[83] Two cell substrates were tested. In case of plain cellulose, decreasing the amount of $CoCl_2$ in the cellulose matrix decreases both hysteresis effect and the RH range, over which the material can be used. Acetylation of the cellulose is found to result in an extension of the effective RH ranges, an increase in reproducibility, and a decrease in hysteresis.

A fiber optic sensor for humidity has been developed by Zhu et al.[84] The sensor utilizes a fluorescent dye entrapped within a perfluorinated polymer matrix. Fluorescence intensity increases strongly and linearly with increasing water vapor partial pressure even though the lifetime of fluorescence is simultaneously lowered (!). The response time of the optrode is approximately 1 s and the presence of CO_2 has no detectable effect. Apparently, the immobilized dye, rhodamine 6G, associates with water to form a complex with higher absorptivity. The dependence of the lifetime on emission wavelength revealed the coexistence of multiple excited states for the water-dye system.

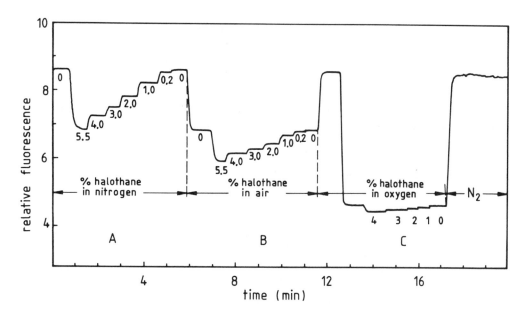

FIGURE 16. Response time, signal change, and reproducibility of a fluorosensor for halothane. (From Wolfbeis, O. S., Posch, H. E., and Kroneis, H. K., *Anal. Chem.*, 57, 2556, 1985. With permission.)

RH changes are known to affect properties such as the reflectivity of thin films of silver or other metals deposited onto glass. A 600-µm PCS fiber was stripped of its coating for 1 to 2 cm at the end, and then plated with 100 nm silver or nickel using vapor deposition techniques.[85] The reflectivity was found to vary by up to 15% over the range 10 to 90% RH, and the effect to be fully reversible. The effect is probably due to changes in metal conductivity and, thence, reflectivity. The response time is within a few seconds. Effects of wavelength variation and type of metal coating were also studied. However, the sensor lacks long-term stability, has a high temperature coefficient, and cannot be operated in a dual wavelength system. Primary, optical type, phase transition humidity sensors which, so far, have not been coupled to fibers, have been reviewed.[86]

XII. MISCELLANEOUS GASES AND VAPORS

Chlorine, bromine, and ozone have absorptions in the visible range that can be used to monitor these species by remote spectrometry, using LEDs as light sources. Since the molar absorbances are rather small, long sample cuvettes have to be employed. The absorption maximum of chlorine is in the UV, but the band extends far into the visible. The absorption maxima of bromine and ozone are at 410 and about 600 nm, respectively, and the bands are very broad. The molar absorbance of the ozone 600-nm band ($4.9 \cdot 10^{-25} m^2$ atm) is about 200-fold weaker than its UV absorbance. However, the blue band matches the emission of yellow LEDs. Assuming the length of the gas cell to be 20 cm, the lower detection limit of ozone is in the order of 1%. A fiberless single beam instrument with a yellow-orange LED as the source and an Si photodiode as a detector has been constructed to monitor ozone in ozone generators for disinfection of drinking water.[87] In a 10-cm optical cell, $5 \cdot 10^{21}$ molecules ozone could be detected in 1 m^3.

Chlorine dioxide, a species employed for paper bleaching, has a weak UV absorption (800 $cm^{-1} M^{-1}$ at 360 nm). Hence, direct fiber monitoring of ClO_2 is impossible with standard fibers. However, in view of the breadth of the band, it was possible[88] to measure, via fibers, the absorption at 470 nm and relate this to the ClO_2 concentration after referencing the signal to the absorption at 540 nm. The bromine spectrum is very broad and extends from 250 to 600 nm. Using a blue LED as a light source, bromine can be monitored by absorptiometry, with a lower limit of detection in the 1000 ppm range in 20-cm cells, and a resolution of about 10 ppm.

The bulk photoluminescence of the aforementioned Schottky-type photodiodes constructed from Pd and CdS is not only sensitive to molecular hydrogen, but also to oxygen, methyl iodide, SO_2, and ammonia,[6] and to iodine.[89] A new class of optrodes with particular sensitivity toward chlorinated hydrocarbons relies on electrical spark or radio frequency excitation (via helium plasma) of the analyte species at the fiber tip which collects its emission.[90] The new probe concept offers the potential for multipoint real-time field sampling and mapping of groundwater contamination plumes or airborne chemical contamination. Figure 17 shows two representative experimental arrangements.

Sulfur *tri*oxide is a secondary pollutant that is made partially responsible for the effect called acid rain and its consequences. Sensing SO_3 at very low levels is therefore highly desirable in environmental sciences. It has a second harmonic (of the μ_3 band) at 3.61 μm, and a combination band at 4.09 μm, but this has not been exploited for sensing purposes yet.

A mechanically and optically simple gas cell for on-line fiber optic absorptiometric monitoring of gases such as UF_6, fluorine, ClF_3, and PuF_6 has been described.[91] With a simple optical pathlength of 7.6 cm, 2 m of fused silica fibers, and UV detection using the mercury lamp emission lines at 254 and 365 nm, detection limits in the 0.01 to 30 torr concentration range were reported. All calibration curves were linear.

Hydrogen chloride (which is produced in the combustion of pvc) conceivably can be sensed via changes in the pH of an internal buffer similar to the CO_2 and SO_2 sensing schemes. Since the pH of the internal buffer can be kept very low, interferences by other acidic gases hardly will be observed. The low pH of the internal buffer, however, will also keep the lower detection limit rather high. HCl also has an IR absorption that makes it a candidate for direct spectrometric determination with plain fibers. The center wavelength of the fundamental band is at 3.465 μm, the strength is 102 $atm^{-1}cm^{-2}$ for $H^{35}Cl$. The low absorbance of the band makes the direct spectrometric assay not very sensitive. Detection limits in 20-cm cells are likely to be in the order of 50 ppm. NIR absorption bands exist at around 1.2 μm.

The fluorescence of Langmuir-Blodgett films of tetraphenylporphyrine (TPP) mixed with a rigid fatty acid was found to be quenched by 1 to 10 ppm of NO_2, HCl, and chlorine.[92] An optical fiber/prism method was used to assess the film response which was reversible for NO_2. Other gases quenched irreversibly, but the effect was made reversible by exposing the film to ammonia.

Hydrogen cyanide has been detected fiber optically by oxidation of the atmospheric sample with chloramine-T impregnated on XAD-7 resin placed in front of the fiber bundle[93] to yield cyanogen chloride. This reacts with picoline and barbituric acid to yield a color change with a maximum absorbance change at around 530 to 565 nm. In air, 1 ppm HCN was detectable within 1 min. Since the method is not reversible, it cannot be used to monitor HCN in air. Nonetheless, the probe is considered to be useful for detection of HCN below its toxic level in a short time. Another detection method is based on the selective reaction of HCN with a picrate dye, which is coated as a film on an optical waveguide.[94]

The strong fluorescence of rubrene and α-naphthoflavone is quenched by traces of iodine and bromine. When a fiber rod with a layer of a 20-ppm rubrene solution in polystyrene attached to its end is immersed into an aqueous solution of bromine or iodine, the halogen diffuses into the polymer membrane. It causes a reduction in the fluorescence intensity to around one-third in the case of iodine.[95] Some metal ions, notably Fe and Cu, also quench rubrene (which is argon laser excitable), but cannot pass into the polymer layer and thus do not interfere.

Gaseous inhibitors of the enzyme acetylcholinesterase (AChE) are used as so-called nerve agents and pesticides. They can be detected in air by virtue of their inhibition of the enzyme:[96]

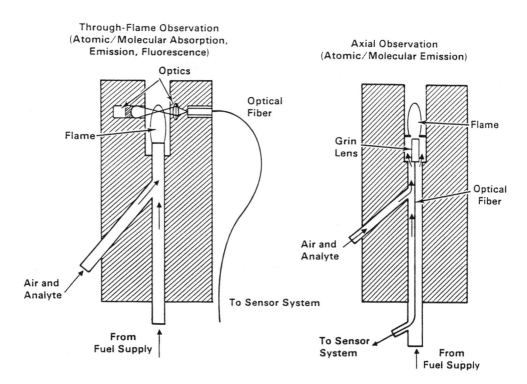

FIGURE 17. Conceptual designs of two spectrochemical emission probes based on flame excitation. (From Griffin, J. W., Olsen, K. B., Matson, B. S., and Nelson, D. A., *Proc. SPIE*, 990, 55, 1989. With permission.)

when AChE is immobilized on polymer beads and is fully active, it can convert a synthetic substrate into a product having a different color. If, however, the enzyme is inactivated, the color will no longer be formed. Figure 18 shows the arrangement used for on-line monitoring of AChE inhibitors in gas. A red dye (the enzyme substrate) contained in reservoir 2 is mixed with buffer from reservoir 1 and is brought into contact with the gas stream to be tested. It passes a column filled with beads containing immobilized AChE where the substrate is converted into a blue product. The fiber optic "sees" the blue dye flowing through the cell. If the enzyme is inactivated, the substrate will no longer be converted into the blue dye, and absorption decreases. The effect has also been used to detect traces of pesticides.[97] The effect of 20 nM of paraoxon (a potent pesticide and inhibitor of AChE) added to the buffer in reservoir 1 is shown in Figure 19.

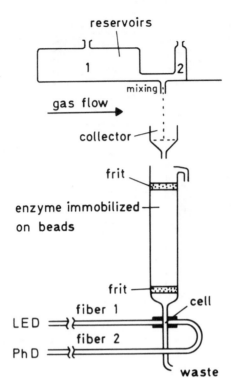

FIGURE 18. Schematic of a remote sensor for gaseous inhibitors of acetylcholinesterase. Reservoirs 1 and 2, respectively, contain buffer and enzyme substrate which, in the enzyme reactor, is converted into a blue product whose absorption is measured via fibers, using a yellow LED as a light source and a photodiode (PhD) as a detector. Note that the whole system is fielf-suited in that it does not need a power supply except for the central light source and detection system which is battery powered. (From Reference 96.)

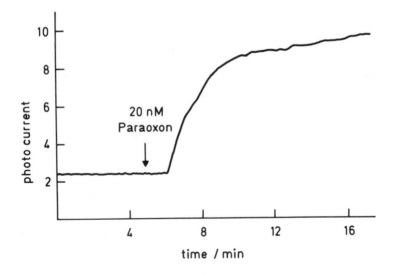

FIGURE 19. Effect of 20 nM paraoxone (a phosphorus type inhibitor of the enzyme acetylcholinesterase) on photocurrent of the sensing device shown in Figure 18. (From Reference 96.)

REFERENCES

1. **Stevens, R. K. and Herget, W. F.,** *Analytical Measurements Applied to Air Pollution Measurements,* Ann Arbor Science, Ann Arbor, MI, 1974.
2. **Butler, M. A.,** Optical fiber hydrogen sensor, *Appl. Phys. Lett.,* 45, 1007, 1984.
3. **Butler, M. A. and Ginley, D. S.,** Hydrogen sensing with palladium-coated optical fibers, *J. Appl. Phys.,* 64, 3706, 1988.
4. **Farahi, F., Akhavan-Leilabady, P., Jones, J. D. C., and Jackson, D. A.,** Interferometric fibre-opric hydrogen sensor, *J. Phys. E,* 20, 432, 1987.
5. **Nishizawa, K. and Yamazaki, T.,** Jap. Kokai 60,209,149; through Chem Abstr. 105:34808, 1986.
6. **Carpenter, M. K., Van Ryswyck, H., and Ellis, A. B.,** Photoluminescent response of palladium-cadmium sufide and palladium-graded sulfoseelenide Schottky diodes to molecular hydrogen, University of Wisconsin Rep. UWIS/DC/TR-85/1, 1985; through Chem Abstr. 104:196121, 1986; also see Eur. Patent Appl. 291,149, 1988.
7. **Guiliani, J. F. and Jarvis, N. I.,** *Sensors Actuators,* 6, 107, 1984; also see U.S. Patent Appl. 709,251, 1985.
8. **Farahi, F., Leilabady, P. A., Jones, J. D. C., and Jackson, D. A.,** Optical-fibre flammable gas sensor, *J. Phys. E,* 20, 435, 1987 and references cited therein.
9. **Stueflotten, S., Christensen, T., Iversen, S., Hellvik, J. O., Almas, K., Wien, T., and Graav, A.,** An infrared fiber optic gas detection system, *Proc. SPIE,* 514, 87, 1984.
10. **Inaba, H., Chan, K., and Ito, H.,** All-optical remote gas sensor system over a 20-km range based on low-loss optical fibers in the near infrared, *Proc SPIE,* 514, 211, 1984.
11. **Chan, K., Ito, H., and Inaba, H.,** All-optical-fiber-based remote sensing system for near IR absorption of low-level methane gas, *J. Lightwave Technol.,* 5, 1706, 1987.
12. **Siato, M., Takizawa, M., Ikegawa, K., and Takami, H.,** *J. Appl. Phys.,* 63, 269, 1988.
13. **Tai, H., Tanaka, H., and Yoshino, T.,** Fiber optic evanescent wave methane gas sensor using optical absorption for the 3.392-μm line of a He–Ne laser, *Opt. Lett.,* 12, 437, 1987.
14. **White, K. O. and Watkins, W. R.,** Erbium laser as a remote sensor for methane, *Appl. Opt.,* 14, 2812, 1975.
15. **Alarcon, M. C., Ito, H., and Inaba, H.,** All optical remote sensing of city gas through CH_4 gas absorption employing a low-loss optical fiber link, *Appl. Phys.,* B43, 79, 1987.
16. **Chan, H., Ito, H., Inaba, H., and Furuya, T.,** 10 km long fiber optic remote sensing of CH_4 gas by near infrared absorption, *Appl. Phys.,* B38, 11, 1985.
17. **Dakin, J. P., Wade, C. A., Pinchbeck, D., and Wykes, J. S.,** A novel optical fiber methane sensor, *Int. J. Opt. Sens.,* 2, 261, 1987.
18. **Samson, P. J., Stuart, A. D., and Inglis, H.,** Fiber optic gas sensing using Raman spectrometry, paper presented at the 14th Australian Conf. Fourier Transform Spectrosc., Brisbane, December 1989.
19. **Samson, P. J. and Stuart, A. D.,** Coal mine methane sensing by optical fibers, in Proc. Australian Conf. Optical Fibre Technology, Hobart, December 4 to 7, 1988, 113.
20. **Sachse, G. W., Hill, G. F., Wade, L. O., and Perry, M. G.,** Fast-response, high-precision carbon monoxide sensor using a tunable diode laser absorption technique, *J. Geophys. Res.,* 92, 2071, 1987.
21. **Manuccia, T. J. and Eden, J. G.,** U.S. Patent 4,509,522, 1985.
22. **Stuart, A. D. and Samson, P. J.,** Optrode sensors for carbon monoxide and relative humidity, in Proc. 13th Australian Conf. Optical Fibre Technology, Hobart, December 4 to 7, 1988, 117.
23. **Goldstein, M., Anderson, T., Meadows, J. H., and Taylor, D.,** PCT Int. Pat. WO 88/05911, 1988.
24. **Lübbers, D. W. and Optiz, N.,** The pCO_2/pO_2 optrode: a new probe for measurement of pCO_2 and pO_2 of gases and liquids, *Z. Naturforsch.,* 30C, 532, 1975.
25. **Lübbers, D. W., Hannebauer, F., and Optiz, N.,** The pCO_2 optode: Fluorescence photometric device to measure transcutaneous pCO_2, *Birth Defects Orig. Artic. Ser.,* 15, 123, 1979.
26. **Optiz, N. and Lübbers, D. W.,** Compact CO_2 gas analyzer with favourable signal-to-noise ratio and resolution using special fluorescence sensors (optodes), in *Proceeds. Intl. Symp. Oxygen Transport to Tissue,* Vol. 4, Bruley, D., Bicher, H. I., and Reneau, D., Eds., Plenum Press, New York, 1984, 757.
27. **Zhujun, Z. and Seitz, W. R.,** A carbon dioxide sensor based on fluorescence, *Anal. Chim. Acta,* 160, 305, 1984.
28. **Heitzmann, H. A.,** U.S. Patent 4,557,900, 1985.
29. **Marsoner, H., Kroneis, H., and Wolfbeis, O. S.,** Eur. Patent Appl. 105,870.
30. **Munkholm, Ch., Walt, D. M., and Milanovich, F. P.,** A fiber optic sensor for carbon dioxide, *Talanta,* 35, 109, 1988.
31. **Wolfbeis, O. S., Weis, L., Leiner, M. J. P., and Ziegler, W.,** Fiber optic fluorosensor for oxygen and carbon dioxide, *Anal. Chem.,* 60, 2028, 1988.
32. **Vurek, G. G., Peterson, J. I., Goldstein, S. W., and Severinghaus, J. W.,** Fiber optic carbon dioxide pressure sensor, *Fed. Proc. Fed. Am. Soc. Exp. Biol.,* 41, 1484, 1982; see also U.S. Patent Appl. 470,920, 1983.

33. **Hirschfeld, T., Miller, F., Thomas, S., Miller, H., Milanovich, F., and Gaver, R.,** Laser fiber optic optrode for real time in vivo blood carbon dioxide level monitoring, *J. Lightwave Technol.*, 5, 1027, 1987.

34. **Leiner, M. J. P., Weis, L., and Wolfbeis, O. S.,** Chemically sensitive optrode membranes for use in blood gas analysis with disposable kits, Unpublished results, 1986-1989.

35. **Kawabata, Y., Kamichika, T., Imasaka, T., and Ishibashi, N.,** Fiber optic sensor for carbon dioxide with pH indicator dispersed in a poly(ethylene glycol) membrane, *Anal. Chim. Acta*, 219, 223, 1989.

36. **Leader, M. J. and Kamiya, T.,** Eur. Patent Appl. 283,116, 1988.

37. **Lopez, M. E.,** Selectivity of the potentiometric carbon dioxide gas-sensing electrode, *Anal. Chem.*, 56, 2360, 1984.

38. **Jensen, M. A. and Rechnitz, G. A.,** Response time characteristics of the pCO_2 electrode, *Anal. Chem.*, 51, 1972, 1979.

39. **Oehler, O., Seifert, M., Cliffe, S., and Mosbach, K.,** Detection of gases produced by biological systems with an enzyme photoacoustic sensor, *Infrared Phys.*, 25, 319, 1985.

40. **Inaba, F.,** Method and apparatus for measuring ammonia gas concentrations, Jap. Kokai 60 86,447, 1985; through Chem. Abstr. 103: 98095, 1985.

41. **Inaba, F.,** Method and apparatus for determination of ammonia gas, Jap. Lokai 59,183,348, 1984; through Chem Abstr. 102: 124798, 1985.

42. **Simhony, S. and Katzir, A.,** Remote monitoring of ammonia using a carbon dioxide laser and infrared fibers, *Appl. Phys. Lett.*, 47, 1341, 1985.

43. **David, D. J., Wilson, M. C., and Ruffin, D. S.,** Direct measurement of ammonia in air, *Anal. Lett.*, 9, 389, 1976.

44. **Smock, P. L., Ororfino, T. A., Wooten, G. W., and Spencer, W. S.,** Vapor phase deterrmination of blood ammonia by optical waveguide technique, *Anal. Chem.*, 51, 505, 1979.

45. **Giuliani, J. F., Wohltjen, H., and Jarvis, N. L.,** Reversible optical waveguide sensor for ammonia, *Opt. Lett.*, 8, 54, 1983.

46. **Giuliani, J. F. and Barrett, T. W.,** The effect of ammonia ions on the absorption and fluorescence of an oxazine dye, *Spectrosc. Lett.*, 16, 555, 1983.

47. **Blyler, L. L., Cohen, L. G., Lieberman, R. A., and MacChesney, J. B.,** Eur. Patent Appl., 292,207, 1988.

48. **Arnold, M. A. and Ostler, T. J.,** Fiber optic gas sensing probe, *Anal. Chem.*, 58, 1137, 1986.

49. **Wolfbeis, O. S. and Posch, H. E.,** Fibre optic fluorescing sensor for ammonia, *Anal. Chim. Acta*, 185, 321, 1986.

50. **Rhines, T. D., and Arnold, M. A.,** Simplex optimization of a fiber optic ammonia sensor based on multiple indicators, *Anal. Chem.*, 60, 76, 1988.

51. **Cagler, P. and Narayanaswamy, N.,** Ammonia-sensitive fibre optic probe utilizing an immobilized spectro-photometric indicator, *Analyst*, 112, 1285, 1987.

52. **Shahriari, M. R., Zhou, Q., and Sigel, G. H.,** Porous optical fibers for high-sensitivity ammonia vapor sensors, *Opt. Lett.*, 13, 407, 1988.

53. **Zhou, Q., Kritz, D., Bonnell, L., and Sigel, G. H.,** Porous plastic optical fiber sensor for ammonia measurement, *Appl. Opt.*, 28, 2022, 1989.

54. **Berman, R. J. and Burgess, L. W.,** Renewable fiber optic based ammonia sensor, *Proc. SPIE*, 1172, 206, 1990.

55. **Wolfbeis, O. S. and Li, H.,** Optical urea sensor with an ammonium optode as the transducer, *Anal. Chem. Acta,* in press, 1991.

56. **Kobayashi, T., Hirama, M., and Inaba, H.,** Remote monitoring of NO_2 molecules by differential absorption using an optical fiber link, *Appl. Opt.*, 20, 3279, 1981.

57. **Munir, Q., Weber, H. P., and Bättig, R.,** Resonant photoacoustic gas spectrometer fiber sensor, Proc. Opt. Fiber Sensor Conference OFS '84, Stuttgart, September 5 to 7, 1984, VDE Publ., Berlin, 1984, 81.

58. **McAndrew, Preses, J. M., and Weston, R. E.,** Infrared fluorescence from NO2 excited at 400-500 nm, *J. Chem. Phys.*, 90, 4772, 1989.

59. **Hirschfeld, T.,** Remote analysis using fiber optics, paper presented at the 13th Cong. Int Committee for Optics, Sapporo, Japan, August 20 to 24, 1984.

60. **Narayanaswamy, R and Sevilla, F.,** Optosensing of hydrogen sulphide through paper impregnated with lead acetate, *Fresenius Z. Anal. Chem.*, 329, 789, 1988.

61. **Narayanaswamy, R and Sevilla, F.,** Flow-cell studies with immobilized reagents for the development of an optical fibre sulphide sensor, *Analyst*, 111, 1085, 1986.

62. **Martinez, A., Moreno, M. C., and Camara, C.,** Sulfide detection using an optical fiber system, *Anal. Chem.*, 58, 1877, 1986.

63. **Wolfbeis, O. S.,** Fiber optic fluorosensors in analytical and clinical chemistry, in *Molecular Luminescence Spectrometry. Methods and Applications*, Vol. 2, Schulman, S. G., John Wiley & Sons, New York, 1988, 240.

64. **Trettnak, W. and Wolfbeis, O. S.,** Kinetic titration of sulfide with heavy metal ions using a highly sensitive fluorescent indicator, *Fresenius Z. Anal. Chem.*, 326, 547, 1987.

65. **Helm, D. A. and Zolner, W. J.,** U.S. Patent 3,845,309, 1974.

66. **Wolfbeis, O. S. and Sharma, A.,** Fibre optic fluorosensor for sulphur dioxide, *Anal. Chim. Acta*, 208, 53, 1988.

67. **Sharma, A. and Wolfbeis, O. S.,** Fiber optic fluorosensor for sulfur dioxide based on energy transfer and exciplex quenching, *Proc. SPIE*, 990, 116, 1989.

68. **Wolfbeis, O. S.,** Unpublished results, 1987.

69. **Tiefenthaler, K. and Lukosz, W.,** Integrated optical humidity and gas sensors, in Proc. Opt. Fiber Sensor Conference OFS '84, Stuttgart, September 5 to 7, 1984, VDE Publ., Berlin, 1984, 215.

70. **Posch, H. E., Wolfbeis, O. S., and Pusterhofer, J.,** Optical and fibre-optic sensors for vapours of polar solvents, *Talanta*, 35, 89, 1988.

71. **Dickert, F. L., Lehmann, E. H., Schreiner, S. K., Kimmel, H., and Mages, G. R.,** Substituted 3,3-diphenylphthalides as optochemical sensors for polar solvent vapors, *Anal. Chem.*, 60, 1377, 1988.

72. **Dickert, M., Vonend, M., Kimmel, H., and Mages, G. R.,** Dyes of the triphenylmethane type as sensor materials for solvent vapors, *Fresenius Z. Anal. Chem.*, 333, 615, 1989.

73. **Dickert, F. L., Schreiner, S. K., Mages, G. R., and Kimmel, H.,** A dipping sensor for organic solvents in waste water, *Anal. Chem.*, 61, 2306, 1989.

74. **Wolfbeis, O. S., Posch, H. E., and Kroneis, H. K.,** Fiber optical fluorosensor for determination of halothane and/or oxygen, *Anal. Chem.*, 57, 2556, 1985.

75. **Nylander, C., Liedberg, B., and Lind, T.,** Gas detection by means of surface plasmon resonance, *Sensors Actuators*, 3, 79, 1982/83.

76. **Russell, A. P. and Fletcher, K. S.,** Optical sensor for the determination of moisture, *Anal. Chim. Acta*, 170, 209, 1985.

77. **Ballantine, D. S. and Wohltjen, H.,** Optical humidity sensor, *Anal. Chem.*, 58, 2883, 1986.

78. **Tokyo Shibaura Electric Co.,** Jap. Kakai 81,112,636, 1981; through Chem. Abstr. 96:87,570, 1982.

79. **Zhou, Q., Shahriari, M. R., Kritz, D., and Sigel, G. H.,** Porous fiber optic sensor for high-sensitivity humidity measurements, *Anal. Chem.*, 60, 2317, 1988.

80. **Zhou, Q., Shahriari, M. R., and Sigel, G. H.,** The effect of temperature on the response of a porous fiber optic humidity sensor, *Proc. SPIE*, 990, 153, 1989.

81. **Posch, H. E. and Wolfbeis, O. S.,** Fibre-optic humidity sensor based on fluorescence quenching, *Sensors Actuators*, 17, 77, 1988.

82. **Omron Tateishi Electronic Co.,** Jap. Kokai 59 19,843, 1984; through Chem. Abstr. 101: 83,210, 1984.

83. **Boltinghouse, F. and Abel, K.,** Development of an optical relative humidity sensor, *Anal. Chem.*, 61, 1863, 1989.

84. **Zhu, C., Bright, F. V., Wyatt, W. A., and Hieftje, G. M.,** A new fluoresence sensor for quantification of atmospheric humidity, Proc. Electrochem. Soc. 87-9 (Proc. Symp. Chem. Sens.) 476, 1987.

85. **Stuart, A. D. and Grazier, P. E.,** A fibre optic relative humidity sensor, *Int. J. Optoelectron.*, 3, 177, 1988.

86. **Elliott, S. B.,** Primary, optical type, phase transition humidity sensors, Moisture Humidity, Proc. Int. Symp., p. 717, 1985;. ISA: Research Triangle Park, NC.

87. **Fowles, M. and Wayne, R. P.,** Ozone monitor using an LED source, *J. Phys. E*, 14, 1143, 1981.

88. **Boisde, G. and Perez, J. J.,** Une nouvelle generation de capteurs: les optodes, *Vie Sci. C. R.*, 5, 303, 1988.

89. **Van Ryswyk, H. and Ellis, A. B.,** Optical coupling of surface chemistry. Photoluminescent properties of a derivated gallium arsenide surface undergoing redox chemistry, *J. Am. Chem. Soc.*, 108, 2454, 1986.

90. **Griffen, J. W., Olsen, K. B., Matson, B. S., and Nelson, D. A.,** Fiber optic spectrochemical emission sensors, *Proc. SPIE*, 990, 55, 1989.

91. **Saturday, K. A.,** Absorption cell with fiber optics for concentration measurments in a flowing gas stream, *Anal. Chem.*, 55, 2459, 1983.

92. **Beswick, R. B. and Pitt, C. W.,** Optical detection of toxic gases using fluorescent porphyrin Langmuir-Blodgett films, *J. Coll. Interface Sci.*, 124, 146, 1988.

93. **Bentley, A. E. and Alder, J. F.,** Optical fiber sensor for detection of hydrogen cyanide in air, *Anal. Chim. Acta*, in press, 1990.

94. **Orofino, T. A., Dand, D. J., and Hardy, E. E.,** A technique for workstation monitoring utilizing optical waveguides, paper presented at the 4th Joint Conf. Sensing Environmental Pollutants, New Orleans, March 23 to 26, 1977.

95. **Hirschfeld, T. and Daton, T.,** Remote fiber fluorimetry: specifc analyte optrodes, lecture presented at the Pittsburgh Conf., 1982, Atlantic City, NJ, March 1982.

96. **Koller, E. and Wolfbeis, O. S.,** Remote fiber optic detection of inhibitors of acetylcholinesterase, in Proc. NATO Conf. A Forward Look Into the Detection and Characterization of Chemical and Biological Species, Salamanca, Spain, April 1 to 5, 1989, 121.

97. **Wolfbeis, O. S. and Koller, E.,** Fiber optic detection of pesticides in drinking water, in Proc. Intl. Workshop on Biosensors: Applications in Medicine, Environmental Protection, and Process Control; Braunschweig (FRG), May 21 to 23, 1989; Verlag Chemie, 1989, 221.

Chapter 12

ENVIRONMENTAL MONITORING APPLICATIONS OF FIBER OPTIC CHEMICAL SENSORS (FOCS)

Stanley M. Klainer, Kisholoy Goswami, Dileep K. Dandge,
Stephen J. Simon, Nelson R. Herron, DeLyle Eastwood, and
Lawrence A. Eccles

TABLE OF CONTENTS

I. INTRODUCTION

A. SCOPE

The impacts of industrialization and its associated environmental pollution have expanded to virtually all communities and nations; consequently, environmental protection is receiving global attention. International agreements on controlling chlorofluorocarbon releases and acidic deposition are two leading examples of environmental concern at the international level.

On the national scale, many industrialized countries have implemented comprehensive regulatory programs to protect the citizens and ecological systems within their borders. The scope of such national programs in the U.S. is exemplified by the selected environmental protection regulations described in Table 1. These federal regulations, in turn, represent minimum environmental protection standards that are often superseded by more stringent state and local regulations.

Each national and international protection issue ultimately necessitates environmental

TABLE 1
Selected Environmental Regulatory Programs and Issues
in the United States

Regulatory Act	Environmental Issue
Comprehensive Environmental Response, Compensation and Liability Act/Superfund Amendments and Reauthorization Act (CERCLA/SARA)[a]	Remediation of abandoned hazardous waste disposal sites
Resource Conservation and Recovery Act (RCRA)	Active hazardous waste storage, treatment, and disposal sites
	Municipal and industrial solid waste treatment and disposal sites
	Uncontrolled releases from fuel and chemical storage tanks
Clean Water Act	Waste water discharges
Safe Drinking Water Act	Safe drinking water standards
	Contamination of ground water from deep waste and brine injection
Clean Air Act	Outdoor air pollution
Occupational Safety and Health Act	Indoor air pollution in the work place
Surface Mining Reclamation and Enforcement Act	Reclamation of land from surface mining operations

[a] Commonly known as Superfund.

monitoring. The scope of the monitoring efforts ranges from quick response, single parameter, localized measurement tasks to planned, multiparameter, long-term programs of regional, national, or international scale. Assessing the nature and quantity of chemicals in the environment is often central to the mandated monitoring programs, and many of the techniques and instruments used in these assessments have been developed in response to specific regulatory requirements. The application of fiber optic chemical sensors (FOCS) to environmental monitoring is a significant example of technology driven by regulation.

B. MAGNITUDE

The magnitude of the environmental monitoring efforts that respond to the CERCLA/SARA and RCRA (Table 1) regulatory programs is enormous. For these programs alone, monitoring is estimated to cost several hundred million dollars per year.[1] There are estimated to be more than 28,000 hazardous waste sites in the U.S. that are uncontrolled or abandoned; billions of dollars are needed to assess and investigate these sites, to remediate them and to monitor them on the long-term (30 year) basis required by law.[2] Currently, there are about 1000 sites on the Superfund National Priority List for cleanup action, and many additional sites are proposed for inclusion. In addition, monitoring is now required for many of the 225,000 municipal and industrial solid waste sites (landfills, storage lagoons, and waste sites as defined in RCRA Subtitle D), because they have been shown to contain hazardous materials.[3] California has recently enacted legislation requiring such monitoring at permanent and temporary waste sites.[4]

The U.S. Environmental Protection Agency (EPA) estimates that there are approximately 1.4 million underground fuel storage tanks in the U.S. and that about 50,000 new tanks are installed annually.[5] EPA surveys indicate that about 35% of these tanks leak into ground water supplies.[6] About 50% of the drinking water in the U.S. is obtained from underground sources that potentially could be contaminated by leakage from underground tanks. The magnitude of risk to the population from such sources is leading to the formulation of very stringent regulations under RCRA Subtitle I which currently defines an unacceptable leak rate as 0.05 ga/h or greater.[5] Many states have implemented even more stringent regulations accompanied by fines that can reach $10,000 per day per tank for a first offense and $25,000 per day per tank for subsequent offenses.[7] Obviously, such regulations will cause severe economic pressure on small businesses and independent gasoline stations if inexpensive, reliable monitoring systems are not available.

C. TARGET COMPOUNDS

Several U.S. regulatory agencies have prepared lists of hazardous substances and have included these lists in regulations to control the use, transport, and disposal of these materials.[8-12] The toxic contaminants that appear on these lists include heavy metal ions, radioactive wastes, and a wide variety of organic compounds ranging from chloro- and bromo-hydrocarbons to aromatic hydrocarbons to phenols, and organophosphates.[13] The organic chlorides, as a group, dominate the ten hazardous compounds most frequently found in contaminated groundwater at hazardous waste storage, treatment, and disposal facilities.[14-19] In addition, many organic industrial chemicals find their way into the soil and groundwater, and some of these are known or suspected to be extremely dangerous. Trichloroethylene (TCE) heads this compendium. TCE is of particular concern because it is known to form the carcinogen vinyl chloride in water.[20] Moreover, it has been estimated that about 23 million people in the U.S. alone are exposed each year to TCE levels ranging from 500 ppb to 5 ppm even though the safe exposure limit is considered to be less than 5 ppb.[17] Furthermore, there is increasing concern about the total concentrations of other, less hazardous, organic compounds present in the soil and groundwater.

D. DISTINCTIONS BETWEEN TYPES OF MEASUREMENTS

One of the key issues in environmental monitoring is the conceptual difference between the diagnostic and the monitoring tasks. The objective of diagnostic evaluation is to define and describe; the objective of monitoring is to watch closely the behavior of a system that has already been characterized. In concept, the transition from diagnosis to monitoring is recognized, but in practice the most prevalent approach to monitoring is to continue to employ the original, more complex and costly, diagnostic methods. The use of diagnostic methodology for the monitoring task has impeded the development of new concepts specifically directed toward monitoring.

1. The Diagnostic Task

The diagnostic task usually requires making a broad range of measurements with limited repetition. The contaminants, their sources, and their toxic limits must be determined. These determinations require long-term exploratory investigations that are not based on any assumptions; consequently, the analytical equipment used must be capable of measuring the anticipated and the unexpected constituents. This need for versatility has led to the development of sophisticated sampling devices and state-of-the-art analytical equipment such as the gas chromatograph-mass spectrometer (GC-MS). In the exploratory phase of diagnostic investigation, the need for versatility overshadows the cost and complexity of the data gathering process; consequently, this phase is conducive to developing new ideas, methodologies, and instruments. In response to diagnostic needs, complex GC-MS, MS-MS, and other state-of-the-art hyphenated systems have become commonplace in the analytical laboratory, as have the highly trained, costly analysts needed to operate these systems.

2. The Monitoring Task[21-35]

The monitoring task, on the other hand, usually requires frequent, focused, preselected, repetitive measurements. The effectiveness of the monitoring task has been limited by the use of expensive, time-consuming diagnostic methodologies. Effective monitoring sacrifices analytical versatility for dedicated, faster, and more economical systems.

Most environmental analytical measurements are for monitoring rather then diagnostic purposes; consequently, most of the fiscal and manpower resources are directed toward these efforts. Although many of the measurements required for monitoring are simple and highly repetitive, versatile diagnostic instruments and analytical requirements are often applied. This approach is wasteful because the additional information obtained usually adds very little to the requisite monitoring information.

Environmental investigators are aware that the use of diagnostic analytical requirements for the monitoring task usually produce null or nondetectable results. This fact has led to an accelerated effort with the EPA to better define the monitoring task and to field more cost-effective systems. The current lack of suitable monitoring systems inevitably forces environmental investigators to compromise the scope or quality of the requisite measurements. They are often forced to accept statistically inferred conclusions rather than the more certain deterministic information.

E. NEED FOR NEW TECHNOLOGY

The scope and magnitude of the environmental monitoring required for reasonable protection of human health and natural resources appears to be beyond the technical, fiscal, and manpower resources of even the high technology industrial nations. The ideal monitoring system would provide *in situ* determination of the levels of potential contaminants at low concentrations. Instrumentation would be inexpensive to install and maintain, would be capable of automatic operation, and would give reliable results when used by operators who have only modest technical training.

The sensitivity, specificity, and versatility of optical methodologies have made spectroscopy a very popular technology on which to base potential monitoring systems. Unfortunately, the very characteristics that make spectroscopy desirable also make it costly and complex. Fiber optic chemical sensors overcome these objections by making inexpensive sensors primarily responsible for the specificity of the measurement and partially accountable for its sensitivity. Furthermore, because FOCS are specific to a particular species or class of compounds, the spectrometer can be greatly simplified. The potential for good monitoring capabilities at a reasonable price give FOCS the capacity to help resolve the environmental monitoring dilemma.

Practical economics are the first prerequisite of a successful monitoring program. Researchers and regulators can define how many loci are to be monitored for each situation; however, the financial ability to obtain the requisite measurements will determine who can comply. In the U.S. existing monitoring requirements have surpassed the practical limits of technology and funding. Expensive monitoring regulations can force business closures and/or cause severe financial hardships.

The economics of existing and anticipated environmental monitoring requirements involve several components: (1) the cost of the fixed equipment, (2) the cost of the sensor, (3) the expense of placing the sensor at the sampling site, (4) the skill level required to make the requisite measurements and to interpret and record the resulting data, (5) the set-up and measurement times, and (6) the cost of maintaining, repairing, and upgrading the instrumentation.

The rationale to develop the FOCS as an alternative to available analytical systems for environmental monitoring is based on these economical considerations, on the ability of FOCS to perform *in situ* measurements, and on the number of individual species that FOCS can measure. FOCS offer users the technology necessary to perform numerous, repetitive analytical measurements at reasonable cost. The FOCS approach can also have advantages in selected screening uses. Diagnostic environmental measurements are not a *forte* of the FOCS approach and should not hence be considered alone for that application.

II. MAKING AND TESTING THE FOCS

The FOCS system consists of four basic elements: an illuminator, the FOCS itself, a spectral sorter and detector, and a data processor and readout. Block diagrams of typical systems are given in Chapter 6.

For environmental applications, it is desirable that the system be portable by one person and, preferably, that it be battery operated. To meet these requirements a typical FOCS system incorporates the following design features: (1) a low-powered, small halogen, tungsten, or mercury lamp is used for illumination; (2) nonfluorescent, narrow-band filters are used for spectral separation; (3) low-powered, solid state electronics are employed for detection and data processing; and (4) miniature digital readout devices display the data. The instrumentation used for environmental measurements is discussed in detail later in this chapter.

Whereas the instrumentation portion of the system is readily available, the techniques for making FOCS are still being investigated. Researchers differ in their preferences for FOCS design and chemistry, in the methods they use for investigating the sensor, and in the test procedures they employ. It is not the intent of this chapter to critique the different approaches or to endorse a preferred methodology. The purpose is to summarize the information available on environmental applications of FOCS so that readers can match the technology to their own applications.

A. DESIGN CONSIDERATIONS

Although, in the final analysis, a FOCS is a comparatively simple device, its design is complex because it involves so many disciplines: chemistry, fiber optics, immobilization, membranes, and optical spectroscopy. Each discipline contributes to the overall development plan and must be integrated to produce a reliable, rugged, long-lived system that meets the needs for environmental monitoring. The starting point, therefore, is a set of specifications, and the end result is a field-hardened device.

1. Specifications for Environmental Use

Environmental FOCS are being developed in response to a set of specifications usually defined by the particular regulatory jurisdiction, or for example, lead and trihalomethane levels in drinking water are regulated. Because monitoring is an emerging requirement, performance criteria for *in situ* monitoring devices are virtually nonexistent. Most of the regulatory specifications that do exist for continuous monitoring devices actually apply to air and water sampling; consequently, few analogies can be drawn with *in situ* chemical analysis. *In situ* water quality monitoring devices in wide use are the conductivity, pH, and dissolved oxygen electrochemical sensors. Performance and operational specifications for these *in situ* devices have been formally established.[36]

2. Possible Approaches to FOCS Development

The development of a FOCS is, in many instances, very much like other research projects. The literature search for appropriate chemical reactions is followed by an evaluation of the available knowledge and the selection of suitable techniques. The one difference in the initial approach is that, in many instances, the needed information is in the older literature, written when wet chemistry was the only method of analysis and absorption (colorimetric) and fluorometric methods were dominant. These are the techniques that are often best suited for chemical measurements using the fiber optics.

3. Logistics of FOCS Development

The selection of the appropriate chemical reaction is the key to a successful FOCS. Everything else is complementary. The chemistry must have the specificity, sensitivity, and stability to make the appropriate environmental analysis. When the reaction has been selected, a compatible FOCS design is developed. Next, the method for measuring the interaction of the chemistry with the environmental targets of interest is chosen and appropriate support instrumentation is selected.

B. FOCS CONFIGURATION

The physical design of a FOCS is based on two major considerations: the actual FOCS configuration and the selection of construction materials. The design must be compatible with the requisite environmental measurements and the materials must be suitable for use in the designated surroundings. A FOCS must be able to interact with a target pollutant in a known way to give a measurable quantity that can be related to the information desired. This interaction is governed by the selected chemistry and, therefore, the physical design must support the chosen chemical systems.

FOCS designs can either be based on commercial fiber optics or on the construction of a device that behaves like a fiber optic. This section presents the evolution of several FOCS, starting with a basic fiber optic. The discussion will be limited to designs for which feasibility has been adequately demonstrated.

1. Tip-Coated FOCS

To date, the most widely used FOCS design has been to put a reagent that specifically and sensitively interacts with the analyte of interest at the tip of the fiber (Figure 1). Fibers in the 100 to 600 µm diameter range are used most often. These have very small surface areas at their tips (7×10^{-5} to 3×10^{-3} cm^2) and it is difficult to provide enough chemistry at the end of the fiber for reliable measurements, i.e., good signal-to-noise ratios.

Several approaches have been used to place enough chemistry at the tip of the fiber to overcome the surface area limitations. These approaches include: (1) applying surface amplification techniques[37,38] that use a unique type of covalent immobilization (Figure 2); (2) imbedding the chemistry into a membrane[39,40] or polymer[41] placed at the tip of the fiber (Figure 3); (3) attaching large surface area, porous glass fibers,[42-44] or porous glass beads[45] to the tip;[34,35] and (4) using very sensitive chemistry, i.e., fluorescence reactions.[46,47] In special situations, where the species to be measured is a volatile, reservoir FOCS (liquid reagents) with gas-permeable membranes have been used (Figure 4).[48,49]

To use the reservoir cell effectively, it is necessary to collect as much of the signal as possible. One way of doing this is to employ nonimaging optics (NIO), specifically the Winston cone. The NIO technology is based on research into solar concentrators.[50] Rather than follow the historical path of imaging optics, which depend on spatial coherence of transmitted rays, the NIO is optimized for power throughput. The device currently being used expands the beam from a 400 µm fiber with a 0.22 NA to 2 mm in diameter with a 0.05 NA.

2. Evanescent-Wave FOCS

The sides of the fiber, for all practical purposes, cannot be used because the light does not reach the outer surface of the clad, or, in cases where the clad is removed, the light comes out the sides and cannot be collected effectively. It has been found, however, that when the sensing reagents are coated on the side of a fiber optic whose clad has been removed, a measurable effect occurs in the evanescent wave region. This is treated in Chapters 4, 5, and 20.

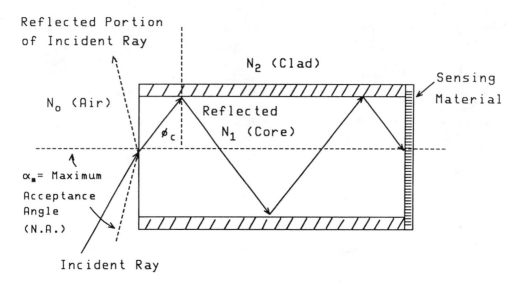

FIGURE 1. Tip-coated FOCS.

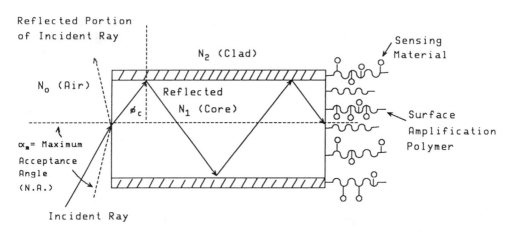

FIGURE 2. Surface-amplified FOCS.

3. Side-Coated FOCS[51]

Coating the sides of the fiber with sensing reagents can be very effective if all of the sensing chemistry can react with the target species and if the resultant changes in light characteristics can be measured. The technique is accomplished by making a multilayered FOCS in which the reacting chemistry is "sandwiched" between the core and the clad (Figure 5). The approach is to make the refractive index, N_3, of the reacting medium greater than N_1 and N_2 while keeping N_1 greater than N_2. Thus, because N_3 is greater than N_1 the light passes through N_3 and is sent back to the core when it reaches N_2. Because N_1 is also of smaller refractive index than N_3, it, too, acts as a clad and redirects the light to N_3. This reaction occurs over the length of the core. Any reactions or interactions that do occur in the reactive layer are enhanced, because the light passes through this medium many times and the signal response is much greater than with a tip-coated or evanescent wave FOCS. The output of the FOCS is a light annulus, dark in the center with a bright ring. In case of planar waveguides, it is a bright thin layer.[52,53]

It is essential that the clad, in addition to having the proper N_2, be sufficiently porous so

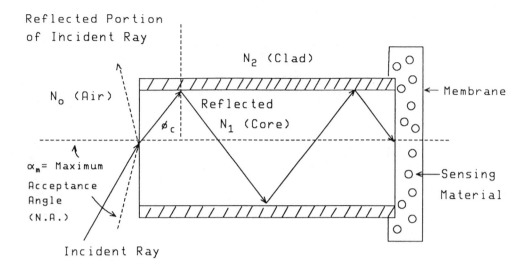

FIGURE 3. FOCS with chemistry imbedded in a membrane or porous glass.

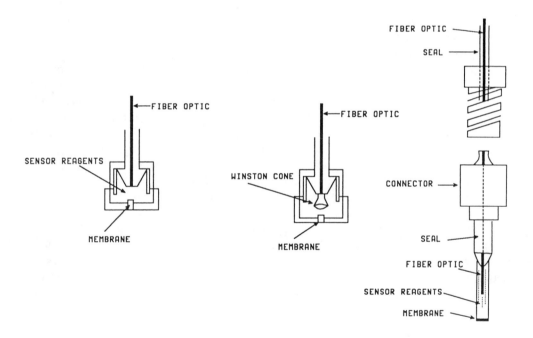

FIGURE 4. Three reservoir FOCS designs.

that the species to be monitored has adequate access to the sensor chemistry. Clads that have a controlled pore size can also be used as protective or gas-permeable membranes. By covalently immobilizing the reactants to the core, any chance of losing the sensing material by "leakage" through the porous polymer clad is obviated. In a similar manner, the clad/membrane is most effective when it is immobilized to the sensing chemistry. Although there are no published examples of a FOCS based on this concept to date, its feasibility is demonstrated in the referenced pending patent.[51]

4. Refractive-Index FOCS

The refractive index FOCS[54] consists of a bare fiber optic core with a thin clad of an organic

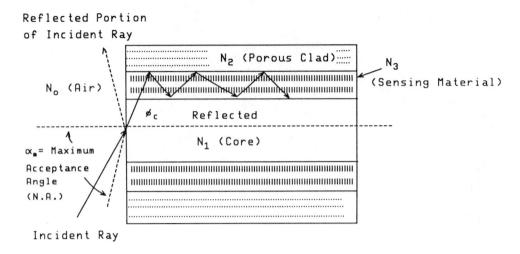

FIGURE 5. Side-coated FOCS.

or inorganic compound attached to its outer circumference near the distal end. In one design, the sensor has a fluorescent tip that is formed with an immobilized dye. An excitation signal is transmitted through the fiber tip, and the fluorescence emission is used as a constant-intensity light source. This emission is detected as the return signal. It is also possible to put the excitation source and the detector at opposite ends of the fiber. A change in the refractive index of the medium surrounding the fiber alters the transmission characteristics and results in a variation in the amount of light that reaches the detector (Figure 6). Results indicate the ability to quantitate over a wide dynamic range of eight orders of magnitude (ppb to %).

If the coating on the side of the fiber is closely matched to the index of refraction of the target molecule, specific sensitive measurements are possible. In order to work properly, however, the side coat must have two characteristics: it must closely match the refractive index of the target, and it must have a high specific affinity with the target species of interest. The latter is necessary because (1) the fiber optic index-matched system is not an accurate refractometer and cannot distinguish between compounds with close refractive indices and (2) the refractive index in a mixture is an average of the species present, and without preselection of the species of interest the data would be uninterpretable. Figure 7 shows data taken with a refractive index based FOCS for trichloroethylene (TCE);[55] this is an environmental sensor currently under evaluation. This FOCS has rhodamine B on the tip. The bare core is side coated with a hydrophobic compound that has a preferential affinity toward TCE. What is important to note in Figure 7 is the preferential response of this sensor to TCE vapor, in comparison to its responses to the vapors of dichloromethane, chloroform, and carbon tetra-chloride. In the case of solid-phase sensors, satisfactory measurements of the reflected return light become difficult. The intensity of reflected light can be more easily determined in those cases, where a wavelength shift occurs due to an immobilized, colored reagent phase. An example of a reflectance-based sensor is an ammonia detector.[56]

C. SPECIALIZED MEMBRANES AND MATERIALS

Sensing reagents, types of measurement, and immobilization/surface amplification are discussed in other chapters. For environmental FOCS, however, there are other critical considerations such as membranes and the refractive indices of the core and clad.

1. Membranes

Membranes are used with FOCS for (1) containing the reaction chemistry or the fluorescent

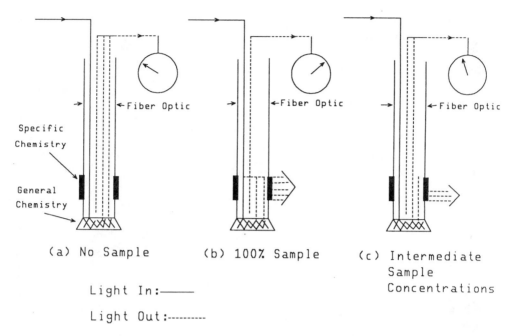

(a) No Sample   (b) 100% Sample   (c) Intermediate
                                        Sample
                                        Concentrations

Light In:———

Light Out:---------

FIGURE 6. How a refractive-index FOCS works.

dyes, (2) protecting this chemistry from undesirable species in the environment being moni-tored, and (3) selective permeation of analytes. Commercially available polymeric membranes are used (as manufactured or with some modification depending on the application) with the reservoir cells (Figure 4). In general, polymers such as cellophane, polytetrafluoroethylene, polyorganosiloxanes, and perfluorinated ionomers have been used as membranes. When coated fibers are used, however, these polymers do not work because of the small size of the substrate, i.e., commercial membranes do not perform well in the micro domain. To overcome this problem, membranes are formed directly on the fiber.

a. Thin-Film Membranes

For the solid state FOCS, the membrane is usually put on the fiber optic from polymer solutions. This method has been found to be unreliable because of pinholes, poor attachment and variations in membrane thickness. Gas-permeable and liquid-impermeable membranes, which are often used in FOCS applications, have been applied this way. Silicone membranes are the most gas permeable but lack mechanical strength. A block copolymer of polycarbonate (hard block) and polydimethylsiloxane (soft block) has been used with a pH sensor[57] to obtain a pinhole-free membrane that is permeable to CO_2 but acts as a barrier to the protons in solution. In a membrane of this type, however, the overall gas permeability is reduced due to the presence of the hard polycarbonate blocks. The substantially increased mechanical strength, however, can be used to improve gas permeability by employing thinner membranes.

Bright et al.[58] have described the use of a thin-film Nafion membrane for the immobiliza-tion of the fluorophore Rhodamine G. The membrane is evaporated on a microscope cover slide and then glued to the distal end of the fiber. The Nafion, which has the chemical structure shown in Figure 8, acts as a host matrix for the detection chemistry. The permeation of cations such as Cr^{3+}, Fe^{3+}, Cu^{2+}, and NH_4^+ quench the Rhodamine G fluorescence, and ions such as H^+, Li^+, Na^+, K^+, Ba^{2+}, Ca^{2+}, Mn^{2+}, Zn^{2+}, and Mg^{2+} reverse the quenching caused by the former group of ions. The quenching and its reversal are concentration and ion dependent phenomena and thus quantitation of both types of ions is possible.

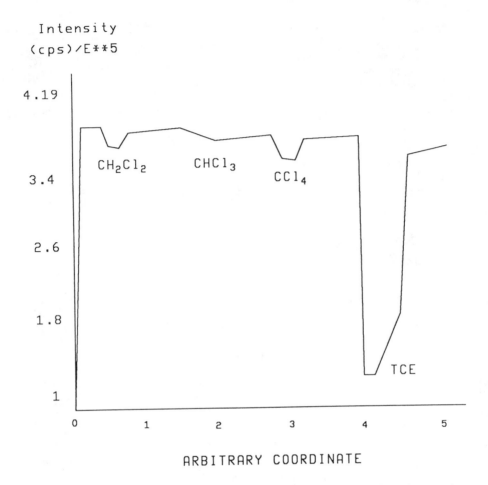

FIGURE 7. Response of the ST&E breadboard TCE$_v$ FOCS (computer trace of actual spectrum).

b. Graft Membranes

The need for making durable membranes with designer properties has led to grafting the membrane directly onto the fiber. The approach is to attach a second polymer to the sensing layer (X) as shown in Equation 1. The second polymer (Y) is then covalently bonded to (X), provided that it has active groups where attachment can take place.

$$\begin{array}{ll} -XXXXXXXXX +Y & -XXXXXXXXXYYYYYYYYY \\ -XXXXXXXXX \rightarrow & -XXXXXXXXXYYYYYYYYY \\ -XXXXXXXXX & -XXXXXXXXXYYYYYYYYY \end{array} \qquad (1)$$

It is also possible to attach the polymer for making the membrane to the immobilization polymer or to the clad if active sites are available. The differences between the thin-film and graft membranes are shown in Figure 9.

The graft material must possess four properties to be useful: (1) it should contain chemically reactive groups that allow it to be covalently bonded to sensing material or to any other selected surface, (2) it should be immiscible with the sensing layer, (3) it should have permeability and selectivity to the target molecule of interest, and (4) it should protect the sensing layer from the normally present interferences in the environment.

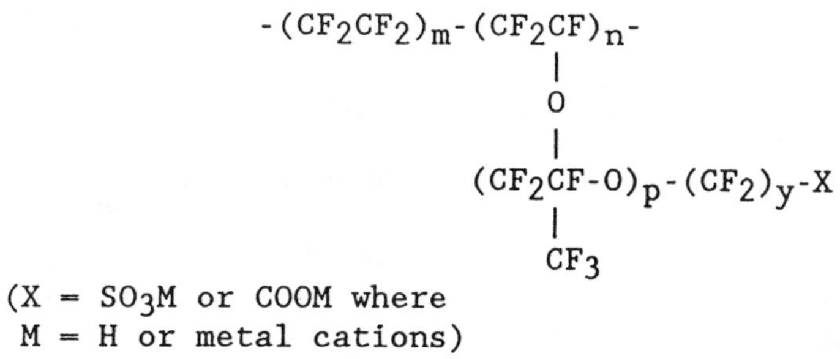

$$-(CF_2CF_2)_m-(CF_2CF)_n-$$
$$|$$
$$O$$
$$|$$
$$(CF_2CF-O)_p-(CF_2)_y-X$$
$$|$$
$$CF_3$$

(X = SO$_3$M or COOM where
M = H or metal cations)

FIGURE 8. Structure of Nafion.

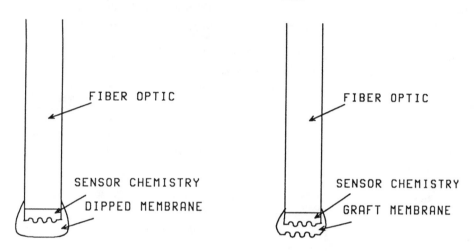

FIGURE 9. Membrane designs.

2. Special Cores and Clads[77]

The FOCS reported in the literature use standard optical fibers as the substrate. This means that the number of usable clad materials is restricted because the selection of refractive indices for the core is limited. For optimum FOCS performance, this impediment must be overcome. This can be accomplished by making the FOCS separate from the fiber and then attaching it to the distal end of the fiber optic. The following discussion applies to side-coated FOCS.[51]

a. Clad

The clad that forms the sensing material is chosen first, so that it is possible to (1) choose the best clad/analyte interaction for the desired measurement, (2) select material(s) that will optimize the clad/analyte interaction, and (3) pick a clad that is stable and that will work properly under the prescribed environmental condition. Organic and inorganic compounds, polymers, and metals may be used as species-specific clads.

In some cases, the clad itself cannot be the sensing material. This is true when (1) the reactive layer does not have the appropriate refractive index, (2) "large" reaction volumes are needed, (3) the sensing reagents are not suitable for deposition as a clad, or (4) the reagents are not stable in the environment where they are to be used.

The clad is attached to the core by vapor deposition, plating, coating, or immobilization. This approach offers an infinite selection of clad materials (polymers, inorganics, organics, ceramics, polymers with reactant(s) covalently attached to, "trapped" in or dissolved in them, etc.); the approach also offers a wide selection of reactants and a good choice of physical properties.

b. Core

After the clad has been chosen, a core can be selected. The appropriate core will have a refractive index that best matches the clad, and it will be composed of a material that does not react with the sensing reagent. The material selected will be stable and will work properly under the prescribed operational scenario and environmental conditions. It will also give optimum light transmission at the desired wavelength. The core can be of the standard extruded type, or it can be a single crystal, a polymer or glass rod, or a rod made from crystalline compounds formed in a mold under high pressure.

In this design the sensor is a custom fiber optic that is attached to the tip of a conventional fiber. The sensor is 1 to 5 mm long; this very short light path means that the core material does not have to follow the normal criteria for good light transmission. The consequent advantages are an infinite selection of core materials, stability and ruggedness, and a wide range of selectable refractive indices and of thermal and light transmission properties.

D. INSTRUMENTATION

Many instruments have been built for use with FOCS. One laboratory fiber optic spectrometer and three units for *in situ* testing merit discussion here for their environmental applications. A detailed discussion of FOCS instrumentation appears in Chapter 6.

1. Breadboard Systems

Two breadboard spectrometer systems have been used to obtain initial environmental measurements. One is a very versatile laboratory instrument for evaluating FOCS designs and chemistries used in initial FOCS characterization and performance. The other is a van-transportable unit that has been used to make initial *in situ* measurements.

a. Breadboard Laboratory Spectrometer

A block diagram for a typical laboratory fiber optic spectrometer is shown in Figure 10. This is basically a high-sensitivity Raman system that has good resolution and excellent stray-light rejection. This design is amenable to fluorescence and refraction measurements.

In order to cover a variety of excitation requirements, three illumination sources are used. The first is an air-cooled argon ion laser operated with a prism wavelength selector. The second illuminator uses a xenon lamp and the third a tungsten lamp. These lamps continuously cover the ultraviolet and visible excitation wavelength ranges using a single monochromator to generate monochromatic light. The monochromator is equipped with a stepping motor and is computer controlled.

Multimode optical fibers that have core sizes of 125 to 600 μm are most often used for environmental FOCS. Some work has also been done using the smaller, single-mode fibers.[23,59] The use of silica, glass, and plastic cores is permissible, provided that the spectral response is nearly "flat" over the wavelength region of interest. In addition, it has been found that the fibers that have larger numerical apertures are usually preferable because they provide optimum illumination and signal collection efficiencies.

A device is needed to focus the illuminating light into the fiber and to direct the returning information-carrying light signal from the fiber into the spectral sorter. The configuration of this device depends on whether a one- or two-fiber system is used.

If two fibers are used, then two optical systems are needed, one to focus the illuminating light into the end of the fiber and the other to direct the diverging returning light into the spectral sorter. In most situations, it is desirable to focus the irradiating light so it underfills the fiber. A special coupler makes it possible to use the same fiber for illumination and

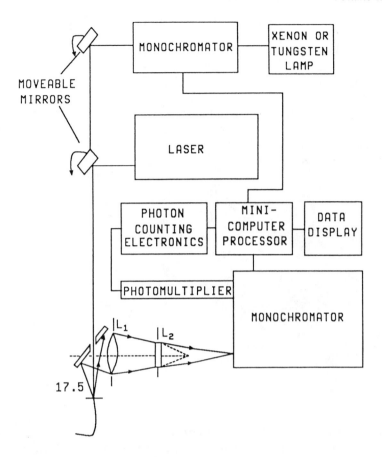

FIGURE 10. Laboratory fiber optic spectrometer.

collection. This design not only eliminates problems caused by fiber alignment variations and modal distortions within the fiber, it maximizes the overlap between the illumination and observation volumes and their subtended solid angle, and thus increases sensitivity and precision. To use a single fiber requires that the outgoing and returning beam be separated at the fiber end. In the laboratory instrument, a perforated mirror that has a small hole in it is used to separate the highly focused incoming beam from the divergent returning beam (Figure 11). This approach is efficient and versatile but is sensitive to optical alignment. The perforated mirror is a must for research where the excitation and returning light wavelengths may vary depending on the sensing reagents. For single-wavelength applications, particularly in portable instruments, a dichroic mirror can be substituted for the perforated mirror to avoid some of the optical alignment problems.

The spectral sorter for the spectrometer is a single monochromator with a prefilter. This instrument uses f/4 illumination from a photographic objective and is scanned by a digital stepping scanner. The monochromator, in this case, is coupled to a quantum counting photomultiplier tube connected to a high-voltage power supply and a discriminator and photon counter. The output of the photon counter is connected to the data system and then to a recording device. The monochromator is computer controlled and can be set up to track the monochromator in front of the two lamp sources for excitation vs. emission measurements.

The output of the photon counter is fed into a system voltmeter controlled by a timer that digitizes the output and transfers it to a personal computer. The system is set up so that the data from the computer is fed to a graphics processor with a hard-copy unit or plotted out directly on a XY plotter.

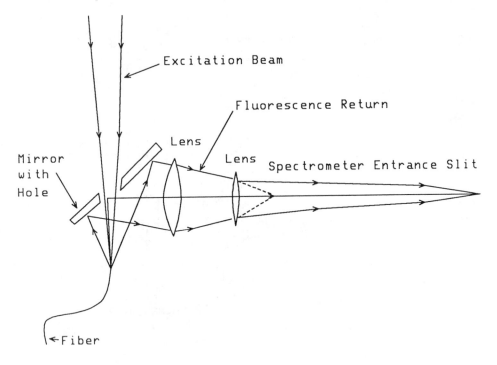

FIGURE 11.   Single fiber optic coupler.

b. Breadboard Van-Transportable Field Spectrometer

Figure 12 is a schematic representation of the first fiber optic spectrometer that was used for environmental field work.[49] This instrument was only used with fluorescence-based FOCS. The system was mounted in a temperature-controlled van. The excitation source was an air-cooled, argon-ion laser. The laser beam (514.5 nm) was expanded and then reduced in intensity to 1 to 10 µW by a series of neutral-density filters. It was then directed through a dichroic mirror and was focused into an optical fiber. The returning fluorescence was reflected with efficiencies greater than 85% by the dichroic mirror and was directed into one of the three photomultiplier tubes set up to accommodate a particular FOCS. Spectral sorting was performed by bandpass filters that were placed in front of the photomultiplier tubes. The output of these tubes was electronically processed and subsequently was recorded on a strip chart. An automatic shutter exposed the FOCS to the laser excitation for a preset duration and sequence to minimize photobleaching.

2. Prototype Portable Spectrometers

One of the biggest advantages of FOCS is their ability to perform *in situ* analyses. These analyses require practical, portable field spectrometers. Two such instruments, a fluorometer and a spectrometer, have been used in this capacity. Although both of these accomplish the same end goal, they are different in concept.

a. Portable Fiber Optic Fluorometer[60,61]

A portable fluorometer, which incorporates several innovations that have led to its reduced size and weight, is shown in Figure 13. The existing portable model has dimensions of 41 × 23 × 19 cm and weighs 10.8 kg and requires an external power supply (2.25 kg). It uses an incandescent lamp, instead of a laser, as the illumination source, and a photodiode detector in place of the photomultiplier tube. Furthermore, the optical system is internally connected by

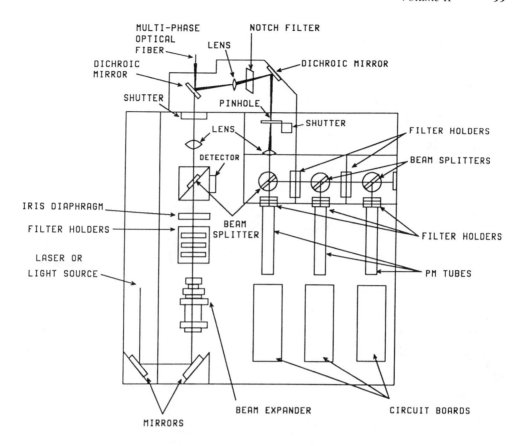

FIGURE 12. Diagram of a breadboard van-transportable fiber optic spectrometer.

using 630 μm fibers to minimize alignment problems. A 400 μm fiber connects the spectrometer to the FOCS. Sensors of 100 to 600 μm in diameter or larger can be used with this device. Three x,y,z translation stages are provided for optical alignment. In addition, a fiber switch permits an external light source, i.e., a laser, to be used, if desired.

In operation, the light from the incandescent lamp is conditioned by the illuminator and is sent to the fiber switch, which directs it into the optical splitter. An electronic shutter controls analysis time and thus reduces or eliminates photobleaching of the fluorophore. The light entering the optical splitter is filtered so that only a single wavelength band is transmitted. The selected wavelength band is then divided into two beams by a dichroic mirror set at a 45° angle to the incident beam. The mirror is designed to reflect the incoming light and to transmit light at the wavelength of the light containing the analytical information. Thus, when the entering light hits the dichroic mirror it is reflected into the FOCS. The dichroic mirror, however, is not perfect; a small amount of the illuminating light passes through the mirror into the reference channel where it is measured by a photodiode. The fluorescence signal, returning from the sample through the fiber, is at a specific wavelength. It passes through the dichroic mirror into the signal channel, where it is purified by a narrow band filter and is detected by a photodiode. Provisions are made for measuring and displaying the individual reference and signal channel outputs, as well as the ratio of the two.

b. Portable Fiber Optic Spectrometer[62]

A portable fiber optic spectrometer (Figure 14) is designed to measure fluorescence or change in refractive index. Although there are many similarities between the portable

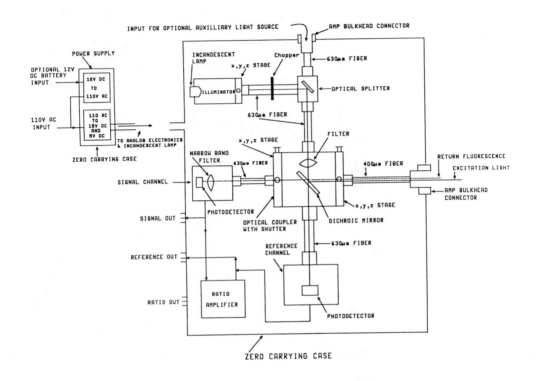

FIGURE 13.   Diagram of a portable remote fiber fluorometer.

fluorometer and the portable spectrometer, the design philosophy of the two instruments is quite different. In particular, the spectrometer is designed to incorporate as many commercial parts as possible. It is intended to be line or battery operated, and is expected to be the precursor to an inexpensive, commercial system. The portable spectrometer uses a miniature optical bench, a highly efficient lamp housing, special high gain amplifiers, audio and visual alarms, and a high-capacity battery-charging circuit. The optical path is similar to that of the portable fluorometer.

The light source is a miniature tungsten halogen lamp whose light output is adjustable up to 840 lm. Two lenses, one at the lamp housing and one in front of the optical fiber, are adjustable with X-Y positioners. These lenses are used for the critical alignment of the light path. A short-wavelength, optical bandpass filter placed in front of the light source furnishes a pure beam of excitation light. A dichroic beam splitter mounted at 45° midway between the lamp and the fiber is optimized to pass the short-wavelength excitation light and to reflect the longer wavelength light signal returning from the optical fiber. At a right angle to the beam splitter, directly in front of the second lens, is a long-wavelength bandpass filter. This lens focuses the returning light directly onto a sensitive photodiode. The photodiode responds to the wavelength range between 300 and 1600 nm. Its sensitivity is 0.45 pA/pW at a peak wavelength of 740 nm. Response is 0.5 μsec. The return signal from the fiber optic passes through a preamplifier, then through a signal data acquisition system and an amplifier.

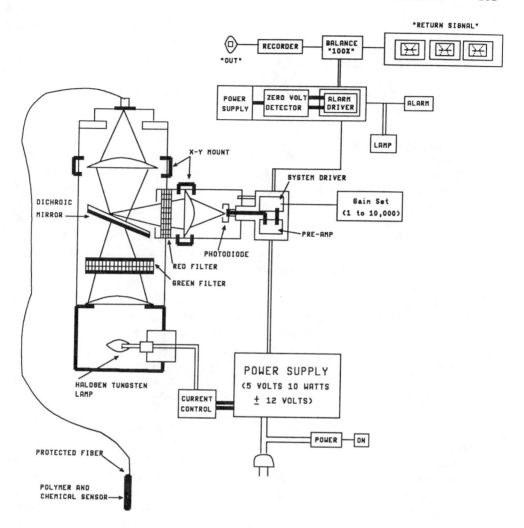

FIGURE 14. Portable fiber optic spectrometer.

Power for the present spectrometer is supplied by either battery (12 V) or AC (110 to 120 V). Its size is 46 × 36 × 8 cm with or without batteries; it weighs 8 kg without batteries and 12 kg with batteries. The minimum battery life is 2 h under full continuous operation, and the battery pack is constantly in a charge mode when the AC mode is in use.

3. Systems Integration

Integration of the FOCS with the spectrometer is straightforward, provided that the instrumentation design is optimized for a particular sensor configuration. Once the type of measurement has been selected, i.e., fluorescence, absorption, or refractive index, then the following parameters must fit: (1) the size of the spot of the irradiation light must underfill the fiber; (2) all of the optical components must be made of materials that *do not* fluoresce at the operational wavelength(s) of the FOCS; (3) system noise must be considerably lower than the minimum signal that is to be detected; (4) the dynamic range of the detection system must be compatible with the highest and lowest signals expected from the FOCS; and (5) the bandwidth of the spectral sorter must be selected to give optimum signal output but must *not* allow interfering spectral information to reach the detector. Fortunately, all of these criteria can be met using accepted spectrometer design and components.

E. CALIBRATION AND TEST

A well-designed calibration and test strategy is needed to confirm that the FOCS developed for a particular environmental application is suited for its intended purpose. The objective is to characterize the FOCS and to establish its performance specifications. The parameters of importance are listed below.[63,64]

Accuracy is the degree of agreement between a measured value and the true or expected value of the quantity of concern. Accuracy is determined by measuring a sufficient number of samples, which have been characterized by one or more independent means, under a variety of conditions so that a statistical analysis of the data can be made.

Bias is systematic error inherent in a method or caused by some artifact or idiosyncrasy of the measurement system. Temperature effects and extraction inefficiencies are examples of this first kind of bias. Blanks, contamination, mechanical losses, and calibration errors are examples of the latter kinds. Bias may be positive and negative, and several kinds can exist concurrently, so that net bias is all that can be evaluated except under special conditions.

Cross-sensitivity is a quantitative measure of the response obtained for an undesired constituent (interferent) as compared to that for a constituent of interest.

Detection limit is the smallest concentration or amount of some component of interest that can be measured by a single measurement with a stated level of confidence. The detection limit is established by continually diluting a standard solution until no response significantly above noise is noted (signal to noise of greater then 3:1 is usually required).

The limit of quantitation (LOQ) is the lower limit of concentration or amount of substance that must be present before a method is considered to provide quantitative results. By convention, $LOQ = 10s_0$, where s_0 is the estimate of the standard deviation at the lowest level of measurement.

Linearity is a measure of the correlation between the test method response and the concentration of the analyte. This is assessed by repetitive measurement of a variety of samples at the same and at different concentrations to determine whether the concentration vs. the response signal falls on a straight line and to assess any deviations from this line.

Operational life is the time span during which a FOCS will function in a sample matrix without degradation of performance. Operational life is assessed by making continuous measurements of a target analyte in a simulated environmental sample until unacceptable changes in response are noted.

Precision is the reproducibility of the test method. It is the degree of mutual agreement of independent measurements as the result of repeated application of the process under specified conditions. It is concerned with the closeness of results.

Reproducibility among sensors is a measure of the similarity of the response of individual FOCS to a specific analyte. Reproducibility is estimated by appraising the response of many sensors to the same set of calibration standards.

Sensitivity is the ability of the FOCS to discriminate between samples that are of different concentrations or that contain different amounts of an analyte. Sensitivity is evaluated by determining the smallest change in concentration that will give a significantly different assay response (signal to noise greater than 5:1 is usually required).

Shelf life is the length of time between manufacture and use during which the sensor will operate properly. Shelf life is determined by taking a set of sensors that were made and calibrated at the same time, and then using them in standard solutions, at predetermined time intervals, until they fail to meet the original calibration response.

Specificity is the ability of the FOCS to respond to a desired substance constituent and not to others. It is determined by measuring the target analyte over its expected concentration range in simulated environmental samples which contain anticipated interferents over their projected concentration range.

1. Laboratory Calibration

The key to obtaining the necessary calibration inputs is to use good, representative standards that maintain their integrity throughout the analytical process. Such standards are very difficult to make for the many marginally soluble, volatile materials and gases of concern to the environmentalist. This is particularly true for some of the more prevalent toxic compounds, including the organic halides, hydrocarbons of low molecular weight, and hydrocarbon mixtures such as gasoline. It is not only troublesome to make standards for these compounds; it is also difficult to use these without changing the analyte concentration. The reason for this enigma is the *loss* of the volatiles during sample preparation and use.

The best available calibration procedure for the organic halides is to make up the calibration solutions in carrier-water mixtures. Alcohols such as methanol and glycol have been found to be suitable. Only small amounts of carrier are needed to stabilize the standard solutions. The best carrier-water ratio is not a constant; it is dependent on the halide used. Samples should be stored in bottles that have large head spaces for gas-phase sensing and minimal head spaces for liquid phase sensing. The bottles should be sealed with septa made of nonreactive, organic-halide-impermeable material. The standard bottles should be identical in volume and in the amount of carrier-water mixture present. Only the organic halide concentration should be varied. The purpose of the septum is to allow small samples (negligible in terms of total volume) to be removed for independent analyses without upsetting the composition of the solution or the vapor phase. FOCS (in a protector, i.e., syringe needle) can be calibrated by inserting them through the septum into either the vapor or solution or the standard can be made to flow through a closed test chamber in which the FOCS is placed. In the latter case, the bottle acts as a reservoir, and the calibration sample can be pumped through the septum.

Carrier-water solutions of the organic halides can be made up in four groups: (1) calibration solution (a single component mixture in a carrier-water matrix), (2) simulated environmental compositions, (3) organic-halide-"spiked" collected environmental samples, and (4) typical solutions from contaminated wells and the vadose zone. Initial calibration is done with organic halide calibration solutions of controlled concentrations, to determine the sensitivity and detection limits of the FOCS, its ability to quantify liquid and vapor samples, and its linearity. The same solutions should also be analyzed for independent corroboration of sample composition by using GC-MS. Calibration standards should also be used to evaluate the performance reproduciblity, bias, and precision of the FOCS. The basic approach presented for the organic halides can be extended to other volatiles. Samples that are less difficult to handle can be made up directly in water to the extent of their solubility.

2. Laboratory Tests

The specificity of the FOCS is of prime interest when monitoring is to be done in uncontrolled matrices such as *in situ* environmental samples. It is essential to know what species may cause false alarms and what compounds will cause the FOCS to give erroneous readings, i.e., wrong contaminant concentrations. To obtain this information requires the use of controlled composition environmental samples. Because these are not always available in natural systems, it is best to make simulated solutions. Standard solutions should be made for the target pollutants, and the designated potential interfering compounds should be added individually to these solutions to determine whether FOCS performance changes in their presence. If no interference from the individual compounds is established, then mixtures should be tried to ensure that no combination of the expected background species affects the FOCS.

The performance evaluation of the FOCS should be based on actual environmental samples and should include field simulations. The simulations are established by using samples from test wells. The samples are first measured as collected, and then are "spiked", if necessary, with known amounts of a particular contaminant. This procedure permits the operation of the FOCS to be evaluated in an existing contaminated system to ensure that nothing unanticipated is occurring. Once proper operation has been confirmed, using a sufficient number of sample sets and corroborating independent analyses, FOCS testing in the field can proceed.

F. FIELD EVALUATION AND TEST
1. Calibration
The FOCS should also be calibrated in the field before *in situ* measurements are made. This is done to ensure that the FOCS are performing properly and that they have not lost their calibration during storage. A two-point measurement is usually sufficient, provided that the laboratory calibration data are available. The easiest way to check the FOCS in the field is to use standard solutions which have been independently analyzed at two concentrations for each contaminant to be measured. The optimum number of test solutions needed to ensure the quality of the field measurements depends on the character of the calibration data; the precision, bias, and linearity required; and the similarity between the individual sensors. Small vials of standard solutions, sealed by septa, are convenient for use in the field. The sensors to be used for *in situ* analyses should be selected on the basis of their similarity and their ability to maintain calibration for an appropriate duration.

2. Field Testing and Evaluation
Field testing and evaluation of a FOCS for an environmental application can be approached in many ways. Generally it is important to know in advance the approximate field conditions (temperature, humidity, air particulates, cable length required, power requirements, health and safety requirements, etc.) and to be certain that the entire system is capable of performing under these conditions. Most prototype equipment is not field hardened, so the actual field test conditions need to be selected carefully. The objectives of the field test, which may vary from a simple demonstration of field operation to a complete comparative evaluation of existing accepted methods, will include accuracy, precision, representativeness, and comparability assessments. A relatively complete field evaluation of a chloroform probe for *in situ* ground-water analysis is reported.[65]

III. SELECTED ENVIRONMENTAL APPLICATIONS

A. CHLOROFORM PROBE
Studies have shown that groundwater from 33% of the hazardous waste sites in the U.S. is contaminated with chloroform.[13,14] The combination of a FOCS for chloroform, based on fluorescence detection of a chloroform reaction product,[60,65] and an associated field-portable fiber optic fluorometer[60] has been described in the literature. A preliminary field demonstration of a first-generation chloroform FOCS that uses a high-powered laser and photon-counting techniques has also been published.[65] The portable fiber fluorometer system and the gas-phase chloroform FOCS have been used to characterize a chloroform-contaminated water-well field.[66] This study was able to provide a good linear correlation between gas-phase chloroform concentrations near the water surface in the wells and the observed FOCS signal, and it detected concentrations in the low parts-per-billion range for aqueous chloroform.

1. Current Status
Considerable work has been completed on the chloroform FOCS. This has led to a working device which has been tested in contaminated wells. The results were verified by independent analysis by GC-MS.

a. Background

The chemistry of the chloroform FOCS is based on the work of Fujiwara[67] and other investigators[68-74] who demonstrated that the absorbance of basic pyridine changed when it was exposed to selected organochlorides, due to the formation of a chromophore. This work is a good example of the ways proven chemistry can be adapted to FOCS use. The reaction undergoes several internal steps, including ring opening and rearrangement, to produce a product. This chromophore, when excited at 535 nm, emits fluorescence at 597 nm. The fluorescence has been linearly related to the quantity of organochlorides in the parts-per-million concentration level.[70]

The Fujiwara reaction chemistry has a number of shortcomings, such as loss of pyridine due to volatilization, inability to maintain pyridine pH, variations in water content, and loss of dye color, which must be considered in the design of the FOCS or the fluorometer. Modifications and improvements reported in the literature[67,123] are concerned with stabilizing the chromophore, using substituted pyridines, etc. The Fujiwara chemistry remains somewhat unpredictable for use under field conditions because the chemistry is complex and the process is diffusion related.

The Fujiwara reaction has been evaluated for FOCS use with one- and two-phase systems. Initially, the two-phase approach was utilized because it compensates for the fact that pyridine and water are immiscible and because it provides a mechanism for replenishing the reagents continuously. The original reaction described by Fujiwara[67] used 10 M KOH to produce the reactive dichlorocarbene[75] intermediate from the chloroform. A single-phase chemistry incorporating tetrapropylammonium hydroxide as the base has since been developed.

The Fujiwara reaction will work for several, but not all, organochlorides. The presence of multiple chlorines is not a sufficient criterion for reaction, and these organo chlorides that respond do so with varying sensitivities. The method has been found to be especially suitable for chloroform or trichloroethylene (TCE).

The organic chloride FOCS is a probe because it is not a reversible chemical system but it integrates the amount of organic chloride vs. time, although selective photobleaching may be used, with caution, to reinitiate the process.[74]

b. Chloroform FOCS Description

The original chloroform FOCS design incorporated a capillary well with the fiber protruding into the capillary tube, to collect the fluorescent signal. The strong base (10 M KOH) attacked the epoxy used to seal the FOCS pieces, and therefore, an air gap was placed between the base layer and the FOCS end seal to prevent this degradation. This modification required that a long stub of fiber extend into the capillary (Figure 15).

The two-sided reagent plug geometry in the original FOCS necessitated a relatively large (5 μl) reagent volume. The loading process was inherently irreproducible and was subject to a high failure rate. The distance from the end of the collection fiber to the reactive pyridine/air interface varied greatly, resulting in an irregular fluorescence collection efficiency at the onset of the experiment.

With the single-phase system, the epoxy used to construct the FOCS is relatively unreactive toward this chemistry. This stability has allowed several significant improvements in the design and operation of the chloroform FOCS. The air gap was removed (Figure 15), which allowed the FOCS to be charged with reagent simply by inserting a microliter syringe to the base of the FOCS and injecting the reagent.

The single surface in the new design results in a stable reagent plug when using only 2 μl of reagent. Reducing the volume of reagent, in turn, permits the FOCS to be constructed with only a short length of fiber extending into the capillary and yields a reproducible distance from

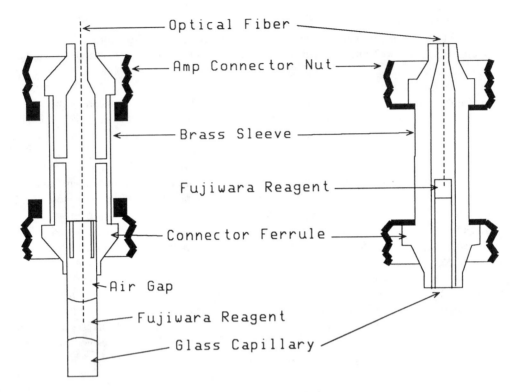

FIGURE 15.   Comparison of original and final gas-phase chloroform FOCS design.

the end of the collection fiber to the pyridine/air interface. The result is an order of magnitude of improvement in the fluorescence collection efficiency from a fluorophore at the interface.

Because the FOCS can be loaded by filling the capillary well from the bottom, it is no longer necessary to observe the charging process. The fragile glass capillary, therefore, can be placed inside a protective cover. In the latest version of this FOCS design, the glass capillary is embedded in an epoxy-filled metal ferrule assembly, resulting in a nearly indestructible FOCS.

Chloroform was selected as the organochloride for the initial laboratory characterization and field tests because: (1) the Fujiwara reaction has high relative sensitivity to chloroform; (2) its relatively high vapor pressure makes it a good candidate for headspace analysis, and (3) chloroform, as noted previously, is commonly found on hazardous waste sites. Furthermore, a well-characterized site with a chloroform groundwater plume was readily accessible.

The current design of the organochloride FOCS is suitable for use with one- or two-phase chemistries, pyridine or substituted pyridines, several bases, with or without a membrane, and with or without an air gap.

c. Laboratory Calibration

The organic chloride FOCS design is a gas-phase, diffusion-limited device. The behavior of a capillary permeation tube is given by the following equation: $R = D * (A_c/L_c) * \Delta C$, where R is the rate of diffusion through the restrictive orifice; D is the gas-phase diffusion constant of the species of interest; A_c and L_c are the cross-sectional area and length of the capillary diffusion device, respectively; and ΔC is the concentration difference between the ends of the diffusion path.[76] Because the analyte concentration is low, the great excess of reagents effectively scavenges the chloroform at the reagent meniscus. The ΔC term is reduced to the bulk chloroform concentration, and the observed rate of change of the fluorescence signal is proportional to the gaseous chloroform concentration.

The common water-well sampling protocol, in which the well is purged of several well volumes of water prior to sampling, lowers the headspace concentrations of volatile analytes.[49,65,66] Because the capillary chloroform FOCS is a gas-phase sampling device that has a relatively slow response time, FOCS calibration and site analysis procedures that depend on ambient gas concentrations have been implemented. A headspace calibration procedure was developed based on a modification of the method of Dietz and Singley.[77] A container, half filled with water, is spiked with a sample of chloroform in methanol. Assuming equilabrium, the gas-phase and aqueous phase concentrations of chloroform can be conveniently determined by using the Henry's Law distribution coefficient of 0.15 at 22°C.[66,78,79] The bottle is sealed during the 30-min equilibration period while the system is magnetically stirred. The septum seal is quickly replaced with the FOCS, and the fluorescence is monitored for 20 min.

d. Results

A calibration of the field-portable FOCS system with the chloroform FOCS was obtained from 4 to 50 ng/ml of gaseous chloroform by using the headspace standard system. These values correspond to aqueous concentrations of 30 to 400 ng/ml (parts per billion). The linear regression through the data yields the intercepts and slopes shown in Figure 16.

The data were collected at 1-min intervals, and the shutter was opened seconds before each reading to minimize photobleaching of the fluorescent material. The lowest concentration of chloroform used in the calibration tests was 4.5 ng/ml in the gas phase over an aqueous concentration of 30 ng/ml.

The intra-FOCS and inter-FOCS response reproducibility were compared for two FOCS at a gas-phase chloroform concentration of 15 ng/ml. One FOCS showed an average maximum slope of 20 mV/min over five repetitions, and the second showed an average response of 21 mV/min over four repetitions. Each FOCS demonstrated a relative standard deviation of 10%.

The chloroform FOCS is a probe (an integrating detector) and the rate of change of the signal is proportional to the concentration of gaseous chloroform. These data are reported in differential form, and they are smoothed with a five-point moving average noise filter. At a gaseous chloroform concentration of 12 ng/ml, the average maximum slope for three experiments at this concentration was 12 mV/min, and the average variance for these three runs was 4 mV/min. The relatively high variance of the experiments reflects the absorption of both the excitation and emitted light by the fluorophor.

e. Field Test

The fluorometer system[60,61] equipped with the chloroform FOCS was tested in a chloroform-contaminated well at water depths to 20 ft. A stainless steel gas sampling tube was attached to the FOCS so that the headspace concentration of chloroform was independently monitored. Two conductance electrodes were attached to the system to ensure reproducible positioning of the FOCS relative to the water surface.

The results of the FOCS field experiments showed a good linear correlation between the maximum slope and the gas-phase concentration of chloroform in the well headspace samples. The headspace samples were withdrawn from the stainless steel sampling tube after pumping in excess of ten times the tube volume (10 ml) of headspace gases. The pumping was done after the FOCS experiment was completed. These data are shown in Figure 17. The lowest detected concentration was 1.2 ng/ml of chloroform in the gas phase.

At the conclusion of the experiments, water samples were drawn and analyzed on a GC-MS. The analysis was done using the EPA Method 624 purge and trap method for volatiles.[76] The lowest aqueous chloroform concentration was determined to be 12 ng/ml (12 ppb) and the values ranged up to 750 ng/ml (750 ppb). The aqueous-phase data show a weaker linear

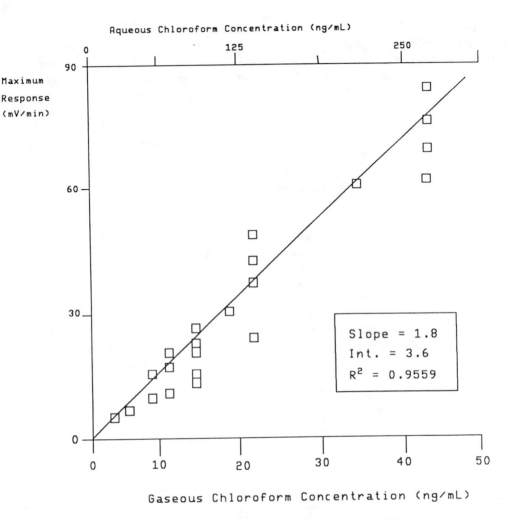

FIGURE 16. Calibration curve for the gas-phase chloroform FOCS.

correlation between the FOCS response and the aqueous chloroform concentration. This probably represents a variation in the transport of chloroform from the aqueous phase to the gas phase, not an inconsistency in the FOCS response. Additionally, if gas-phase concentration is plotted vs. aqueous concentration of chloroform as is done during the verification stage of calibration, the slope obtained is 0.02. This amounts to a 90% negative bias in gas-phase chloroform concentrations, relative to those predicted by Henry's Law of thermodynamics.

2. Research in Progress

The existing gas-phase chloroform FOCS and the single-phase Fujiwara reagent show several serious deficiencies as environmental monitoring tools for groundwater contamination. Under field conditions, transport of chloroform from the condensed phase to the gas phase was demonstrated to exhibit a large negative bias relative to Henry's Law equilibration, which indicates that the aqueous and gas phases are not in equilibrium.[66] Also, the reagent has a very low viscosity, which results in convective mixing of the reaction product from the surface layer with the bulk reagent. This effect results in a highly irregular coupling of the fluorescence with the collection fiber and introduces noise into the data. Furthermore, the reagent is hygroscopic, and it often absorbs enough water from the atmosphere to completely quench the reaction.

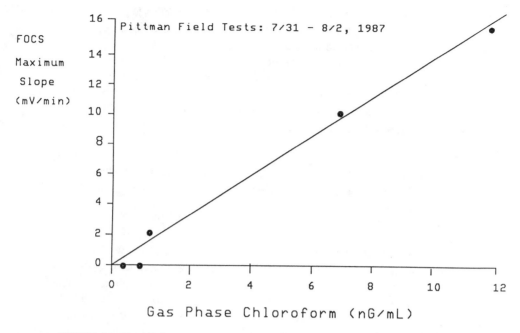

FIGURE 17. Plot of FOCS response vs. gaseous chloroform concentration in water wells.

a. Fujiwara Chemistry

These problems have led to additional research toward developing a *in situ* probe and an improved reagent to enable the probe to be placed directly in the water. The first efforts to increase the viscosity and to reduce the volatility of the reagent were performed while developing the reservoir gas-phase FOCS. The original reagent was found to perform equally well with up to 25% of the more viscous and less volatile dimethylsufoxide (DMSO) included. This new reagent, however, was still very hygroscopic and suffered the quenching problem of the original reagent. Polymers are being added to the reagent that should simultaneously allow the viscosity to be modified and limit the solubility of water in the reagent.

Recent experiments have examined a different chemical approach.[126,133] This research is built upon the work of Bower and Ramage concerning the annulation of 2-aminomethylpyridine by formic acid and phosphorus oxychloride.[134] A method has been developed which is based on the reactions of both dichloroacetylene and the dichlorocarbene with 2-aminomethyl pyridine. The main problem with this chemistry is the complexity of the reaction mechanism pathway necessary to generate the requisite fluorophor quantitatively.

b. Optics

In order to most efficiently collect the fluorescence from the solution, the analyte must diffuse into a very small conical region of fixed acceptance angle very close to the fiber. For the single phase Fujiwara reagent, this transport was observed to be very inconsistent. One method of resolving this problem is to use a fiber of lower numerical aperture. The cone of optimum coupling will be lower, both from the excitation source and the fluorescent material. The present investigations focus on the use of the reservoir cell with nonimaging optical (NIO) lens or the Winston Cone to improve light collection (Figure 4).

Preliminary results show that for a fixed fluorophore with equal illumination intensities, the returned fluorescence in the cone of fixed acceptance angle returned by the NIO lens, optimized for fiber optic communications, is equal to or greater than that returned by the bare fiber. The volume of the acceptance cone for the current NIO lens is significantly greater than that of the bare fiber with the apex extending far beyond the physical boundary of any real cell.[82] For practical sensor cells the divergence angle of this NIO lens can be tailored to optimize light collection vs. the volume of the cell. For a fluorescence sensor the collected signal can, therefore, be expected to be much greater.

An additional advantage of this type of sensor is that the wider aperture allows the experimenter to use a larger piece of membrane. A larger piece of membrane performs an averaging of variations in chemical transport due to micro-inhomogeneities in the membrane. Also, the larger convergence cone of the NIO lens will illuminate a larger area of the membrane, yielding a faster system response.

c. Data Collection

Computerized data acquisition at 30-sec intervals is expected to increase the density of data points sufficiently so that a quadratic nine- or eleven-point Savisky-Golay[83] fit will be possible. This will permit mathematical removal of turbulence artifacts and will give the improved signal fit and noise reduction characteristics of these filters.

B. GENERAL HYDROCARBON SENSORS

Very little work is reported on the use of FOCS for the detection of hydrocarbons. In most cases, where fiber optics are used, their purpose is to get the exciting light to the sample and the return signal to a spectral sorter so that wavelength information can be used for identification.[84,85] In addition, total internal optical scattering is reported for simple alkanes.[86] Successful detection of hydrocarbons by using organophilic compounds on single and multimode fibers has been reported.[23,59] This could improve the ability to perform vadose zone and groundwater analyses.[89] The use of blue thermal paper has been shown to work for a variety of polar solvents,[88] but this chemistry does not respond to hydrocarbons or organic chlorides. Recently, the use of FOCS to detect gasoline[89] and jet fuel has been demonstrated.

C. GASOLINE SENSOR

The gasoline sensor can be used to monitor leaking underground storage tanks, spills, and the spaces between the walls of double-liner storage tanks. This capability is urgently needed because the contamination of drinking and groundwater by gasoline leaking from underground storage tanks presents a considerable health hazard in the U.S. As this directly relates to the availability of potable water, the need to monitor is acute. Early detection of leaking gasoline is imperative since it would not only protect the Nation's water supplies, but would also prevent costly cleanup operations and avoid heavy government penalties and fines. The EPA underground storage tanks (UST) program mandates that these tanks be monitored.[5] The monitor, however, must be reliable, inexpensive, easy to operate, and able to reversibly detect and quantify gasoline as a vapor, a liquid, or water emulsion. Several attempts to develop a gasoline sensor that meets these minimum requirements[90-93] have been unsuccessful. A FOCS that meets these criteria is described below.

1. General Description

The sensor consists of a fluorescent dye attached to the tip of the fiber, which acts as a monochromatic light source, and a 2-cm long coating of a refractive, index-matched material, with selective high affinity for gasoline, immobilized on the surface of a fiber core at its distal end. Figure 6 shows how the sensor works. In the absence of gasoline, the returning light has a high intensity because air (or water) has a smaller refractive index than the core and thus acts as a clad. Under these circumstances, all of the light is propagated through the fiber. When gasoline is present, it is attracted to the high affinity material coated on the surface of the core, and there is an increase in refractive index at the surface of the core. As the refractive index of the clad changes and approaches that of the core, light leaks out of the sides. The return signal decreases proportionately to the amount of gasoline present, due to the reduction in the fluorescent light, until it goes to zero when the refractive index of the clad exceeds that of the core. This loss of light signal can be shown to be directly proportional to gasoline concentration. From Figure 18, which shows the response of the FOCS to gasoline vapor and liquid, it can be seen that liquid gasoline does extinguish the signal and that the FOCS is *completely reversible*. The reversible nature of this sensor indicates that only a physical interaction is taking place between the coating and the gasoline. The fact that laboratory data show that this FOCS responds to gasoline and only slightly to kerosene and jet fuel substantiates the specific affinity of the coating to key gasoline components.

This gasoline FOCS: (1) responds to liquids, vapors, dissolved species, and emulsions; (2) has wide dynamic range, i.e., percent to parts per billion; and (3) operates in either the alarm or quantitation modes. Furthermore, this technology can be extended to the measure of other species. The FOCS is designed to detect gasoline in groundwater, vadose zone, and surface water. Its first major applications will be to monitor leaking from underground storage tanks and large spills.

2. Laboratory Test

The gasoline FOCS has been tested in the laboratory against liquid gasoline and the volatile gasoline components. Several brands of unleaded regular and high test gasoline have been measured in the vapor state. All samples were made up and measured at 20°C. The results obtained for six regular unleaded gasolines between the concentrations of 1 and 50 $\mu l/l$ are summarized in Figure 19. The term $V_0-\Delta V/V_0$ is the normalized response when V_0 is the reading with no gasoline and ΔV is the change observed when various concentrations (microliters per liter) of gasoline are present. These measurements were taken with gasoline samples that were allowed to come to equilibrium with the air in 4-l bottles for 3 h. For those results where the response was not linear over the concentrations measured, all of the gasoline did not vaporize. In these situations, first ultra-small droplets were observed in the test chamber and then liquid remained on the bottom of the bottle at the higher concentrations.

Figure 20 shows the response of four of the main constituents of gasoline from 10 to 60 $\mu l/l$. Figure 21 gives the response of this compound from 0.1 to 600 $\mu l/l$. By changing the coating the low concentration response to benzene can be enhanced (Figure 22).

D. ENVIRONMENTAL APPLICATION OF pH, CO_2, AND O_2
1. pH Sensor

The increased interest in the effects of acid rain has made the measurement of pH an important water quality parameter. For the quantitative description of ocean chemistry, measurement of seawater pH is an essential component. The strongly pH-dependent oceanic processes include mineral solubility, bioavailability, dissolution kinetics, and redox kinetics. Fiber optic sensors are practical for measuring pH in sea water, and this has recently been demonstrated in Chesapeake Bay, MD.[93] For a detailed description of other fiber optic pH sensors, see Chapter 8.

2. Oxygen Sensor

Dissolved oxygen is necessary to support all life in the marine environment. It is, therefore, the most important water quality parameter. Fiber optic technology is very useful for measuring and monitoring dissolved oxygen and its minute, real-time changes in the marine ecosystem. This early warning capability will be useful in predicting the ability of the ocean to support life. Figure 23 shows the response of an oxygen FOCS in seawater. The Winkler titration method for oxygen was used for an independent analysis. The results of the two systems are comparable.[94,95] A number of fiber optic chemical sensors for oxygen measurements have been reported in the literature. Chapter 10 contains a thorough discussion of these sensors.

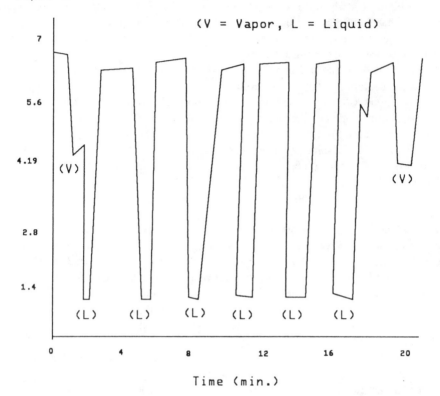

FIGURE 18.   Reversibility of gasoline FOCS (computer trace of actual spectrum).

3. Carbon Dioxide Sensor

The increase in levels of carbon dioxide on a global scale has warranted the need for improved methods of oceanic and atmospheric gas analysis and monitoring. According to the recent climate model predictions, the volumes of the so-called "greenhouse" gases in the Earth's atmosphere are increasing continuously. This will result in a global warming trend by as much as 2 to 3°C in the next 50 years. Because the Earth is covered mostly by water, an *in situ* measurement capability for dissolved carbon dioxide in the ocean is an important step toward prediction and possible alteration of the Earth's climate. FOCS have the potential to satisfy this need. A carbon dioxide FOCS was run in seawater[94] in parallel with the oxygen FOCS (Figure 24). There was, however, no independent CO_2 measurement available and, therefore, the data cannot be validated. A detailed description of CO_2 sensors appears in Chapter 11.

E. OTHER ENVIRONMENTAL SENSORS

Many of the sensors such as ammonia (Chapter 11) and benzo(a)pyrene (Chapter 18) listed elsewhere in this book have environmental sensors are discussed below.

1. Aluminum

Aluminum was once thought to be innocuous; however, mounting evidence indicates that it is involved in numerous biochemical processes. In aqueous solutions the aluminum(III) ion

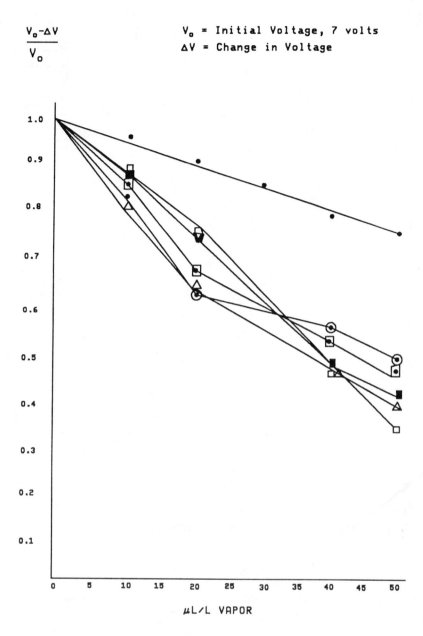

FIGURE 19.   Response of FOCS to regular unleaded gasoline vapors (3-h equilibrium).

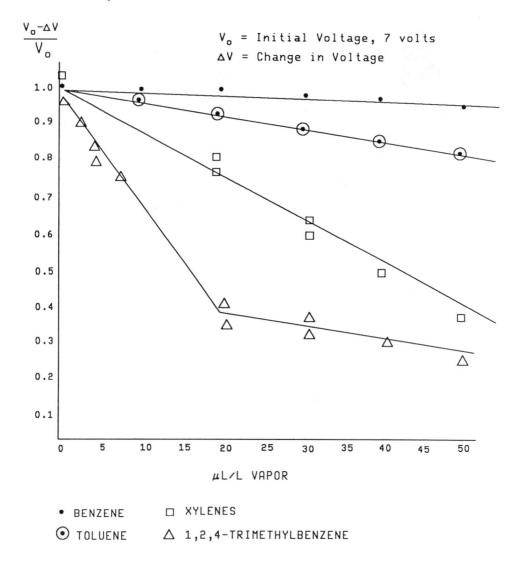

FIGURE 20.   Response of FOCS to 10 to 50 µl/l vapor of four individual gasoline constituents (3-h equilibrium).

hydrolyzes surrounding water molecules forming hydroxyl complexes which decreases the pH of the solution. It forms many inorganic complexes in soil and water.

The physiological mechanism of aluminum toxicity in plants has been summarized by Foy et al.[96] Hutchinson[97] presents a historical overview of the subject and points out the increasing impact of aluminum on aquatic ecosystems. Increased aluminum concentrations and the corresponding reductions in fish populations have been attributed to the mobilization of aluminum from edaphic to aquatic environments due to acidic deposition.[9] The implication of aluminum in several human disorders such as Alzheimer's Disease and Senile Dementia has been of public concern and scientific controversy in recent years.[99] The effects of aluminum on biological and chemical systems is highly dependent upon the form in which the metal occurs. The effects of aluminum in the environment are as numerous as its forms.[100] Plaguing the toxicologist and chemist is a lack of analytical methods for isolating and quantitating the various species of aluminum. More refined monitoring methods are needed to understand the sources, transport, and chemistry of aluminum in the environment.

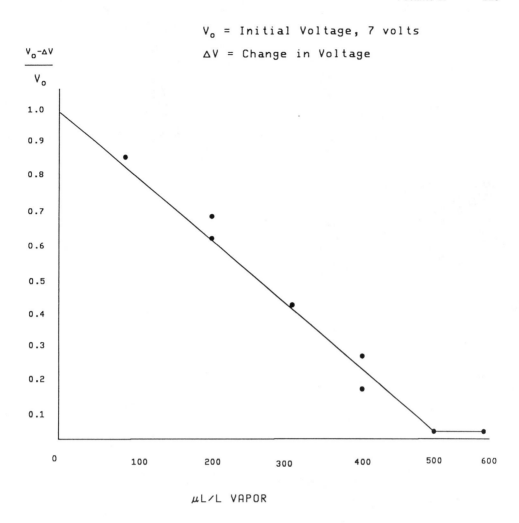

FIGURE 21. Response of a gasoline FOCS to 100 to 600 μl/l bebzene vapor (3-h equilibrium).

Fiber optic chemical technology has been employed[101] for end-point determination in the direct complexometric titration of Al^{3+}. Morin [2-(2,4-dihydroxyphenyl)-3,5,7-trihydroxy-4H-benzopyran-4-one] and DCTA [(diaminocyclohexane-tetra-acetic acid)] have been used as the indicator and complexing agent, respectively. Morin itself is essentially nonfluorescent, but forms an intensely fluorescing chelate with aluminum.[102] This fluorescence intensity has been monitored during titration by using a bifurcated fiber optic light guide.[101] The excitation (420 nm) and emission (500 nm) maxima of the complex are both in the visible region. The end point of the titration is characterized by a sharp drop in the fluorescence intensity. The dynamic range of detecting aluminum (III) is 1 to 800 ppm at pH 4.6

2. Cyanide

Detection of cyanide is very important because of its severe toxicity. Uncontrolled release of cyanide into public waterways poses adverse effects on public health. Development of a real-time fiber optic cyanide sensor is underway.[103] The sensing chemistry is based on the reaction of 1,4-benzoquinone with cyanide, which leads to the formation of a fluorescent complex with an excitation peak at 400 nm and an emission peak at 500 nm.

The sensing material is immobilized on the side of a short (6 in.) fiber optic core. The core

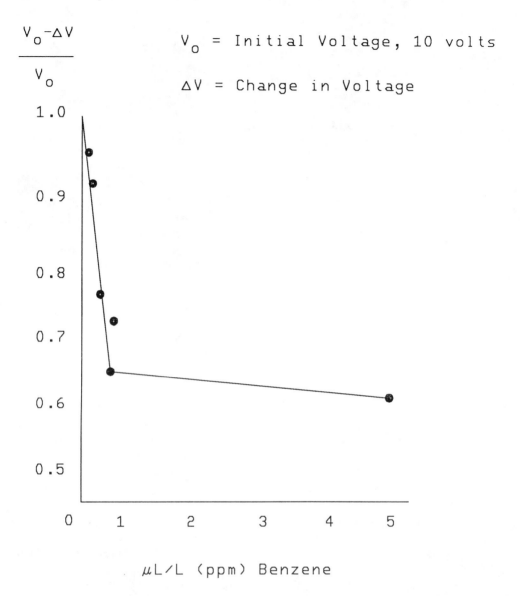

FIGURE 22.   Response of FOCS from 0.05 to 5 µl/l benzene vapor (90 s equilibrium).

is then coupled to the distal end of a long fiber optic cable. This configuration permits the use of disposable probes. Preliminary studies show that the device is capable of detecting cyanide ion concentration between 500 ppb and 5 ppm.

3. Sulfur Dioxide

Sulfur dioxide (SO_2) is one of the most common and harmful air pollutants. The maximum allowable exposure limit as set by the Occupational Safety and Health Act (OSHA) is 5 ppm in 8 h weight average, but lower values have been reported to be toxic to plants and corrosive to metallic construction materials. Wolfbeis and Sharma[104] have described an SO_2 FOCS based on the dynamic fluorescence quenching of benzo(b)fluoranthene, which is immobilized in a silicone polymer and held on the tip of a bifurcated optical light guide. The detection limit for SO_2 in air has been found to be 84 ppm. In another[105] sensor configuration, Sharma and

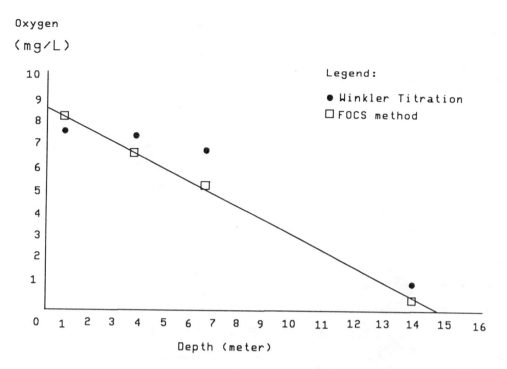

FIGURE 23. Monitoring oxygen in the Chesapeake Bay by FOCS (38° 22.78′ N, 76° 18.83′ W).

Wolfbeis have used the inhibition of the electronic energy transfer from pyrene (donor) to perylene (acceptor) which are both dissolved in a thin layer of silicone polymer for SO_2 detection. The detection limit has been established as 10 ppm.

IV. SUMMARY AND CONCLUSIONS

At the present stage in FOCS development, three questions must be answered:

1. How close have the FOCS come to meeting the monitoring requirements of environmental scientists?
2. Are there any fundamental reasons that all their measurement needs cannot be met?
3. What has to be accomplished before FOCS are an accepted monitoring method?

A. CURRENT STATUS

It has been shown that FOCS are potentially a viable monitoring technique. There are several areas where acceptability has been demonstrated:

Chemistry — It is possible to select chemical reactions or interactions with target compounds that can be measured at the distal end of a fiber optic.

Sensitivity — It is possible to meet the detection sensitivities mandated by environmental regulations.

Specificity — It is possible to make FOCS that are specific to a single species of class of compounds.

Response time — It is possible to make FOCS with reasonable response times (<1 min).

Size — It is possible to make FOCS of reasonable size (100 μm to 10 mm).

Instrumentation — It is possible to make portable systems for use with those FOCS chemistries that are illuminated by and give a return signal in the visible spectral region.

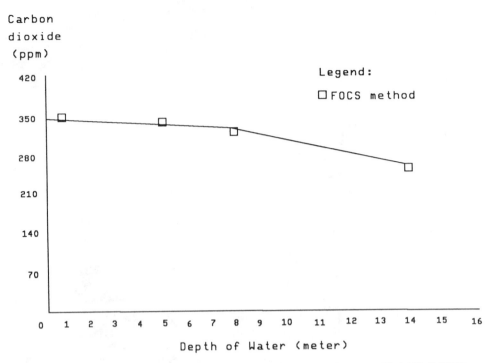

FIGURE 24. Monitoring carbon dioxide in the Chesapeake Bay by FOCS (38° 22.78′ N, 76° 18.83′ W).

B. THEORETICAL LIMITATIONS
There are no fundamental reasons that FOCS cannot meet the requisites for a monitoring system suitable for use by the environmental scientist.

C. AREAS WHERE RESEARCH AND DEVELOPMENT ARE NEEDED
Chemistry — There is a need for many more stable chemistries for monitoring individual species or classes of compounds.

Immobilization — There is a need for more reliable repeatable methods of attaching the specific reagents to the fiber.

Similarity — There is a need to be able to make several FOCS of the same kind whose performance characteristics are similar.

Operational longevity — There is a need to make sensors whose chemistry and antifouling properties are such that they will operate *in situ* for extended time periods (>6 months).

Shelf life — There is a need to be able to store FOCS for reasonable times (>1 year) without degrading their performance.

Calibration — There is a need to develop calibration techniques for characterizing the FOCS and for checking them before and during use.

Instrumentation — There is a need to develop instrumentation, with realistic size, weight, and power specifications, that can illuminate the chemistry and process the return signal below 500 nm (260 to 500 nm).

REFERENCES

1. **Rajagopal, J.,** Conceptual design for ground water quality monitoring strategy, *Environ. Prof.*, 8, 244, 1986.
2. Superfund Strategy, OTA-ITE-252, U.S. Congress Office of Technology Assessment, Washington, DC, April 1985.
3. U.S. Environmental Protection Agency report to Congress, Resource Conservation and Recovery Act, Subtitle D, 1987.
4. State of California, Assembly Bill 3374 (Calderon Bill, 1986) Solid Waste: Disposal Sites: Air Monitoring: Water Pollution Reports, Sacramento, CA.
5. Underground storage tanks, proposed rules, *Fed. Regist.*, (40 CFR Parts 200 and 281), 52, 74, 1987, 12662.
6. Summary of state reports on releases from underground storage tanks, EPA/600-M-86/020, U.S. Environmental Protection Agency, August 1986.
7. **Souder, K.,** State to tackle problem of leaking storage tanks, *Albuquerque (New Mexico) Tribune*, August 30, 1987.
8. U.S. Environmental Protection Agency, Substances found at proposed and final NPL sites, EPA Internal Report, October 1986.
9. List (phase I) of hazardous constituents for ground water monitoring, *Fed. Regist.*, (40 CFR Parts 264 and 270), 52, 131, 1987, 25942.
10. Hazardous materials tables and hazardous materials communication regulation, *Fed. Regist.*, (40 CFR 172.101), 1987.
11. Occupational safety and health standards, *Fed. Regist.*, (29 CFR 1910 Subpart Z), 78, 1974.
12. Identification and listing of hazardous waste, *Fed. Regist.*, (40 CFR 261), 362, 1987.
13. **Plumb, R. H., Jr.,** A comparison of ground water monitoring data from CERCLA and RCRA sites, *Ground Water Monitoring J.*, 7, 94, 1987.
14. **Plumb, R. H., Jr., and Pitchford, A. M.,** Volatile organics scans: Implications for ground water monitoring, in Proc. National Water Well Association/American Petroleum Institute, Conference on Petroleum Hydrocarbons and Organic Chemicals in Ground Water, Houston, 1985.
15. **Arneth, J., Milde, G., Kerndorff, H., and Schleyer, R.,** Waste deposit influences on ground water quality as a tool for waste type and site selection for final storage quality, Proc. Swiss Workshop on Land Disposal, Gerzinsee, Switzerland, 1988.
16. **Strobel, K. and Grummt, T.,** Aliphatic and aromatic halocarbons as potential mutagens in drinking water. I. Halogenated methanes, *Toxicol. Environ. Chem.*, 13, 205, 1987.
17. **Cothern, C., Coniglio, W., and Marcus, W.,** Estimating risk to human health, *Environ. Sci. Technol.*, 20, 111, 1986.
18. **Strobel, K. and Grummt, T.,** Aliphatic and aromatic halocarbons as potential mutagens in drinking water. III. Halogenated ethanes and ethenes, *Toxicol. Environ. Chem.*, 15, 101, 1987.
19. **Montiel, A. and Rauzy, S.,** Halogenated organic compounds (volatile and nonvolatile) as parameters for the evaluation of drinking water quality, *Toxicol. Environ. Chem.*, 10, 315, 1985.
20. **Vogel, T. and McCarty, P.,** Biotransformation of tetrachloroethylene to trichloroethylene, dichloroethylene, vinyl chloride, and carbon dioxide under methanogenic conditions, *Appl. Environ. Microbiol.*, 49, 1080, 1985.
21. **Angel, S., Daley, P., and Kulp, T.,** Optical chemical sensors for environmental monitoring, presented at DOE Contractor's Meeting on Chemical Toxicity, Monterey, CA, June 1987.
22. **Hirschfeld, T., Deaton, T., Milanovich, F., and Klainer, S.,** The feasibility of using fiber optics for monitoring ground water contaminants, *Opt. Eng.*, 22, 527, 1983.
23. **Kawahara, F., Fiutem, R., Silvus, H., Newman, F., and Frazar, J.,** Development of a novel method for monitoring oils in water, *Anal. Chim. Acta*, 151, 315, 1983.
24. **Klainer, S., Milanovich, F., and Eccles, L.,** The remote detection of ground water contaminants using fiber optics and optrodes, presented at the National Water Quality Symp., Austin, TX, April 1985.
25. **Klainer, S., Eccles, L., and Simon, S.,** The role of fiber optic chemical sensors (FOCS) in hydrologic monitoring, presented at the 1987 American Geophysical Union Spring Meeting, Baltimore, May 1987.
26. **Klainer, S., Koutsandreas, J., and Eccles, L.,** Monitoring ground water and soil contamination by remote fiber spectroscopy, presented at the 5th Int. Conf. Chemistry for Protection of the Environment, Torino, Italy, September 1987.
27. **Klainer, S., Schwarz, W., and Milanovich, F.,** The use of fiber optic chemical sensors for monitoring specific parameters and species in aqueous systems, *Oceans 86 Proc.*, 3, 828, 1986.
28. **Narayanaswamy, R., Russell, D., and Sevilla, F., III,** Optical-fibre sensing of fluoride ions in a flow-stream, *Talanta*, 35, 83, 1988.
29. **Hirschfeld, T., Deaton, T., Milanovich, F., and Klainer, S.,** The feasibility of using fiber optics for monitoring ground water contaminants, EPA Topical Report (UCID 19774), Las Vegas, 1984.

30. **Klainer, S.,** Monitoring ground water and soil contamination by remote fiber spectroscopy, presented at the 1986 ASTM Conf., Cocoa Beach, FL, February 1986.
31. **Klainer, S., Hirschfeld, T., and Milanovich, F.,** Unique applications of laser-induced fluorescence to environmental analysis, presented at the 25th Rocky Mountain Conf., Denver, 1983.
32. **Milanovich, F., Hirschfeld, T., and Roe, J.,** Remote sensing over optical fibers: air pollution sensing optrodes, presented at 1983 Laser Conf., San Francisco, December 1983.
33. **Milanovich, F., Hirschfeld, T., Roe, J., and Aziz, N.,** Air pollution sensing optrodes for optical fibers, presented at Pittsburgh Conf., Atlantic City, March 1984.
34. **Shahriari, M., Sigel, G., and Stokes, G.,** Porous fiber optic chemical sensors, presented at the Fiber Optics and Chemical Sensors Session of the 1st Int. Symp., Field Screening Methods for Hazardous Waste Site Investigation, Las Vegas, October 1988.
35. **Klainer, S., Milanovich, F., Miller, H., and Hirschfeld, T.,** Fiber optrodes for the remote quantification of air and water contaminants, presented at the 26th Rocky Mountain Conf., Denver, 1984.
36. U.S. Geological Survey, National Handbook of Recommended Methods for Water-Data Acquisition, Chapter 5, Chemical and Physical Quality of Water and Sediment, U.S. Department of Interior, 1977.
37. **Klainer, S., Walt, D., and Gottlieb, A.,** Amplification of signals from optical fibers, Patent pending, 1986.
38. **Munkholm, C., Walt, D., Milanovich, F., and Klainer, S.,** Polymer modification of fiber optic chemical sensors as a method of enhancing fluorescence signal for pH measurement, *Anal. Chem.*, 58, 1427, 1986.
39. **Lübbers, D.,** Method and arrangement for measuring the concentration of gases, U.S. Patent 4,003,707, 1977.
40. **Duportail, G. and Weinreb. A.,** Photochemical changes of fluorescent probes in membranes and their effect on the observed fluorescence values, *Biochim. Biophys. Acta*, 736, 171, 1983.
41. **Hsu, L. and Heitzmann, H.,** Dye containing silicon polymer composition, U.S. Patent 4,712,865, 1987.
42. **Finger, S. M., Macedo, P. B., Barkatt, A., Hojaji, H., Laberge, Mohr, R., and Penafiel, M.,** Porous glass fiber optic sensors for field screening of hazardous waste sites, presented at the Fiber Optics and Chemical Sensors Session of the 1st Int. Symp., Field Screening Methods for Hazardous Waste Site Investigation, Las Vegas, October 1988.
43. **Macedo, P., Litovitz, T., Lagakos, N., Mohr, R., and Meister, R.,** Optical sensing apparatus and method, U.S. Patent 4,342,907, 1982.
44. **Macedo, P., Litovitz, T., Lagakos, N., Mohr, R., and Meister, R.,** Optical sensing apparatus and method, U.S. Patent 4,443,700, 1984.
45. **Heitzmann, H. and Kroneis, H.,** U.S. Patent 4,557,900, 1985.
46. **Wolfbeis, O.,** The development of fibre-optic sensors by immobilization of fluorescent probes, in *Proc. NATO Meeting on Immobilized Biol. Compounds*, Guilbault, G. G. and Mascini, M., Eds., Reidel, Dordrecht, 1988, 219.
47. **Wolfbeis, O.,** Fiber optical fluorosensors in analytical and clinical chemistry, in *Molecular Luminescence Spectroscopy: Methods & Applications, 2*, Schulman, S. G., Ed., John Wiley & Sons, New York, 1988, 3.
48. **Herron, N., Cardenas, D., Hankins, W., Curtis, J., Simon, S., and Eccles, L.,** Modification, calibration, and field test of a chloroform specific fiber optic chemical sensor, Rep. No. EPA/600/X-87/416, EPA Environmental Monitoring Systems Laboratory, November 1987.
49. **Klainer, S., Koutsandreas, J., and Eccles, L.,** Monitoring ground water and soil contamination by remote fiber spectroscopy, *ASTM Proc.*, 963, 370, 1988.
50. **Welford, W. T. and Winston, R. A.,** *The Optics of Nonimaging Concentrators: Light and Solar Concentrators*, Academic Press, New York, 1978.
51. **Klainer, S.,** Fiber optics which are inherent chemical sensors, U.S. Patent, 4, 846, 548, 1989.
52. **Lukosz, W. and Tiefenthaler, K.,** Sensitivity of integrated optical grating and prism couplers as (bio)chemical sensors, *Sens. Appl.*, 15, 273, 1988.
53. **Nellen, Ph. M., Tiefenthaler, K., and Lukosz, W.,** Integrated optical input grating couplers as biochemical sensors, *Sens. Appl.*, 15, 285, 1988.
54. **Le Goullon, D., Goswami, K., Klainer, S., and Milanovich, F.,** Fiber optic refractive index sensor, U.S. Patent, 4, 929, 049, 1990
55. **Goswami, K. and Klainer, S.,** The development of fiber optic chemical sensors for water monitoring, 1st quarterly report, Contract No. 417-88, AWWA Research Foundation, November 1988.
56. **Giuliani, J., Wohltjen, H., and Jarvis, N.,** Reversible optical waveguide sensor for ammonia vapors, *Opt. Lett.*, 8, 54, 1983.
57. **Munkholm, C., Walt, D., and Milanovich, F.,** A fiber optic sensor for CO_2 measurement, *Talanta*, 35, 109, 1988.
58. **Bright, F., Poirier, G., and Hieftje, G.,** A new ion sensor based on fiber optic, *Talanta*, 35, 113, 1988.
59. **Boiarski, A., Kingsley, S., Kawahara, F., Roesler, J., and O'Herron, R.,** Status report on design and evaluation of fiber optics toxic chemical analyzer, Draft Final Report, Contract No. 68-03-3224 (Work Assignment 2-06), USEPA, August 1987.

60. **Milanovich, F., Daley, P., Klainer, S., and Eccles, L.,** Remote detection of organochlorides with a fiber optic based sensor. II. A dedicated portable fluorimeter, *Anal. Instrum.*, 15, 347, 1986.
61. Available from Douglas Instruments Company, Palo Alto, CA.
62. Available from ST&E, Inc., Livermore, CA.
63. E13189, Vol. 1401 of annual Book of ASTM Standards, to be published 1989.
64. **Taylor, J.,** *Quality Assurance of Chemical Measurements,* Lewis Publishers, Chelsea, MI, 1987.
65. **Milanovich, F. P., Garvis, D. G., Angel, S. M., Klainer, S. M., and Eccles, L. A.,** Remote detection of organochlorides with a fiber optic based system, *Anal. Instrum.*, 15, 137, 1986.
66. **Herron, N. R., Simon, S. J., and Eccles, L. A.,** Remote detection of organochlorides with a fiber optic based system. III. Field test of a chloroform FOCS, *Anal. Instrum.*, submitted, 1988.
67. **Fujiwara, K.,** Ueber eine neue sehr empfindliche reaktion zum chloroformnachweis, *Sitzungsber. Abh. Naturforsch. Ges. Rostock.*, 6, 33, 1916.
68. **Lugg, G.,** Fujiware reaction and determination of carbon tetrachloride, chloroform, tetrachloroethane, and trichloroethylene in air, *Anal. Chem.*, 38, 1532, 1966.
69. **Taha, A., El-Rabbat, N., and El-Kommos, M.,** Novel modification of the Fujiwara reaction, *J. Pharm. Belg.*, 35, 107, 1980.
70. **Okumura, K., Kawada, K., and Uno, T.,** Fluorimetric determination of chloroform in drinking water, *Analyst*, 107, 1498, 1982.
71. **Mantel, M., Molco, M., and Stiller, M.,** Improved spectrophotometric method for the determination of small amounts of chloroform, *Anal. Chem.*, 35, 1737, 1963.
72. **Daroga, R. P. and Pollard, A. G.,** Colorimetric method for the determination of minute quantities carbon tetrachloride and chloroform in air and in soil, *J. Soc. Chem. Ind.*, 60, 218, 1941.
73. **Hunold, G. A. and Schuhlein, B. L.,** Die Pyridin-alkali-Reaktion als Grundlage zur Bestimmung chlorierter aliphatischer Kohlenwasserstoffe, *Z. Anal. Chem.*, 179, 81, 1961.
74. **Milanovich, F., Hirschfeld, T., Miller, H., Garvis, D., Anderson, W., Miller, F., and Klainer, S. M.,** The feasibility of using fiber optics for monitoring ground water contaminants II. Organic chloride optrode, EPA 600/X-85-044, April 1985.
75. **Angel, S. M., Daley, P. F., Langry, K. C., Albert, R., Kulp, T. J., and Camins, I.,** The feasibility of using fiber optics for monitoring ground water contaminants. VI. Mechanistic evaluation of the Fujiwara reaction for the detection of organic chlorides, EPA 600/X-87/467, December 1987.
76. **Palmes, E. D. and Gunnison, A. F.,** Personal monitoring device for gaseous contaminants, *Am. Ind. Hyg. Assoc. J.*, 78, 1973.
77. **Dietz, E. A., Jr. and Singley, K. F.,** Determination of chlorinated hydrocarbons in water by headspace gas chromatography, *Anal. Chem.*, 51, 1809, 1979.
78. **Dilling, W. L.,** Interphase transfer processes. II. Evaporation rates of chloromethanes, ethanes, ethylenes, propanes and propylenes from dilute aqueous solutions. Comparisons with theoretical predictions, *Environ. Sci. Technol.*, 11, 405, 1977.
79. **Mackey, D. and Shiu, W. Y.,** A critical review of Henry's Law Constants for chemicals of environmental interest, *Phys. Chem. Ref. Data*, 10, 1175, 1981.
80. **Angel, S. M., Daley, P. F., and Kulp, T. J.,** Optical chemical sensors for environmental monitoring, DOE Contractors Meeting on Chemical Toxicity, June 1987.
81. **Bower, J. D. and Ramage, G. R.,** Heterocyclic systems related to pyrocoline, part 1:(2:3a) Diazaindene, *J. Chem. Soc.*, 2834, 1955.
82. **Herron, N. R., Whitehead, D. W., and Miller, V. J.,** Evolution of a FOCS for monitoring dissolved volatiles, SPIE, 990, paper 29, 1988, in press.
83. **Savitsky, A. and Golay, M. E.,** Smoothing and differentiation of data by simplified least squares procedures, *Anal. Chem.*, 36, 1627, 1964.
84. **Chan, K., Ito, H., and Inaba, H.,** All-optical-fiber-based remote sensing system for near infrared absorption of low-level methane gas, *J. Lightwave Technol.*, LT-5, 1706, 1987.
85. **Farahi, F., Leilabady, P., Jones, J., and Jackson, D.,** Optical-fibre flammable gas sensor, *J. Phys. E*, 20, 435, 1987.
86. **Guiliani, J. F. and Jarvis, N. L.,** Detection of simple alkanes at a liquid gas interface by total optical scattering, *Sens. Act.*, 6, 107, 1984.
87. **Marrin, D. L. and Kerfoot, H. B.,** Soil-gas surveying techniques, *Environ. Sci. Technol.*, 22, 740, 1988.
88. **Posch, H., Wolfbeis, O., and Pusterhofer, J.,** Optical and fibre-optic sensors for vapours of polar solvents, *Talanta*, 35, 89, 1988.
89. **Klainer, S. M., Dandge, D. K., Goswami, K., Eccles, L. A., and Simon, S. J.,** A fiber optic chemical sensor (FOCS) for monitoring gasoline, EPA 600/X-88/259 NV, 1988.
90. Hydrocarbon Sensor System (Product Brochure), Westinghouse Bio-Analytic System Company, Madison, PA, 1988.

91. Pollulert (Product Brochure), Mallory Electronics Group, Indianapolis, 1988.

92. Leak·X (Product Brochure), Leak·X Corp., Englewood Cliffs, NJ, 1988.

93. Hydrostatic Tank Monitor (Product Brochure), Owens Corning Fiberglass Corp., Toledo, OH, 1988.

94. **Klainer, S., Goswami, K., and Kennedy, J.,** Fiber optic chemical sensors (FOCS) for the measurement of pH, CO_2 and O_2 in sea water, Annual Report, National Oceanic and Atmospheric Administration (NOAA), Contract No. 50 DMNA-7-00162, September 1988.

95. **Goswami, K., Klainer, S. M., and Tokar, J. M.,** Fiber optic chemical sensor for the measurement of partial pressure of oxygen, Proc. SPIE, 990, 111, 1989.

96. **Foy, C. D., Chaney, R. L., and White, M. C.,** The physiology of metal toxicity in plants, *Annu. Rev. Plant Physiol.*, 29, 511, 1978.

97. **Hutchinson, T. C.,** A historical perspective on the role of aluminum in toxicity of acidic soils and lake waters, in Proc. Int. Conf. Heavy Metals in the Environment, Heidelberg, West Germany, 1983, 201.

98. **Baker, J. P. and Schofield, C. L.,** Aluminum toxicity to fish in acidic waters, *Water Air Soil Pollut.*, 18, 289, 1982.

99. **Ganrot, P. O.,** Metabolism and possible health effects of aluminum, *Environ. Health Perspect.*, 65, 363, 1986.

100. **Lewis, T. E., Ed.,** *The Environmental Chemistry and Toxicology of Aluminum*, Lewis Publishers, Chelsea, MI, in press.

101. **Wolfbeis, O., Schaffar, B., and Chalmers, R.,** Direct complexometric titration of aluminum (III) with DCTA, *Talanta*, 33, 867, 1986.

102. **Saari, L. and Seitz, W.,** Immobilized morin as fluorescence sensor for determination of aluminum (III), *Anal. Chem.*, 55, 667, 1983.

103. **Goswami, K., Klainer, S., and Simon, S.,** The detection of cyanide ion (CN^-) in ground water, Annual Report, Lockheed Engineering and Service Company, Contract No. EM4035B, October 1988.

104. **Wolfbeis, O. S. and Sharma, A.,** Fiber-optic fluorosensor for sulphur dioxide, *Anal. Chim. Acta*, 208, 53, 1988.

105. **Sharma, A. and Wolfbeis, O. S.,** Fiber optic fluorosensor for sulphur dioxide based on energy transfer and exciplex quenching, *Proc. SPIE*, 990, 1989, (in press).

Chapter 13

OPTICAL FIBERS IN TITRIMETRY

Otto S. Wolfbeis

TABLE OF CONTENTS

I. GENERAL ASPECTS

Although not a continuous method, titrimetry is still one of the most widespread tools in analytical chemistry. The major advantage of titrimetry over direct sensor methods is that the process of calibration is very simple, if necessary at all. Most titer solutions are stable for prolonged periods, and titrimetry is therefore the method of choice in many kinds of field tests. Although optically indicated titrations have been known for a long time, their application has been limited to situations where the endpoint can be recognized visually. Moreover, the laborious process of transferring a sample from the titration beaker to the fluorimeter cuvette, doing a measurement, returning the sample to the beaker, and continuing titration has prevented analysts from using the method more often. One solution to avoid this cumbersome procedure is to use small spectrophotometer cells which allow flow-through operation,[1] but this may result in some dead volume and delayed endpoint indication. Visual endpoint determination, on the other hand, is more convenient, but is operator-dependent and limited to strongly fluorescent indicators having visible emission.

The remedy for this situation is the fiber optic light guide[2] which can be dipped into the solution to be titrated. Aside from a fairly simple experimental setup (Figure 1), the fiber method allows the use of UV indicators and measurement of even minute spectral changes, requires only tiny sample volumes, and gives an operator-independent endpoint. When compared with methods in which the sample to be titrated is pumped through a flow-through cell in the photometer,[1] the fiber optic approach has an unique advantage in that the photometer comes to the sample, rather than the sample to the photometer.

Recent work demonstrates that fibers can replace electrodes in various titrimetric procedures including acid-base titrations, argentometry, bromometry and iodometry, complexometry, or redox titrations. There are two ways to perform this. In the first one, an indicator is added to the solution to be titrated, and relative changes in absorption or fluorescence during titration are followed via the fiber which acts as a light pipe. This method can offer considerable cost reductions since plain optical fibers along with indicator solutions are much less expensive than electrodes. The optical and electrochemical detection systems are of comparable price.

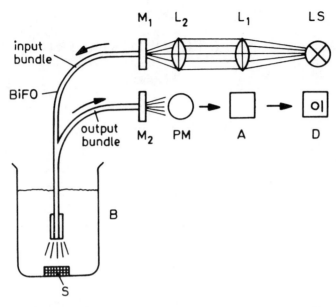

FIGURE 1. Schematic diagram of an experimental arrangement for performing optical fiber titrations. LS, light source; BiFO, bifurcated fiber optic bundle; PM, photomultiplier or photodiode; A, amplifier; D, display; M_1 and M_2, monochromators (or optical filters); L_1 and L_2, lenses; B, beaker; S, stirrer.

In the second method, the analyte-sensitive indicator is immobilized at the fiber end to give an analyte-selective optrode, which is immersed into the solution during titration. A typical example would be the use of a pH optrode to monitor the course of an acid-base titration. However, since most optrodes have a certain response time, the use of analyte-selective optrodes results in longer titration times than by using dissolved indicators. It may be predicted that the endpoints of the majority of standard titration methods can be determined by fiber optical methods with the same precision as with electrochemical methods, although the full potential of fiber optical titrimetry is not fully exploited yet.

A manifold of indicators for endpoint determination in titrimetry has been described,[3] most of them based on changes in the color of an added indicator solution.[4] Fluorescent indicators seem to be less popular but provide a more expanded range of analytical possibilities than absorptiometric indicators. Thus, acid-base indicators, redox indicators, chelating reagents, and adsorption indicators are usually suitable for both absorptiometry and fluorimetry, but quenching and lifetime titrations are confined to the latter method.

An experimental setup for performing fiber-optical titration is shown in Figure 1. Light from a light source (LS) passes lenses L_1 and L_2 as well as an interference monochromator M_1 and then is coupled into the end of the input bundle of a bifurcated fiber optic (BiFO). The common end is immersed into the titration beaker containing analyte plus indicator (if necessary). Fluorescence is collected by the strands of the output bundle and guided to a photomultiplier after it has spectrally been isolated from scattered light by the emission monochromator M_2. The signal is amplified in A and displayed in D.

A fiber optical titration system can be relatively inexpensive. The light source can be a halogen tungsten lamp or, even cheaper, an LED.[5,6] The latter are available for wavelengths above 460 nm only, but cover the rest of the visible spectrum as well as the near infrared. Another promising light source is the semiconductor laser which emits at wavelengths above 600 nm. It is the cheapest laser light source at present for use in electron spectroscopy. Cut-off filters and plastic fibers for use in titrimetry are inexpensive likewise, and costly photomultipliers can be replaced by cheaper photodiodes.

Since fibers can be made smaller than any electrodes, the use of typically 1.0 to 0.1 mm thick fibers (both in a bundle or a single strand device) enables the titration of very small sample volumes. As a result, the low-cost opto-electronic equipment required for fiber optical titrations is likely to result in reasonably priced fiber titration equipment.

An interesting feature of fiber optic titrations is remote titration. Unlike electrodes, which can be used over fairly short distances only, fiber titrations can be performed over distances of typically 2 to 100 m, even in hostile environment. A major field of application is the titration of radioactive materials or the titration in radioactive environment. Glass fibers are resistant to radiation doses in the order of 10^7 rad or more.

In order to restrict the adverse effect of ambient light entering the output fiber bundle, titration should be performed in a brown- or black-teflon-covered beaker. However, some background light may be tolerated because it can be subtracted electronically, when the instrument is run in the pulsed mode, and because it is the relative change in intensity that is observed during a titration, not an absolute value.

II. ACID-BASE TITRATIONS

An opto-electronic titration unit, which can be used for monitoring the course of acid-base titrations via an optical lightguide, has been described in some detail.[5] A blue light emitting diode (LED) was used as a light source, and an inexpensive photodiode as a detector to monitor changes an the fluorescence of an added indicator. Figure 2 shows the overlap of the absorption spectra of the frequently used pH indicator 1-hydroxypyrene-3,6,8-trisulfonate (HPTS) and fluorescein with the emission of the blue LED.

The advantages of the method include simplicity, broad applicability, the ease of automation, and the possibility of performing assays with sample volumes as small as 0.5 ml. Both strong acids and strong bases can be titrated with a precision of better than 0.4%. Figure 3 shows typical plots obtained when HPTS is used as an indicator. A voltage change of about 100 mV is observed at the endpoint. The precision is not affected when methylene blue (an intensely colored and redox-active dye) is present, or when the solution is made turbid by adding silver nitrate plus potassium chloride.

Aside from strong acids and bases, acetic acid as well as hydrofluoric acid were successfully titrated using HPTS as the indicator. The precision was ±0.5%, provided the ratio of the concentrations of acid or base to indicator was greater than 100. Otherwise, the titration of the indicator which itself is a weak acid will cause interferences. Hydrofluoric acid is a particularly challenging example as it cannot be titrated using glass electrodes. In the fiber optic method, and using plastic fibers, a sharp endpoint is observed. Concentrations down to 1×10^{-3} have been determined with excellent precision. Typical plots for the titration of acetic acid and HF are given in Figure 4.

Phosphoric acid with its two pK_as of 2.16 and 7.21 can be titrated using an indicator having pK_as in the same region. Both steps are easily recognizable in the titration plot, provided the proper wavelength of excitation is applied (Figure 5). Sodium dihydrogen phosphate can be titrated with 0.1 M NaOH using HPTS as the indicator (Figure 4). As a result of the buffer capacity of the hydrogen phosphate formed during titration, the curve is characterized by a steady increase in intensity, followed by an almost constant final value when the endpoint is reached. Similarly, hydrogen phosphate has been titrated with HCl, but fluorescein was found to be a better indicator in this case. The same is true for the titrimetric determination of ammonia (Figure 5).

Titrations have also been performed in very small volumes, typically 0.5 ml, which may be useful in the microtitration of biophosphates such as adenosine phosphates and their congeners, and in studies on the binding of proteins (with their intrinsic fluorescence) to either drugs or other proteins, and in investigating host-guest complexes.

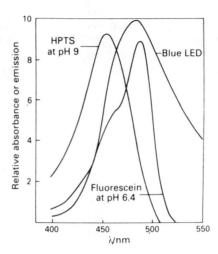

FIGURE 2. Comparison of the absorption spectra of 1-hydroxypyrene-3,6,8-trisulfonate and fluorescein with the emission spectrum of the Siemens SLB 5410 blue LED.

Another LED-based titration device was introduced[6] that is based on absorptiometric measurement of a pH-sensitive dye (phenol red). It incorporates a green LED source and solid state electronics for the detection of light transmission at about 565 nm and is said to have a very good and reproducible response from pH 7 to 9. The big advantage of such a system is the extremely quick response time since the indicator is in bulk solution rather than immobilized on a solid support.

Instead of adding a fluorescent indicator to the solution, one of the fiber optic pH sensors described in Chapter 6 may as well be utilized for this purpose. These, hovever, have certain limitations: the immobilized indicator may be stripped off by strong acid or base, the response time is usually longer than 20 sec, and the pK_a of most known pH sensors is not suitable for titration of, e.g., ammonia or dibasic phosphates. The latter fact, however, appears to be only a matter of time.

Acid-base titrations have been performed[7] with a miniaturized optical system consisting of an LED light source and a CdS photocell or a photo-IC. The device, when immersed into a test solution, can indicate its absorbance. If the photocell surface is coated with a poly(vinyl chloride) (PVC) membrane incorporating an acid-base indicator, the probe can be used as a membrane photosensor for endpoint detection in titrations. In order to match the absorption profile of the test solution, it was suggested to use a multicolor LED whose emission color is governed via a simple additional circuit. Although the system was not coupled to a fiber optic system, it easily could be so.

The measurement of both the intrinsic color of the analyte solution and the color of the PVC film whose color responds to the chemical parameter of interest in an otherwise colorless test solution were investigated.[7] Thus, acid-base titrations were conducted with methyl orange as the indicator and a green LED as the light source. When the endpoint is reached, there is a sharp drop in the resistance of the photocell. A series of chelatometric titrations has also been performed and will be discussed in Section 4.

III. ARGENTOMETRY

The endpoint of argentometric titrations is usually indicated by either electrochemical or optical methods. In the latter, advantage is taken of color changes as they occur with surface-adsorbed indicators such as dichlorofluorescein when the endpoint is reached. Notwithstand-

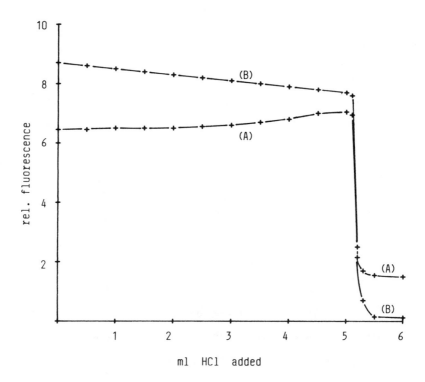

FIGURE 3. Typical curve as obtained in the fiber-optical pH titration of 5 ml 0.1 *M* sodium hydroxide with 0.1 *M* hydrochloric acid. In curve B, 20 mg of silver chloride were added to produce a turbidity which, however, does not affect the sharpness of the endpoint. Indicator: 2 μ*M* HPTS.

ing the widespread use of the Fajans method, it has three decisive advantages: it has to be performed at near-neutral pH, its endpoint is operator-dependent, and it cannot be automated easily.

Based on the finding that the fluorescence of certain quaternized heterocyclics such as 6-methoxyquinoline or acridine is dynamically quenched by halides and silver(I) ion, a new method for the endpoint determination in argentometry has been worked out that allows the fiber optic titration of either halides or silver ion.[8]

A typical example is the titration of chloride or related halides with silver nitrate at pH 2.3 in the presence of acridine. Other indicators such as quinine and phenanthridine may also be used. The analytical principle in this case is dynamic fluorescence quenching of the indicator by both halide and silver ion. Figure 6 shows the resulting plot obtained by following, with a plain two-armed fiber bundle, the fluorescence intensity changes of acridinium cation, when 0.1 N AgNO$_3$ is added to 0.01 KCl. At the beginning fluorescence is low due to dynamic quenching by chloride. As the titration proceeds, the concentration of chloride becomes smaller and reaches a minimum at the endpoint. At this stage practically all chloride is precipitated (as AgCl), and fluorescence quenching is not observed. Addition of excess silver(I) which acts as a quencher too, results in a decrease in fluorescence again.

The endpoint is clearly evident in the plots, even in 2 N nitric acid. In contrast to the popular Fajans method, where a *color change* is observed, there is an *intensity change* evident in this method. The standard deviation in the titration of 0.01 *N* chloride was ±1.8%, and 0.001 *M* halides were titrated with 0.01 *N* AgNO$_3$ with practically the same precision. Other ions that have successfully been determined by this method include bromide, iodide, isothiocyanate, sulfide, and silver(I) ion. The endpoint is not affected by the turbidity of the solution, and surface potential effects as observed gith Ag/AgCl electrodes were not found.

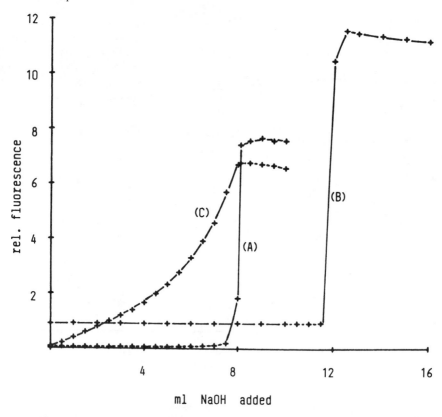

FIGURE 4. Optical signal change in the fiber optic titration of (A) 5 ml acetic acid (0.16 *M*); (B) 5 ml hydrofluoric acid (0.24 *M*); (C) 5 ml monobasic phosphate (0.16 *M*) in 45 ml water with 0.1 *N* sodium hydroxide Indicator: 2 μ*M* HPTS.

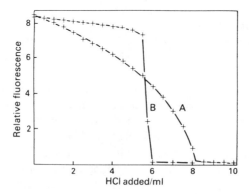

FIGURE 5. Optical signal change obtained in the titration of: A, 5 ml of hydrogen phosphate (0.16 *M*), and B, 5 ml of ammonia (0.12 *M*) in 45 ml of water with HCl (0.1 *M*) using fluorescein as an indicator (2 μ*M*).

Instead of adding the halide-sensitive indicator to the solution and following its intensity changes with a plain fiber, the fiber-optic halide sensors described in Chapter 7 may be utilized to follow the course of argentometric titration. While this may save the addition of indicator to each sample, the less efficient quenching of immobilized dye makes the shape of the titration curve less characteristic and, consequently, the endpoint less sharp.

Sulfide has been kinetically titrated[9] with heavy metal ions such as Pb(II) or Ag(I) based

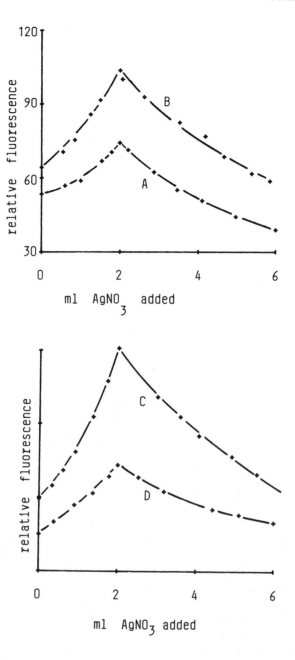

FIGURE 6. Fiber-optical titration of 20 ml 0.01 N potassium chloride with 0.1 N silver nitrate in aqueous solution of pH about 2.3 using acridine as an indicator. Excitation at 410 nm, emission taken at 480 nm. Acridine content: (A) 1.6 mg; (B) 2.9 mg; (C) 5.4 mg; (D) 9.9 mg.

on the finding that hydrogen sulfide anion adds to acridinium ion at alkaline pH. The resulting reaction product is nonfluorescent. In addition, hydrogen sulfide acts as a dynamic quencher of the acridinium fluorescence. At the beginning, the fluorescence of added acridine is low because of both "static" and dynamic quenching, but as the hydrogen sulfide disappears because of precipitation, its intensity increases and reaches at maximum at the endpoint, as shown in Figure 7. Sulfide or hydrogensulfide have been determined this way in concentrations from 1 to 10 mM with an average error of ±0.5%.

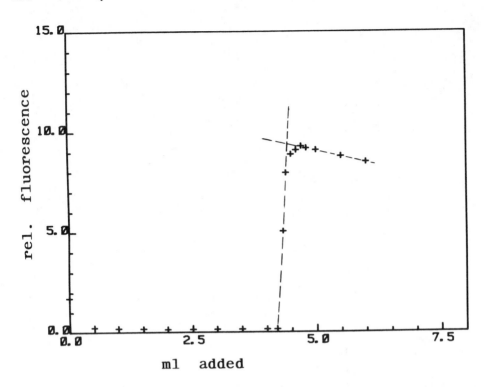

FIGURE 7. Plot of relative fluorescence intensity of acridinium cation (33.5 μM) during titration of 50 ml of sodium sulfide (8.9 mM) with silver(I) titer solution (0.2 M). Excitation wavelength 410 nm, emission taken at 480 nm.

IV. COMPLEXOMETRY

In this technique, changes in the color of a metal-complexing ligand are exploited to recognize the endpoint. Numerous methods and indicators are known[3,4] and await their adaption to fiber titrimetry. This also may contribute to the development of new analytical concepts for problems for which no solution existed so far.

Thus, the old problem of direct complexometric determination of aluminum(III) with 1,2-diaminocyclohexane tetra-acetic acid (DCTA) has successfully been solved by fiber-optical titration using morin as an endpoint indicator.[10] The fluorescent Al(III)-morin complex is slowly destroyed by addition of EDTA titer solution and reaches a low and constant value at the endpoint (Figure 8). Aluminium in concentrations from 1 to 200 ppm may be determined by this method with good precision, even in colored or turbid solution, although the slow complexation kinetics at levels below 20 ppm is annoying. Interestingly, the point of equivalence predicted by stoichiometry is different from the experimental value, resulting in an empirical but fully reproducible relation between DCTA consumption and Al(III) concentration.

Matsuo et al.[7] have used the photosensor described before to titrate Zn(II) and Mg(II) with ethylenediamine tetra-acetic acid (EDTA) in the presence of Eriochrome Black T (EBT). Figure 9 shows how the optical transmission of a mixture of Fe(III) and Cu(II) (each 0.025 M) changes on titration with 0.05 M EDTA in the presence of EBT. The endpoint can be found be correctly from the intersections of the extrapolated linear portions of the curve. Phototitrations have also been performed using a PVC membrane containing EBT. A decrease in the illuminance is observed as a result of the reaction of Zn(II) with EBT dissolved in the membrane. The shape of the curve is more complex in this case due to slow diffusion of metal ion, and ligand in and into the membrane. It was stated that the nature of the metal-dye-ligand interactions at the membrane surface requires further investigation.

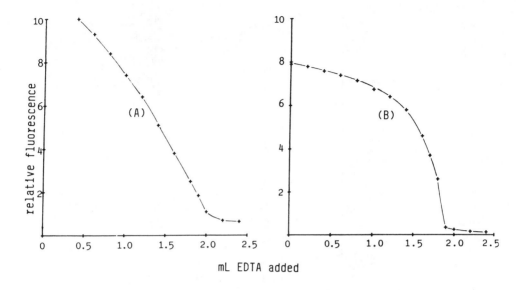

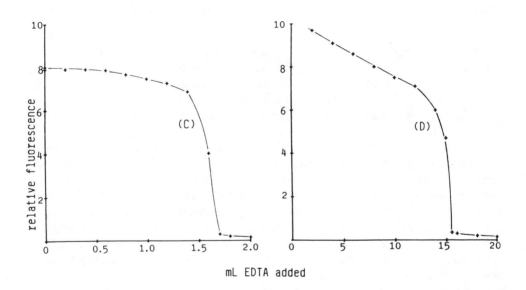

FIGURE 8. Titration of aluminum(III) with EDTA using morin (10 μM) as an indicator. A, 0.96 ppm Al(III) with 0.001 M EDTA; B, 9.6 ppm Al(III) with 0.01 M EDTA; C, 96 ppm Al(III) with 0.1 M EDTA; D, 960 Al(III) with 0.1 M EDTA. Excitation at 420 nm, emission at 490 nm. The concentration range over which titration can be performed is very large, but at less than 20 ppm Al(III) the titration kinetics is very slow.

Saari and Seitz[11] have applied a sensor based on immobilized calcein to titrate Cu(II) with EDTA. Initially, fluorescence is very low because of total fluorescence quenching by Cu(II). At the endpoint, there is a sharp increase in fluorescence when Cu(II) is pulled away from calcein because it binds more strongly to EDTA.

V. OTHER METHODS

Cerimetry is a frequently used method in oxidation-reduction (redox) titrations, because it offfers remarkable advantages over permanganatometry. It has the potential of being followed by fluorescence, since cerium(III) is fluorescent, whereas cerium(IV) is not.[12] The rather broad excitation and emission bands have peaks at 275 and 375 nm, respectively. Glass light guides

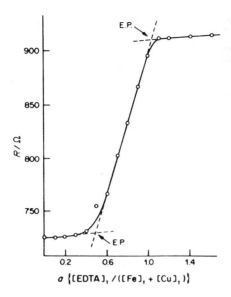

FIGURE 9. Phototitration of a mixture of Fe(III) and Cu(II) with EDTA using a photosensor membrane containing Eriochrome Black T and a red LED as a light source.

with their UV absorption shift the broad excitation maximum to 350 nm, and fluorescence is best observed at 400 nm. Since cerimetry is mostly performed in strong acid and elevated temperature, the use of fibers offers particular advantages in following the course of the reaction, because they are resistant to both acid and heat. Various other rare earth metal ions are known to be fluorescent and change their fluorescence intensity after being complexed, e.g., by EDTA. The increase in fluorescence can be utilized to monitor the course of titration.

Dissolved oxygen is frequently determined by the Winkler method which involves addition of excess $MnCl_2$ and KI to the sample solution, followed by phosphoric acid, and back-titration at pHs of the free iodine released from KI by MnO_2. We have developed an alternative procedure by using an oxygen optrode for direct determination of the endpoint.[13] The sample (typical volume 100 ml) is transferred into an airtight container and directly titrated, at a pH of 13 to 14, with about 0.05 to 0.1 N Mn(II) titer solution. The consumption of oxygen is indicated by the oxygen optrode immersed in the sample. The endpoint is characterized by a constant signal level. Other titrants for oxygen are dithionite, ferrous ion (in the presence of copper(II) at slightly acidic pH), and cerium(III).

One alternative to the use of oxygen-sensitive optrodes is to add a water-soluble oxygen indicator — for instance pyrene-butyric acid, anthracene-sulfonate, or a metal-organic ruthenium complex — to the solution and to use plain fibers to monitor changes in the fluorescence. Unfortunately, all good oxygen indicators either require UV excitation or are instable in strong alkali. Therefore, the use of an oxygen optrode seemed preferable. When, in contrast, the amperometric Clark electrode was tried in these experiments, no reproducible results were obtained, obviously because of the formation of surface potentials and diffusional barriers, caused by surface-deposited MnO_2.

The endpoint of a variety of other titrimetric procedures may be determined by fiber optic techniques. The course of bromometric and iodometric titrations, for instance, can be followed using the sensors known for elemental bromine and iodine, or those for bromide or iodide. The assay of dithionite stock solution[14] (using flavine mononucleotide as a fluorescent indicator) has been reported to give a sharp endpoint even under visual observation and may be refined by making use of fibers. In a series of papers, Goto[15,16] has presented various new fluorimetric procedures for redox and quenching titrations in cuvettes which, after being coupled to fiber optics, may be simplified and automated. Fluoride may be titrated with fluorescent aluminium chelates which results in the formation of AlF_6^{3-} and release of nonfluorescent ligand. A variety of fluorogenic reactions has been described by Weisz et al.,[17] and Bishop has shown[18] that 8-quinolinol can be used as a chromogenic titrant for bivalent heavy metal cations.

Aside from the titrations described before, which focus on electrolytes only, the interaction of intrinsically fluorescent biomolecules such as tyrosine or tryptophan and proteins with a second species (a binder) may be monitored via fibers. Such titrations can also be used to detect the number of binding sites of a fluorescently tagged macromolecule, using, for instance, Scatchard plots. Since performing a titration aliquot by aliquot is both tedious and cumulatively inaccurate, an automated titration is considered to be a substantial improvement. A simple automatic titrator (albeit without fibers) has been described and some of the problems encountered while performing titrations have been discussed.[19] Ideally, the process of fluorescence titration may be computer-controlled[20] and then would lend itself to fully automated titration of the tremendous numbers of specics for which phototitration protocols have been described.[21]

REFERENCES

1. **Frans, S. D. and Harris, J. M.,** Re-iterative least squares spectral resolution of organic acid/base mixtures, *Anal. Chem.*, 56, 466, 1984.
2. **Daly, J.C.,** *Fiber Optics*, CRC Press, Boca Raton, FL, 1984.
3. **Bishop, E., Ed.,** *Indicators*, Pergamon Press, Oxford, 1972.
4. **Kraft, G. and Fischer, J.,** *Indikation von Titrationen*, de Gruyter Verlag, Berlin, 1972.
5. **Wolfbeis, O. S., Schaffar, B. P. H. S., and Kaschnitz, E.,** Construction and performance of a fluorimetric acid-base titrator with a blue LED as a light source, *Analyst*, 111, 1331, 1986.
6. **Benaim, N., Grattan, K. T. V., and Palmer, A. W.,** Simple fibre optic pH sensor for use in liquid titrations, *Analyst*, 111, 1095, 1986.
7. **Matsuo, T., Masuda, Y., and Sekido, E.,** Acid-base and chelatometric photo-titrations with photosensors and membrane photosensors, *Talanta*, 33, 665, 1986. In this article, the legend of figures 5 and 6 obviously should be interchanged.
8. **Wolfbeis, O. S. and Hochmuth, P.,** A new method for the endpoint determination in argentometry using halide-sensitive indicators and fiber-optic light guides, *Mikrochim. Acta (Vienna)*, III, 129, 1986.
9. **Trettnak, W. and Wolfbeis, O. S.,** Kinetic titration of sulphide with heavy metal ions using a highly sensitive fluorescent indicator, *Fresenius Z. Anal. Chem.*, 326, 547, 1987.
10. **Wolfbeis, O. S., Schaffar, B. P. H. S., and Chalmers, R. A.,** Direct complexometric titration of aluminium(III) with DCTA, *Talanta*, 33, 867, 1987.
11. **Saari, L. A. and Seitz, W. R.,** Immobilized calcein for metal ion preconcentration, *Anal. Chem.*, 56, 810, 1984.
12. **Kroeger, H. and Barker, J.,** Die Luminescenz von Cer-Verbindungen, *Physica*, 8, 628, 1941.
13. **Greschonig, H. and Wolfbeis, O. S.,** Unpublished results, 1988.
14. **Eberhard, A.,** Anaerobic fluorescence titration of dithionite solutions, *Analyst*, 111, 475, 1986.
15. **Goto, H.,** The use of phosphine, fluorescein and rhodamine B as fluorescence indicators in oxidimetric titration, *J. Chem. Soc. Jpn.*, 59, 1357, 1938.
16. **Goto, H.,** Naphthoflavone as a fluorescence indicator in iodometry and bromometry, *J. Chem. Soc. Jpn.*, 59, 365, 1938; see also *Chem. Abstr.*, 35, 1720 ff, 1941.
17. **Weisz, H., Pantel, S., Dilger, C. M., and Glatz, U.,** Application of fluorescence reactions in the ring oven segment technique, *Mikrochim. Acta (Vienna)*, I, 69, 1984.
18. **Bishop, J. A.,** 8-Quinolinol-5-sulfonic acid as a titrant for bivalent cations, *Anal. Chim. Acta*, 35, 224, 1966.
19. **Igoli, G.L. and Penzer, G. R.,** Automatic fluorescence titrations and their interpretation, *Anal. Biochem.*, 64, 239, 1975.
20. **Kleinfeldt, A. M., Pandiscio, A. A., and Soloman, A. K.,** A computer-controlled titration fluorescence spectrometer, *Anal. Biochem.*, 94, 65, 1979.
21. **Mooibrock, M.,** Methods of Phototitration, Mettler Handbook, Mettler Gmblt., Grevfensce, Switzerland, 1978.

Chapter 14

FIBER OPTIC CHEMICAL SENSORS IN NUCLEAR PLANTS

Gilbert Boisde, Floreal Blanc, Patrick Mauchien, and Jean-Jacques Perez

TABLE OF CONTENTS

I. INTRODUCTION

The use of optical fibers in nuclear environments remained a relatively untouched subject for many years, due to its connection with military or industrial objectives. However, many applications have stimulated growing interest since 1972. At the time, the most explicit investigations dealt with the effects of ionizing radiations (X-ray, γ, electrons, neutrons) on optical fibers and thick samples of vitreous silica from various sources.[1,2] The objectives were mainly concerned with remote data transmission. With the expanding needs of telecommunications, silica fibers, or fibers doped with germanium gradually replaced all others. The main applications, which include nuclear testing, nuclear power plants, and electromagnetic pulse testing, were the driving force of the many worldwide activites in this field, punctuated by numerous specific international conferences.[3-10]

Civilian research on fiber optic sensors in the period from 1975 to 1980 enjoyed privileged applications in two areas: nuclear reactors and spent fuel reprocessing facilities. The former involved the performance of diagnostic measurements for pressurized water reactor (PWR) research programs.[11,12] Sensors for two-phase and vacuum ratio measurements were tested in a nuclear reactor.[13] They measured the refractive index variations observed at the fiber ends between liquids and gases. This type of sensor is mainly used in research laboratories and pilot plants to establish a mapping of the phase states in the boilers and pressurizers and will not be discussed here.

To the best of our knowledge, the earliest experiments on chemical measurements using optical fibers in nuclear installations are related to *in situ* on-line monitoring of plutonium (IV) in gloveboxes. This work, performed by French nuclear investigators in 1974, was not reported at this time. It involved a differential measurement at two wavelengths, the absorption peak (477 nm) and an absorption valley (520 nm) of the optical spectrum. The interference filter photometer ($\Delta\lambda$ = 5 nm) used was the ancestor of the Telephot®, (see Section II.C). A bundle of glass fibers transmitted the data which are related to the Pu (IV) concentration to two photomultipliers via bifurcated fibers. Sealed windows inserted between the inner and outer optical fibers served as barriers to protect from radioactive contamination. The major value of these investigations was to indicate very early the technical problems faced in making optical fibers operational in nuclear environments with particular emphasis given to operating constraints and influence of radiation.

Since 1980, considerable technological progress has been achieved in the nuclear field. Advances specific to chemical measurements are reported here, particularly in spectro-photometry and spectrofluorimetry. The special aspects of operation in a nuclear environment will also be examined.

II. OVERVIEW OF FIBER OPTIC CHEMICAL SENSORS

Four main civilian nuclear branches are concerned with optical fiber applications in chemistry: power reactors, environmental sciences (geology, wastes), iostopic separation, and spent fuel reprocessing plants.

A. NUCLEAR REACTORS

Since reactors using molten salts, molten metals, and gases were mainly of the experimental type, the main applications of optical fibers were essentially related to water chemistry in boiling water reactors (BWR) and PWR. Water testing is essential to prevent corrosion and leakage of radioactive products. A major obstacle to *in situ* measurements is imposed by the operating conditions of these reactors including high pressure, temperature, and neutron activity. Accordingly, the measurements are always taken by samplings at temperatures below 50°C. Analytical robots with a continuous (Technicon) or sequential flux, such as the CEA DIMA® (discrete multi-analyzer)[14] generally performed these functions for most conventional measurements. Typically, lithium, boron, iron, silicates, phosphates, and chlorides were assayed using colorimetric techniques that are well-known in water chemistry.

The typical configuration of an analytical chemical robot incorporating optical fibers is shown in Figure 1. After the pressure is lowered, the sample is taken by simply opening a valve connected to the main stream, and transferred to constant-volume dose distributors. These units consist of vessels fitted with primable siphons. The standard and the blank samples can be conveyed to these instruments by means of a slight vacuum or by pressure applied to the corresponding containers. The transfer of liquid from one container (chemical reactor) to the second (measurement) is carried out by identical measurers. The reagents may be acids or bases. Solvents are added when a liquid-liquid extraction is necessary. The advantage of the colorimetric measurement by optical fibers is evident if the samples are radioactive or if they require solvent which always implies the risk of explosion. The measurement cells, usually of the flow-through cell type, have optical path lengths of 0.5 to 10 cm. Volumes of 0.3 to 0.5 ml are easily obtained for an optical path of 5 cm.

Another application of optical fibers for monitoring reactor waters is the *in situ* measurement of particles in suspension (corrosion products, degradation products of ion exchange resins) in pressurized circuits. Pressurized cells (180 bar) with thick silica windows and optical fibers were accordingly placed in the cooling circuit upstream and downstream from the filtration systems. Optical absorption measurements permitted the determination of concen-

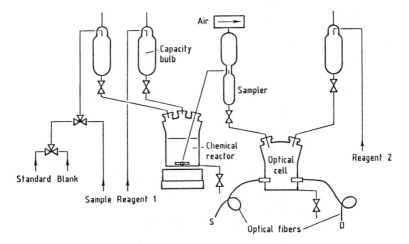

FIGURE 1. Schematic representation of a chemical robot with liquid-liquid extraction for measurement of traces in reactor water or wastes. The optical fiber cell is used to separate the electronic device form solvent zone.

trations in the range of 0.01 to 10 ppm with 50 cm optical path cells. Calibration was performed with magnetite in water suspension.[15] When using a white light source, the cut-off of the near infrared (>750 nm) is mandatory to prevent a temperature rise in the solutions and the production of microbubbles at the interface of the windows. Similarly, the pressurization and cooling conditions of the solutions remained problematic. The sensitivity of 1-m optical path cells[16] is such that measurements of a few tens of parts per billion of chlorides were achieved by the injection of $AgNO_3$ resulting in a precipitation of $AgCl$ and, therefore, some turbidity.

For reasons specific to the operators of nuclear power plants, these applications of chemical measurements using optical fibers were never generalized to our knowledge. The secondary circuit is normally free of radioactive products, and the measurements are taken conventionally. The primary circuits are inaccessible and present severe constraints such as high pressure and temperature and intense neutron or gamma radiation.

B. ENVIRONMENTAL SCIENCES

The monitoring of subsurface waters in nuclear waste repositories was investigated by the Lawrence Berkeley Laboratory jointly with the Lawrence Livermore National Laboratory.[17,18] Particularly, the migration of actinides in the soils and mineral deposits of various types requires extremely sensitive analytical techniques. This also applies to geological exploration for uranium and thorium ores. The laser-induced fluorescence excitation (LIFE) technique uses an argon laser (1 mW) of 488 nm (for uranium). The fluorescence spectrum is dominated by the lines at 527/528 and 520/521 nm and a peak at 530.3 nm. The calibration curve is linear over the concentration range from 10^{-5} to 10^{-8} M (2.7 ppb). The detection limit can be as low as 10^{-14} M for a signal-to-noise ratio of 7. This sensitivity is greater than that afforded by radioactive techniques. The usual drawbacks of the simple fluorescence technique, i.e., the influence of contamination, are largely reduced by the UO_2–CaF_2 co-precipitation method, which concentrates UO_2 (II) by a factor of 50 to 100 and removes the impurities, and by the calcination that destroys the organic contaminants. This method is by far more effective than necessary, because neutral waters have a uranium concentration approaching 5.10^{-10} M.

The extension of this laboratory method to *in situ* measurement in natural waters uses a silica optical fiber with a 100 μm core, 0.2 to 1 km long, with the remote fiber fluorimetry (RFF) technique.[18,19] The optical performance and calibration are related to the Raman spectrum of naphthalene.[20] In an early system,[17] the optode consisted of two fibers with a fixed angle between their ends, in order to define a volume related to the numerical aperture of the transmitting and receiving fibers. The use of a single fiber with a sapphire microsphere serving as a lens at the fiber end[19] helped to detect minute traces of rhodamine 6 G serving as a tracer to calibrate the device. A pierced mirror is used to separate the excitation light produced by the laser from the return fluorescence emission. A capillary placed at the end of the fiber increases the fluorescence signal. The incorporation of insoluble reactants or substances covalently bonded to the end of the fiber is being considered for specific chemical measurements,[21] e.g., for chloride.

These investigations were supplemented by the development of semipermeable membrane optodes. This membrane isolates a reservoir containing an aqueous solution of phosphoric acid, which enhances the fluorescence of the uranyl ion[19] migrating across the membrane. The method makes it easy to obtain sensitivites in the range of 10^{-6} M for the uranyl ion. Some quenchers, such as Fe(III), can be analyzed by the decrease of the uranyl fluorescence signal.

The extension of the method is the laboratory measurement[22] of Pu(III) at concentrations of 10^{-4} M. The energy level of this ion is close to the highly fluorescent terbium (III) ion. Plutonium decreases its fluorescence by energy transfer. This effect is enhanced in the presence of D_2O or of complexing agents such as acetate and trifluoroacetate.[23]

In France, trace amounts of uranium have been determined by three methods. The first is currently restricted to uranium ore processing plants to monitor liquid wastes. The chemical robot called DIMA 12 is schematically shown in Figure 1. The uranium is extracted from the aqueous solution using trioctylphosphine oxide dissolved in an aliphatic hydrocarbon and reacted with dibenzoylmethane to give a chromophore having $\varepsilon = 25{,}000$ at 410 nm. The concentrating extraction with a volume ratio 1 to 50 between aqueous phase and organic phase allows for detection limits of 5.10^{-7} M. The use of optical fibers is necessary due to the presence of solvents. This system, used on-line in a river with the measurement system in a caravan, has the drawback of high solvent consumption.

The second method is the direct measurement using cells with a long optical path.[16] Tested on wastes, its sensitivity is nevertheless inferior, 5.10^{-6} M of uranium (IV) and 10^{-5} M for uranium (VI) being detectable with 1-m cells. This method has been supplanted by the RFF method both for monitoring waters with tracers (fluorescent rare earths) and for traces of uranium. Passive optodes with optical fibers in the form of a ring have also been developed.[24]

The fluorescence excitation by pulsed nitrogen laser operating at 337 nm is possible if the silica fibers are less than 20 m long, due to losses in the UV by Rayleigh diffusion. The fluorescence spectrum of uranium is given in Figure 2. *In situ* operation to monitor liquid wastes and for analyses of natural subsurface waters is under way via remote measurement up to 1.5 km. The detection limit is about 5.10^{-12} M using a spectrofluorometer with a time-gated intensifier and photodiode array.[25]

C. ISOTOPIC SEPARATION

Gaseous diffusion across ceramic or metallic barriers and centrifugation are industrial methods of isotopic separation. The problems raised by the kinetics of the gaseous phases, the corrosion effects of fluorides, and process control led to the development of real-time spectrometric observation. Hexavalent uranium (UF_6) can be determined by fluorescence. In the Los Alamos National Laboratory a technique has been developed for the observation of UF_6 in the centrifuges.[26-28] A central nonrotating column assembly was designed for real-time measurement. A laser pulse is transmitted by a 600 μm core optical fiber and focused in a peripheral region by means of an assembly of lenses and mirrors. The fluorescence image returned is observed outside the centrifuge using an internal telescope and an array of coherent fibers that examine clearly defined optical paths. The UF_6 molecules can be excited within a wide band between 375 and 420 nm with a peak at 392 nm. Filters eliminate scattered excitation light. The lifetime of the fluorescence depends on the pressure and temperature,

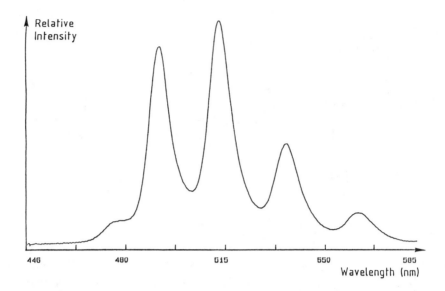

FIGURE 2. Fluorescence spectrum of uranium (VI) traces taken with a photodiode array spectrofluorometer with a time-gated intensifier and a passive optode.

while the intensity is directly proportional to the number of molecules excited by the laser. The gate delays vary from 100 to 900 ns, and the gate width from 20 to 200 ns.

In a similar area, but for the purpose of process control, and absorption cell for measuring the gases UF_6, F_2, ClF_3, and PuF_6 was built by Saturday.[29] The source is a low-pressure mercury lamp having strong lines at 253.7 and 365 nm. Measurements up to 2.5 absorbance units were achieved despite high losses of light due to the measurement cell and to the silica optical fibers used in the UV. The minimum pressures detected are 0.01 torr for PuF_6, 0.05 torr for UF_6, 1 torr for ClF_3, and 5 torr for F_2.

The Chemex process is a process of isotopic enrichment by chemical method. The aqueous phase is a concentrated hydrochloric acid medium. The main species are the uranium valencies (III) and (IV). The monitoring of the species present, i.e., U (III), U (IV), and U (VI), in the different chemical operations led to the development of a remote *in situ* measurement program using optical fibers in an industrial pilot plant.[30] Figure 3 shows the relative absorptions of U (III) and U (IV).

Two types of the Telephot® fiber photometer, described elsewhere,[31] were tested. The first, with four wavelengths, served to perform the simultaneous monitoring of U (III) and U (IV) at the outlet of the electrolyzer and in isotopic exchange, using two measurement points. The high total uranium concentrations (between 1.2 and 1.6 M) required measurements on the sides of the absorption spectrum for U (III) at 746 and 754 nm, at the absorption peak (655 nm) at lower concentration for U (IV), and the choice of a common absorption valley as a reference for both spectra (780 nm). The flow-through measurement cells with 1 mm optical path were made with thick silica windows and poly(vinylidenedifluoride) to withstand the pressure (10 to 15 bar) and resist corrosion. The optical fibers of the PCS 1000 type (core diameter 1 mm) were specially sheathed for industrial operation. The results of real-time measurements without sample transfer, which is necessary because of the simultaneous presence of several oxidation phases, are in the same range as the laboratory measurements. The precision for U (III) is better than 1%.

The second type is a Telephot® fiber photometer with two wavelengths, ideal for trace monitoring. Two measurements were attempted in new conditions at the oxidation outlet, one to monitor oxidation of uranium (III) and the other for traces of U (VI) to evaluate the excess oxidation. The measurement is taken in a concentrated solution of uranium (IV). The optical

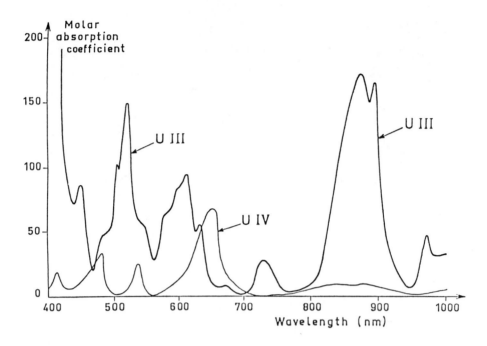

FIGURE 3. Comparative absorption spectra of uranium (III) and (IV) in chloride medium.

fiber measurement cells used an optical path that was easy to provide (0.5 cm). The method already described[31] uses two selective wavelengths of the optical spectrum with the same absorbance (isoabsorbance) for the main species as U (IV). It is first necessary to confirm the Beer-Lambert laws relative to each of the species in the required acidity ranges. The concentrations measured by this method are about 2.10^{-3} M for uranium (III), and 4.10^{-3} M for uranium (VI) in 1.3 to 1.5 M of uranium (IV). A third measurement was also attempted with this type of instrument and a poly(vinylidenedifluoride) optical fibers cells with 50 cm optical path. This was done to monitor traces of uranium (VI) at the extraction outlet in a 6 to 7 N HCl aqueous phase. The detection limit obtained is about 1 mg.l^{-1}.

D. SPENT FUEL REPROCESSING

The separation of plutonium from depleted uranium in nuclear power plant fuels, the complex dissolution operations for these fuels, the isolation of the fission products, the recovery of radioactive isotopes such as molybdenum and americium, the reconditioning of recoverable products, and the disposal of wastes (vitrification, etc.) are operations conducted in specialized chemical units, the so-called spent fuel reprocessing plants. Although the purposes of these units which serve for plutonium recovery and radiochemical separation are varied, the means employed are generally of the same type.

The relative insensitivity of fibers to radiation was exploited to develop units for remote spectrometric measurement, laboratory analysis, and process control.[32] Four sites are mainly concerned by the use of optical fibers in nuclear environments:

- In the U.S., the Savannah River Laboratory (SRP) (E. I. Du Pont de Nemours) associated with the LLNL (Livermore) for the measurement of fluorescence
- In France, the different facilities of the Commissariat à l'Energie Atomique (CEA) based on the work of DERDCA (Division d'Etudes de Retraitement et des Déchets et de Chimie Appliquée) at the Fontenay aux Roses Nuclear Research Center

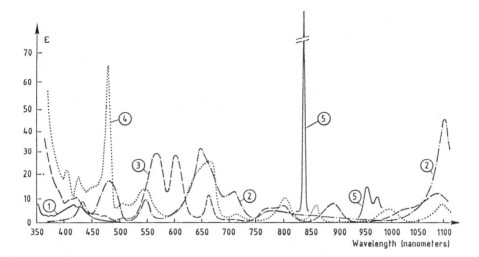

FIGURE 4. Comparative spectra of uranium and plutonium species in nitric solutions. 1 U (VI) - 2 U (IV) - 3 Pu (III) - 4 Pu (IV) 5 Pu (VI). ε is the molar absorption coefficient.

- In Great Britain, the Dounreay Nuclear Power Development Establishment (DNPDE) and UK Atomic Energy Authority (UKAEA)
- In West Germany, the Karlsruhe facility (Kernforschungszentrum)

The basic guidelines and main achievements are discussed, with a distinction between the two techniques that are so far operational, namely absorption and fluorescence.

1. Absorption

The examination of settling and agitation of slurries of wastes in storage tanks employed in nuclear reprocessing plants has been studied at UKAEA with a solid-state fiber optic absorption sensor[33] and at CEA with a granulometer designed with a remote contrifugation unit and optical fibers for a high activity cell. However, the most closely investigated species in the past 10 years are uranium and plutonium, generally in nitric acid medium. The absorption spectra of the species U (IV), U (VI), Pu (III), Pu (IV), and Pu (VI) are compared in Figure 4. The variation in the acid content is an important factor for the chemical analysis and continuous monitoring of these species. Many investigations have therefore been devoted to this influence, on both uranium[34,35] and plutonium[36] for the needs of *in situ* on-line control.

The development of interference filter photometric analyzers incorporating silica optical fibers (49 strands of 200 μm) was the subject of the earliest developments. In the U.S. (SRP), the Du Pont 400 fiber photometer was improved for remote monitoring, at 50 ft, of the elution of plutonium and neptunium in ion exchange columns.[37,38] The nitric acid concentration (7 to 8 M) is adjusted to obtain quantitative results. The peak at 546 nm and a reference valley at 586 nm are used for Pu (VI), the peak at 595 nm and a reference at 656 nm for Np (IV), a species which also has peaks at 720, 980, and 1180 nm. In France, this type of equipment for measurement at two to five wavelengths with a Telephot® fiber photometer is represented by portable instruments for the needs of nuclear inspectors.[39] Thus, uranium (VI) is measured from 1 to 300 g.l⁻¹ to within 3%. For monitoring of uranium losses at the bottom of the liquid/liquid extraction column of the Purex process this photometer was used with 1 m optical path cells or according to the chemical extraction method described above, with a sensitivity of a few mg.l⁻¹ of uranium (VI). In West Germany, the oxidation and electrolytic reduction steps of the uranium and plutonium extraction processes give rise to a special development, a photometer using a dye laser pumped by a nitrogen laser as a light source.[40-42] The distance between the source and the flow-through cells of 1 mm or 1 cm optical path length, is about 100 m, and employs silica fibers with 200 μm cores. A beam splitter in a hot cell provides the zero of the optical line. The plutonium concentrations are obtained with a deviation of about 1 g.l^{-1} in the range 20 to 50 g.l^{-1} for a single-sample analysis. The deviation for a nitric acidity measurement is better than 0.2 M in the range 0.5 to 2.5 M. For on-line measurement, the sensitivity for plutonium is less than 0.14 g.l^{-1}, while it is better than 3 g.l^{-1} for uranium (range 14 to 77 g.l^{-1}).

The need for rigorous evaluations of plutonium using the spectrometric method for measuring Pu (VI) at 830 nm in the presence of neodymium as reference led to the incorporation of optical fibers in commercial spectrometers since 1980.[43] This equipment, which is still operational, required in 1977 the production of a PCS type optical fiber (QS) with a large core diameter (1000 μm, and even 1500 μm experimentally). Adaptors[44] for optical fibers were thus developed for most existing spectrometers (Varian 2300, Beckman DU7, Safas single and double beam, Hewlett Packard). They are marketed in France under the trade name "Fispec" by the Photonetics Society. Concurrent to this development an identical one occurred at the SRP for the equipment of the new photodiode array spectrometers (HP 8451 and HP 8452).[45] The advantage of this type of instrument is to permit real-time measurement with excellent reproducibility. This equipment, used on the OSUR (on-site uranium recycle) prototype, has been designed for the continuous monitoring of the uranium content of the cationic resin by analyzing the concentration in the effluent.[46] Regulation is achieved by modifications of the uranium (VI) concentration of about 25 ppm. 10 ± 0.5 ppm of uranyl nitrate are detectable.

In Great Britain, a fiber optic spectrophotometer system has been developed at DNPDE using modular commercial available units and a bundle (6 mm) of glass (or quartz) fibers wrapped with lead.[47] Computer programs have been used to assist routine analysis of Pu (VI) with a precision of better than 0.5% in glove box and first cycle raffinate samples in highly active conditions, with levels of plutonium as low as 5.10^{-3} g.l^{-1}. Peak stripping routines have also been developed for identifying 0.5% Pu (III) in mixed valence solutions. Americium (III) can be measured at 503 nm in raffinate streams in presence of plutonium which causes a 10% error when the plutonium to americium ratio is 6:1.

The complexity of U and Pu mixtures, the wide range of needs including americium and neptunium, and the many possible interferences in the processes such as hydrazine, hydroxylamine, and fine particles, led the staffs of the SRP in the U.S.[48] and the CEA in France[31] to combine three technologies: optical fibers, photodiode array spectrometers, and multivariable analysis, which requires a modeling of the optical spectra of the different species present. In the U.S.A., these instruments are used for ion exchange columns, and for air-lift samplings of uranyl nitrate and plutonium (III) for recycling. The HP 8451 and HP 8452 spectrometers with a concave holographic lattice and 328-element photodiode arrays were selected along with PCS 600 fibers. The standard deviations obtained are ±0.34% for uranyl nitrate within the 0.2 to 20 g.l^{-1} range, 0.82% for Pu (III) within the 0.05 to 5 g.l^{-1} range, and 0.94% for Pu (IV) within the 0.5 to 5.l^{-1} range. In France, applications concern the control of mixer-settler batteries in the uranium/plutoium partition step of the Purex process[49] and for the equipment of a process for the oxidizing dissolution of plutonium dioxide in nitric medium in the presence of Ag (II) ions[50] with the simultaneous measurement of Pu (VI) at 622 nm, americium at 503 nm, and Ag (II) by calculation using the absorbance at 550 nm. In both cases, the spectrometers are Hewlett Packard instruments. However, for work in the near infrared, e.g., Pu (VI) at 830 nm and neptunium at 980 nm, the CEA has developed[31] an optical fiber and photodiode array spectrophotometer consisting of 1728 photodiodes, called the DTC 1000, which is marketed under the trade name "Spectrofip" by the Photonetics Society. The PCS 1000-type optical fiber used so far is gradually being replaced by the all-silica Polymicro fiber of core diameter 400 μm.

2. Fluorescence

The use of the RFF (remote fiber fluorometry) method for process control was attempted in the U.S. and France. The advantages offered by the sensitivity of this technique are extremely interesting in a radioactive environment, because the samples can be diluted, thereby reducing the specific activity and interferences to an acceptable level.

At the SRP,[51] the uranyl ion is excited at 416 nm by means of an Nd:YAG laser operating at 1060 nm, a third harmonic generator to produce 355-nm light, and a Raman shifter. The maximum fluorescence induced is obtained at 513 nm and measured by means of an intensified photodiode spectrometer. The uranium optode terminates in a ruby placed between the fiber and the solution, and the fluorescence of the ruby serves as a reference. The measurable uranium concentration range is 0.2 to 10 g.l^{-1}. Phosphoric acid enhances the fluorescence intensity, and the addition of two drops of $FeCl_3$ at g.l^{-1} decreases it by quenching its lifetime to 97 µs as opposed to the 197 µs lifetime in pure phosphoric acid medium.

This technique, discussed for environmental measurements,[24,25] has supplanted absorption methods for leak testing of column bottoms or process effluents. With its high selectivity, it offers the advantage of being less sensitive to fission products and darkening of the solvents under radiation, because of the possibility of dilution. Optodes with fibers placed in a ring arrangement with a conical pierced mirror have also been built. The sensitivity (10^{-8} M) with optical fibers, which can be improved with further optimizations, is in the same order of magnitude as with a conventional 1-cm cell using a Fluo 2001 spectrometer with intensified photodiode arrays from Dilor Society.

Another application of RFF is in the continuous analysis of iodine in gaseous effluents,[52] intended to model the operation of the process and to monitor the efficiency of the traps. The method employs the shift in the spectra of I_2^{127} and I_2^{129}, which ensures isotopic selectivity. The excitation of I_2^{127} at 514.5 nm with argon laser gives emission lines at 557 and 570 nm, and the clearly resolved lines of I_2^{129} in mixtures of I_2^{127} and I_2^{129} can be analyzed. The knowledge of the isotopic ratio allows the direct monitoring of I_2^{129} with a detection limit better than 10^{-10} M in the presence of pure and dry air. However, NO_2 quenching is observed at partial pressures above 50 torr.

III. CONCEPTS FOR ON-LINE MEASUREMENT

The widespread research conducted throughout the world shows that optical fibers are now firmly established in nuclear pilot facilities. Their implementation into plants is slow because of the structure of these plants. However, to gain time in data transmission and to operate with fewer staff, plant operators show great interest in remote *in situ* on-line measurement by optical fibers. Figure 5 shows a schematic diagram of spectrometric measurement (RFF) in the main stream or on a derivative stream in a nuclear environment. The light source, consisting of a pulsed nitrogen laser and a dye laser or harmonic analyzer and Raman shifter,[51] can be replaced by a preferably modulated white light source for spectrophotometry and reflectance measurements. The optodes, which have a single optical cable (possibly multifiber) or flow-through cells are easily disconnected to guarantee interchangeability. Optical multiplexing can be achieved outside the radioactive zone. For on-line control, the new concepts introduced by the advent of optical fibers were recently analyzed, and the modeling of optical spectra and the need for an internal reference has been discussed.[31,32]

A. MODELING

The use of multi-element optical spectra largely depends on their complexities. In absorption, the minimum wavelength range for simultaneous measurements is [n + N + (x + 1)], where n is the number of independent species, N the number of interfering functions or species, and x a modeling factor depending on the background noise correction method. The

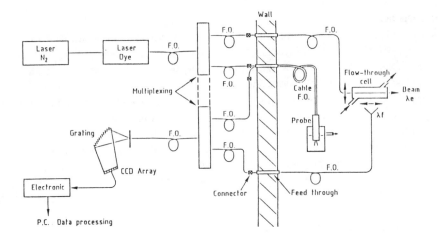

FIGURE 5. Schematic of a remote fiber fluorimetry in nuclear plants. When the losses in line are low, e.g., in visible, a white light source can be sufficient for the remote fiber spectrophotometry in place of the lasers.

simplest instruments therefore have a capacity of two wavelengths (N = 0, x = 0), one for measurement and the second for the reference. This minimum has been adopted for most determinations of uranium[30,38] and even plutonium[42] in process control. A rigorous analysis, attempted for uranium using its complexation constants and the data on the influence of ionic strength[35] proved to be less effective than an empirical determination.[34,39,53] The simple function:

$$A = \sum_{i=0}^{n} \left[M_i \left[a_i [A] + b_i \right] \right]$$

where A is the anion, M the cation, a and b the calibration coefficients for the i species present, has been exploited for different valencies of uranium[30,31] and for uranium/plutonium mixtures.[49] Computer programs with algorithms are usually necessary.[48]

The selectivity of fluorescence does not usually entail this modeling except in the presence of undesirable influences such as pressure and temperature[28] or of quenching processes that must be monitored.[51]

B. INTERNAL REFERENCE

Many proposals have been made to provide an intrinsic reference for measurement by optical fibers. They are summarized as follows:

- A spectral reference in wavelength,[54] e.g., at an absorption valley or on a side of the optical spectrum
- A reference optical line, possibly placed in the active zone,[42] permitting to compensate the losses of the measurement line
- A Raman spectrum from the optical fiber, in correlation with a known product such as the naphthalene[20]
- The addition of a compound at the end of the optode (such as ruby) whose fluorescence serves to determine the excitation intensity[51]
- The fluorescence lifetime,[17] with a preliminary calibration
- The addition of species that favor quenching for certain laboratory analysis,[21] with the determination of quencher influence
- The addition of a standard in solution,[31] such a known concentration of neodymium for the analysis of plutonium (VI)
- The construction of an optode with an associated measurement, used as reference

The choice of this reference is especially important in a nuclear environment because the optical line may itself undergo spectral modifications due to the ionizing radiation.

IV. FIBERS AND INSTRUMENTATION FOR NUCLEAR MEASUREMENTS

A. FIBERS

Two considerations are essential for the use of optical fibers in a nuclear environment: Their special behavior under radiation and their attenuation in the ultraviolet.

Many investigations have been performed as to the influence of radiation on optical fibers (see Section I), chiefly aimed at the transmission of optical signals with its consequences in the presence of high doses received during a short time interval (analysis of transients and of fiber regeneration). The purity of the fiber, its composition, its manufacturing characteristics, the dose rate, its type, and the integrated doses are all criteria that need to be analyzed. Summaries of these different influences have been reported.[55,56] However, little specific research on spectrometry by optical fibers applications has been conducted. The possibilities of on-line monitoring of uranium gave rise to the first published work[57] on bundles of quartz and silica fibers exposed to around 10^9 rad (dose 1.2 10^6 rad.h^{-1}). With the same objective, radiation induced attenuation between wavelengths 0.4 to 2.5 m have also been measured with fibers of various types including plastic, cerium doped glasses,[58] silica,[59] and fluoride glasses.[60]

The main data gathered worldwide are summarized in Figure 6, with the addition of the latest investigations on all-silica fibers with polyimide sheaths (polymicro Technologies, Inc)[61] produced from Heraeus preforms. The optical losses at 0.60 to 0.85 and 2.4 μm of the different commercial fibers are given as a function of the integrated dose. It is obvious from curve 3 that glass fibers are not suitable for use under high radiation. The cerium doping, although effective under radiation, strictly limits the usable spectral zone because it results in high attenuation below 530 nm. Plastic fibers (PMMA and polystyrene) are still valid for integrated doses less than 10^5 to 10^7 rad in a narrow spectral zone. Fluoride glasses are usable in infrared spectrometry for monitoring solvents and powder moisture content.[60] Above 10^7 rad, fracture of the large-diameter PCS fibers, caused by embrittlement of the sheath, makes them useless.[59] For high integrated doses (up to 10^{10} rad), all-silica fibers are generally suitable. The variations in the results observed are related to their purity and their manufacturing quality. Polyimide sheathed fibers of 400 μm core (e.g., from Polymicro) are perfectly satisfactory under radiation[61] for most applications in the visible and the near infrared (<1.8 μm).

Fibers for use in the far ultraviolet, e.g., for absorption measurements of nitrogen oxides or for laser excitation in the UV fluorimetry, require a specific spectral investigation[62] and the evaluation of the losses on various samples made from preforms doped or not doped with OH ions. The spectral variations obtained under radiation (10^8 rad) are shown in Figure 7. It can be seen that OH doped fibers are more advantageous in the ultraviolet,[59] although the presence of these ions is a disavantage in the near infrared, especially at wavelengths close to the corresponding absorption peaks at 0.725, 0.945, 1.23, and 1.39 μm.

B. ASSOCIATED INSTRUMENTATION

Optodes — Most optodes made for nuclear purposes are of the "passive" type in that the fibers act as simple light guides. However, for low integrated doses (environment), specific U (VI), Pu (III), I_2 optodes have been considered.[23] Research underway on pH measurements in acidic or basic media[31] is still too prospective to be used in a plant in the presence of ionizing radiation.

Components — The constraints of a nuclear environment which can include irradiation and a harsh chemical environment necessitate specific systems, for the connectors in stainless

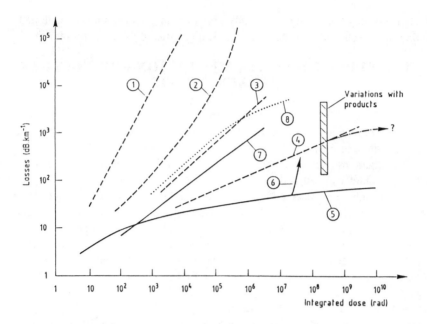

FIGURE 6. Approximate comparison of the radiation-induced attenuation with the integrated dose for different fibers. (A), *dashed* (—) at 0.6 μm: 1 glass fiber, 2 plastic fiber (PMMA), 3 doped cerium glass, 4 SiO_2/F fiber at "drawing" peak, (B), *full line* (——) at 0.85 μm: 5 SiO_2/F fiber, 6 PCS silica fiber with silicone cladding, 7 SiO_2 (Ge)/ SiO_2 or F (Ge) fiber, (C), *dotted* (·····) at 2.4 μm: 8 fluoride glass (ZBLA).

steel, for sealed penetrations, and for the optical cable with, for instance, a coating by heat-shrinkable olefin. The remote positioning of the fibers requires a structure of the fiber ends adapted to technological environment.

Optical multiplexers still require significant development. While industrial (Guided Wave, Inc.) or local products[43] suggest that the problems have been solved for multipoint measurements in absorption techniques, a good knowledge of link budgets is necessary in the area of fluorimetry and in the UV range.[32]

The equipment selected is concerned with absorption and fluorescence. In absorption, the latest developments concern rugged instruments with interference filters for continuous on-line control and photodiode array spectrophotometers for real-time measurements. In the former case, the increase in the measurement range (0 to 80 dBm) by modulation of the light, automatic scale change, and the use of stable photodiodes as detectors are the selection criteria. An instrument of this type should be marketed shortly in France under the name Telephot®. In the later case, the single beam photodiode array spectrometers of the HP 8450/ HP 8452 and Spectrofip (Photonetics) type have demonstrated their capability of dealing with complex problems.[31,48] However, associated computerization remains necessary to check the calibrations and to implement programs with multivariable measurements.

In fluorescence, the trend is also towards laser spectrometry and time-gated detection with intensified photodiode arrays. A recent instrument, the Fluo 2001, has just been marketed for this purpose (Dilor Society) and tested with optical fibers for a nuclear environment.[25]

V. CONCLUSIONS

The use of optical fibers in nuclear environments is far from marginal and becomes operational in Purex process control.[63] In addition to the usual advantages which include in-sensitivity to electrical disturbances and the possibility of miniturization, fibers are relatively insensitive to radiation and corrosive chemical environments and can be adapted to the

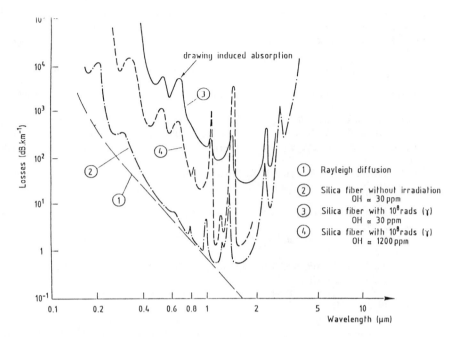

FIGURE 7. Spectral characteristics of silica fibers under γ radiation.

construction of special cells which make them ideal tools in a radioactive environment. The most immediate applications concern environmental protection, including geology, hydrology, liquid and gaseous wastes isotopic separation, spent fuel reprocessing, and nuclear safety, e.g., in nuclear power stations and plants. Developments under way concern both *in situ* process control and trace analysis. The environmental constraints require the development of methods, components, and even specific instruments with a wide dynamic range. Active optodes and the use of ultraviolet wavelengths with optical fibers in the presence of ionizing radiation still offer a wide field for investigation. Major advances can also be anticipated in the field of infrared spectrometry and Raman spectrometry for many remote determinations in nuclear environments.

REFERENCES

1. **Mattern, P. L., Watkins, L. M., Skoog, G. D., Brandon, J. R., and Barsis, E. H.,** The effect of radiation on the absorption and luminescence of fiber optic waveguides and materials, *IEEE Trans. Nucl. Sci.*, NS 21, 81, 1974.
2. **Lyons, P. B.,** Fiber optic applications in hostile environments, Rep. LA-UR-86-1249, Los Alamos National Laboratory, 1986.
3. **Lyons, P. B., Ed.,** Fiber optic in adverse environments, *Proc. S.P.I.E. Int. Soc. Opt. Eng.*, 296, 1981.
4. **Boucher, D., Ed.,** Fiber optic in adverse environments, *Proc. S.P.I.E. Int. Soc. Opt. Eng.*, 404, 1983.
5. **Greenwell, R. A., Ed.,** Fiber optic in adverse environments, *Proc. S.P.I.E. Int. Soc. Opt. Eng.*, 506, 1984.
6. **Greenwell, R. A., Ed.,** Radiation effects in optical materials, *Proc. S.P.I.E. Int. Soc. Opt. Eng.*, 541, 1985.
7. **Greenwell, R. A., Ed.,** Fiber optic in adverse environments, *Proc. S.P.I.E. Int. Soc. Opt. Eng.*, 721, 1986.
8. **Greenwell, R. A., Ed.,** Optical devices in adverse environments, *Proc. S.P.I.E. Int. Soc. Opt. Eng.*, 867, 1987.
9. Fiber optics in the nuclear environment, Proc. Symp. Adelphi, U.S., Publ. Information Gatekeepers, Inc., Boston, 1981.
10. **Paul, D. K., Ed.,** Fiber optic relativity: benign and adverse environments II, E'O Lase Boston (U.S.A.), *Proc. S.P.I.E. Int. Soc. Opt. Eng.*, 992, 1988.
11. **Watson, B. L. and Evans, R. P.,** On optical liquid level detector for high temperature/pressure water environment, *Proc. 8th Water Reactor Safety Research Information Meeting,* Gaithersburg, October 27 to 31, 1980; Rep., NUREG-CP-0023-V, U.S. Nucl. Reg. Comm.; Div. Tech. Inf. Doc. Contr. Wash. (DC)(Ed), 1982.

12. **Veteau, J. M.,** Mesure des aires interfaciales dans les écoulements diphasiques, Rep. CEA - R - 5005, Commissariat Energie Atomique (Fr) Doc. Centre. Etud. Nucl. Saclay. Fr. (Ed) 1979.

13. **Partin, J. K.,** Radiation response of optical fibers in a nuclear reactor, *Proc. S.P.I.E. Int. Soc. Opt. Eng.,* 506, 42, 1984.

14. **Perez, J. J.,** Analyse en ligne industrielle et contrôle des eaux: perspectives des mesures spectrometriques, *Inf. Chim.,* 275, 219, 1986.

15. **Rozenberg, J.,** *Determination de la teneur en suspensions dans les circuits d'eau, Private communication,* 1979.

16. **Boisdé, G. and Boissier, A.,** U.S. Patent 4,188,126, 1980, and U.S. Patent 4,225,232, 1980.

17. **Klainer, S., Hirschfeld, T., Bowman, H., Milanovich, F., Perry, D., and Johnson, D.,** A monitor for detecting nuclear waste leakage in a subsurface repository, *Report Livermore — Berkeley Laboratory* 11981, 1981.

18. **Perry, D. L., Klainer, S. M., Bowman, H. R., Milanovich, F. P., Hirschfeld, T., and Miller, S.,** Detection of ultratraces levels of uranium in aqueous samples by laser-induced fluorescence spectrometry, *Anal. Chem.,* 53, 1048, 1981.

19. **Milanovich, F. P. and Hirschfeld, T.,** Process, product, and waste stream monitoring with fiber optic, *Adv. Instrum.,* 38, 404, 1983.

20. **Hirschfeld, T.,** Remote and *in situ* analysis, *Proc. Int. Conf. Exhibit Philadelphia* (October 21 to 24, ISA 85), Instrum. Soc. Am., 305, 1985.

21. **Hirschfeld, T., Deaton, F., Milanovich, F., and Klainer, S.,** Feasibility of using fiber optics for monitoring ground water contaminants, *Opt. Eng.,* 22, 527, 1983.

22. **Malstrom, R. A. and Hirschfeld, T.,** On line uranium determination using remote fluorimetry, *Anal. Chem. Symp. Ser.,* 19, 25, 1984.

23. **Hirschfeld, T.,** Fiber optic remote analytical techniques in nuclear analysis, *Proc. Int. Conf. Anal. Chem. Nucl. Techn. Karlsruhe* 3,6 June, Abstract Invited Lecture, Kernforschungszentrum Karlsruhe (Ed), 1985.

24. **Boisdé, G., Kirsh, B., Mauchien, P., and Rougeault, S.,** Nouvelle optrode passive pour la spectrofluorimetrie et la spectrometrie Raman, *Proc. Conf. Opto.,* 88, ESI Publications (Ed) Masson, Paris, 1988, 294.

25. **Decambox, P., Kirsch, B., Mauchien, P., and Moulin, C.,** Fluo 2001, nouveau spectrofluorimètre avec détection par barrette de photodiodes intensifiée et pulsée, *Proc. Conf. Opto 88,* ESI Publications (Ed) Masson, Paris, 1988, 357.

26. **Allison, S. W., Magnuson, D. W., and Cates, M. R.,** Use of fiber optics for remote uranium hexafluoride laser-induced fluorescence measurements, *Proc. S.P.I.E. Int. Soc. Opt. Eng.,* 380, 369, 1983.

27. **Cates, M. R., Allison, S. W., Marshall, B., Davies, T. J., Franks, L. A., Nelson, M. A., and Noel, B. W.,** Laser-based data acquisition in gas centrifuge environments using optical fibers, *Proc. S.P.I.E. Int. Soc. Opt. Eng.,* 506, 64, 1984.

28. **Caldwell, S. E., Gentry, R. A., White, R. W., and Allison, S. W.,** Measurement of gas density and temperature profiles in UF6 using laser induced fluorescence, Rep. La-UR-86-3247, Los Alamos National Laboratory, 1986.

29. **Saturday, K. A.,** Absorption cell with fiber optic for concentration measurements in a flowing gas stream, *Anal. Chem.,* 55, 2459, 1983.

30. **Costes, R. M., Boisdé, G., Travert, A., Bonnejean, C., Girard, M., and Devy, P.,** Contrôle non destructif à l'aide de fibres optiques. Application au procédé Chemex, *Proc. Eur. Conf. Ind. Line Spectrogr. Anal.,* Rouen, 19, 21 June, Societé Chimie Industrielle (Ed) Paris, 1985, 5, and Rep. CEA-Conf. 8016.Doc. Centre Etud. Nucl-Saclay Fr (Ed) 1985.

31. **Boisdé, G., Blanc, F., and Perez, J. J.,** Chemical measurements with optical fibers for process control, *Talanta,* 35, 75, 1988.

32. **Boisdé, G. and Perez, J. J.,** Aspects of optical fibers and spectrometric sensors in chemical process and industrial environments, *Proc. S.P.I.E. Int. Soc. Opt. Eng.,* 1012, 58, 1988.

33. **Dakin, J. P., Batchellor, C. R., Pearce, D. A. J., Forrest, C. W., and Wilson, A. P.,** A practical optical fibre absorption sensor for measuring solids concentration in slurries, *Proc. 3rd Conf. Eurosensors,* Sensors and Applications, Cambridge, September 22 to 24, Cavendish Laboratory (Ed) Abstract, 1987, 103.

34. **Bostick, D. T.,** Acid-compensated multiwavelength determination of uranium in process stream, *N.B.S. Spec. Publ.* Washington, DC, 582, 121, 1980.

35. **Corriou, J. P. and Boisdé, G.,** Comparison of numerical and physicochemical models for spectrophotometric monitoring of uranium concentration, *Anal. Chim. Acta,* 190, 255, 1986.

36. **Schmieder, H., Kuhn, E., and Ochsenfeld, W.,** Die Absorptionspektren von Pu (III), Pu (IV), U (IV), and U (VI) in Salpetersäure and Tri-n-butylphosphat-n-Alkan-Lösungen und ihre Anwendung in der automatischen Prozesskontrolle, *Kernforschunszentrum Karlsruhe* - RFA (Ed) Rep. 1306, 1970.

37. **Spencer, W. A., Killeen, T. E., and Herold, T. R.,** Neptunium detector using fiber optic light guides, *Anal. Chem. Nucl. Tecnol. Proc. Conf. Anal. Chem. Energy, Technology 25th,* 205, 1981, Lyon, W. S., Ann Arbor Sci. (Ed).

38. **Van Hare, D. R., Prather, W. S., Boyce, D. A., and Spencer, W. A.,** On line fiber optic photometer, Rep. DP-MS-85-127, E. I. Du Pont de Nemours and Co., Savannah River Laboratory, 1985.

39. **Boisdé, G., Guillot, P., Monier, J., and Perez, J. J.,** Contrôle non destructif de l'uranium en solution à l'aide d'un photomètre portable à fibres optiques, *Commissariat Energie Atomique,* Rep CEA-R-5207, Doc. Cent. Et. Nucl. Saclay, Fr. (Ed) 1983.

40. **Groll, P., Persohn, M., Romer, J., and Schuler, B.,** Erprobung des Lichtleiter-Laserphotometers an Proben der Plutoniumanlage PUTE, *Kernforschungszentrum Karlsruhe (Ed) Rep.* 3843, 1984.

41. **Romer, J., Groll, P., Persohn, M., and Schuler, B.,** Ein Lichtleiter-Laser-photometer für on-line-Messungen in Purex-Prozess, *Kernforschungszentrum Karlsruhe (Ed) Rep.* 38844, 1984.

42. **Groll, P., Romer, J., Röder, L., Persohn, M., and Schlosser, B.,** An optical fibre laser photometer for on-line measurements, *Anal. Chim. Acta,* 190, 265, 1986.

43. **Perez, J. J., Boisdé, G., Goujon, de Beauvivier, M., Chevalier, G., and Issac, M.,** Automatisation de la spectrometrie du plutonium, Analusis 8,344, 1980 (Fr) and Los Alamos National Laboratory (Trad US) Rep. LA-TR-8214, 1982.

44. **Boisdé, G., Linger, C., Chevalier, G., and Perez, J. J.,** Fiber optic couplers for spectrophotometry. Prospects for *in situ* and remote measurement, *Proc. S.P.I.E. Int. Soc. Opt. Eng.,* 403, 53, 1983.

45. **Van Hare, D. R. and Prather, W. S.,** Fiber optic modification of a diode array spectrophotometer, Rep. DP-1714, E. I. Du Pont de Nemours and Co., Savannah River Laboratory, 1986.

46. **Ofalt, A. E. and O'Rourke, P. E.,** Development and performance of on-line uranium analyzers, Rep. DP-MS.85-142, E. I. Du Pont de Nemours and Co., Savannah River Laboratory, 1985.

47. **Brown, M. L., Mills, C. L., and Kyffin, T. W.,** The use of spectrophotometry in FBR reprocessing analysis, *Dounreay Nuclear Power Development Establishment-GB (Ed), Rep.* ND-R-1266 (D), 1985.

48. **O'Rourke, P. E. and Van Hare, D. R.,** On-line fiber optic spectrophotometry Rep. DP-MS-87-100, E. I. Du Pont de Nemours and Co., Savannah River Laboratory, 1987.

49. **Boisdé, G., Deroite, A., Mus, G., and Tachon, M.,** Spectrophotometrie en ligne par fibres optiques. Application à la Séparation uranium-plutonium en usine de retraitement des combustibles irradiés, *Proc. Eur. Conf. Ind. Line Spectrogr. Anal.,* Rouen, June 19 to 21, Société Chim. Indust. Paris Fr (Ed) 7, 1985, and Rep. CEA-Conf. 8017. Doc. Centr. Et. Nucl. Saclay Fr (Ed), 1985.

50. **Lecomte, M., Bourges, J., and Madic, Ch.,** Applications du procédé de dissolution oxydante du bioxyde de plutoium, *Proc. Recod* Soc. Fr. Energ. Nucl. Paris Fr (Ed) Tome 1, 441, 1987.

51. **Malstrom, R. A.,** Uranium determination on-line using remote fiber fluorimetry, Rep. DP-MS, 85-76, E. I. Du Pont de Nemours and Co., Savannah River Laboratory, (Ed), 1985.

52. **Berthoud, T., Drin, N., and Remy, B.,** Analyse en continu de l'iode dans les effluents gazeux des usines de retraitement par spectroscopie laser, *Proc. Eur. Conf. Ind. Line. Spectrogr. Anal.* (Rouen 19-21 June), Société Chim. Indust., Paris (Ed) 20, 1985.

53. **Boisdé, G., Linger, C., and Perez, J. J.,** Developpements récents de la spectrophotometrie par fibres optiques pour le contrôle in-situ de l'uranium (VI) en solution, *Proc. 5th Annu. Symp. Safeguards Nucl. Mat. Manag. Eur. Saf. Res. Dev. Ass.* (19-21 April), Versailles, Com. Eur. Communities, Joint Research Centre, Ispra, Italy, 1983, 203.

54. **Boisdé, G. and Perez, J. J.,** Remote spectrometry with optical fibers, ten years of development and prospects for on-line control, *Proc. S.P.I.E. Int. Soc. Opt. Eng.,* 514, 227, 1984.

55. **Siegel, G. H., Friebele, E. J., and Gingerich, M. E.,** Recent progress in the investigation of radiation resistant optical fibers, *Proc. S.P.I.E. Int. Soc. Opt. Eng.,* 296, 2, 1981.

56. **Friebele, E. J., Long, K. J., Askins, C. G., Gingerich, M. E., Marrone, M. J., and Griscom, D. L.,** Overview of radiation effects in fiber optics, *Proc. S.P.I.E. Int. Soc. Opt. Eng.,* 541, 70, 1985.

57. **Bauer, M. L., Bostick, D. T., and Strain, J. E.,** Evaluation of fiber optics for in-line photometry in hostile environments, *Proc. S.P.I.E. Int. Soc. Opt. Eng.,* 296, 251, 1981.

58. **Boisdé, G. and Perez, J. J.,** Effet du rayonnement γ sur les fibres optiques. Applications à la mesure à distance en milieu irradiant, *Proc. Opto. 82,* ESI Publications, Masson, Paris, 1982, 48.

59. **Boisdé, G., Bonnejean, C., Perez, J. J., Neuman, V., Wurier, B., and Boucher, D.,** Irradiation up to 109 rads on all silica fibers. Comparison with PCS fibers and massive samples between 0.4 and 2.5 μm, *Proc. S.P.I.E. Int. Soc. Opt. Eng.,* 404, 17, 1983.

60. **Abgrall, A., Poulain, M., Boisdé, G., Cardin, V., and Mazé, G.,** Infrared study of γ irradiated fluoride fibers, *Proc. S.P.I.E. Int. Soc. Opt. Eng.,* 618, 63, 1986.

61. **Barnes, C. E., Greenwall, R. A., and Nelson, G. W.,** The effect of fiber coating on the radiation response of fluorosilicate clad, pure silica core step index fibers, *Proc. S.P.I.E. Int. Soc. Opt. Eng.,* 787, 69, 1987.

62. **Boisdé, G., Rougeault, S., and Perez, J. J.,** Spectrometrie à distance dans le domaine de l'ultraviolet, *Proc. Opto. 86,* (May 13 to 15), ESI Publications, Masson, Paris, (Ed) 1986, 71, and Repost CEA-Conf. 8572. Doc. Cent. Et. Nucl. Saclay, Fr, 1986.

63. **Bürck, J., Krämer, and König, W.,** "Lichtleiteradaption des interferenzfilterphotometers Spectran für den Einsatz als in line-Monitor in der Purex-Grozeozkontrolle", Kernforschungszentrum Karloruhe (Ed), Rep. 4672, 1990.

Chapter 15

FIBER OPTIC TECHNIQUES FOR TEMPERATURE SENSING

Kenneth T. V. Grattan

TABLE OF CONTENTS

I. INTRODUCTION

Measurement of temperature is one that is particularly important in most spheres of industrial and scientific activity. The areas of use of thermometers of various types extend widely in the industrial environment for the precise measurement and control of many physical and chemical manufacturing processes. These environments can present many significant difficulties for the determination of temperature. The region to be measured may be moving, extremely hostile, in a position where access is extremely difficult, or where the physical contacting of a sensing probe may even be impossible. As a result, temperature measurement techniques have evolved which rely upon contact and noncontact techniques. In the latter case, for example, radiation monitoring approaches have been used in the industrial measurement of temperature in glass and plastics manufacture, the monitoring or remote furnace temperatures, the observation of temperatures in gas turbines for aircraft use or other power generation purposes, and indeed in many specially designed industrial plants. For contact methods the most familiar approach is the use of the thermocouple or a related technique, but often these devices have short lifetimes due to the hostility of the environment and the observation of the signal received can be difficult in the presence of interference from other electromagnetic noise. In the chemical process industry, temperature monitoring is of particular importance as reactions are influenced to a very great extent by the temperature of the environment. In many cases, due to the highly corrosive nature of chemicals used, it may not be possible to make contact measurements using conventional metal thermocouples, and so alternative means of temperature sensing must be sought to ensure accurate control of the process.

The importance of temperature measurement can be seen simplistically by consideration of the financial aspects of the sensors and devices used worldwide. Temperature is a difficult concept to define, and a considerable body of literature has been built up by national standards organizations in order to present a range of working definitions. By contrast to mass and length, which are common and accurate measurements, the more difficult to measure temperature is of much greater importance. Estimates on the worldwide sales of temperature sensors run approximately $1 billion per year, a figure that could be increased several times when the associated controllers, indicators, and other aspects of the measurement system are added. In spite of the major financial consequences of errors in temperature measurement, the basic principles which underly the measurement process are frequently misunderstood or even ignored. The international temperature scales have been closely defined on the basis of the definitions within the SI Unit standards, but a number of defining fixed points have been used to enable practical experimenters to use temperature scales. It is beyond the scope of this contribution to describe these in detail, but publications from the National Physical Laboratory[1] in the U.K. give a useful overview of the scales.

II. FIBER OPTIC TEMPERATURE MEASUREMENT — HISTORICAL OVERVIEW

It is now almost 25 years since the first concepts of the use of fiber optic techniques for sensor purposes were discussed and what we would now recognize as fiber optic sensors were introduced on the market. In many cases fiber optic techniques simply make more convenient the use of "open air path" optical approaches, through the more convenient channeling of the light to and from the region of measurement. Thus, for example, radiation pyrometry can easily be adapted to fiber optic measurement.[2] However, fiber optic sensors in their own right have been produced in recent years and much research effort has been directed toward the development of this technology. As a result, a considerable body of publications has appeared and a number of commercial products have been advertised by major manufacturers. There is an element of reluctance seen in some aspects of the process and control industries and in biomedical monitoring to take fiber optic sensors seriously. This is somewhat to be expected for a new and rapidly evolving technology, but with the increase in the number of products available, and the greater familiarity of technicians working in these industries with optoelectronics and optical devices, it is expected that some of this initial reluctance can be overcome. In addition, fiber optics for communication purposes are now well established.[3] The U.S., Japan, and most European countries have extensive programs of fiber optic cabling for the main trunk networks of their telecommunications systems.

Many computer data links rely on fiber optics, and there is considerable discussion of the use of fiber optics as communications channels for data into the home. By this means the superior data handling capacity of fiber optics over conventional cabling may be used. All this has resulted in a considerable decrease in the price of fiber optics and their associated optoelectronic devices, i.e., emitters and detectors. With the increase in sophistication of electronic signal processing techniques, there appears to be no good reason why inexpensive fiber optic sensors relying upon sophisticated electronic signal processing cannot be commercialized at reasonable costs. There has been extensive work in many European countries, particularly the U.K. and in the U.S. in the development of fiber optic sensor techniques. The aim is to to provide a sound technological base upon which the development of these fiber optic sensor products can continue.

There is perhaps most diversity in the techniques that are used for temperature measurement as a result of the fact that temperature can fairly readily be transduced into a number of other energy forms, and there has been a wide range of proposed fiber optic sensor devices. Indeed, temperature sensing is perhaps the oldest of the parameters to be approached seriously by fiber optic means, and the fiber optic temperature sensor is perhaps the most successful of the limited commercial range of such products. This largely results from the ease of converting temperature to be a measureable optical characteristic of a material and the fact that often temperature effects in other fiber optic sensors are significant and may lead to confusion of the measurand in these other sensors. As a result, it is important to know the temperature of the environment, either to allow for design of the sensor to eliminate temperature effects or, as is possible now, the measuring of temperature as well and the correction through electronic techniques via stored temperature calibration data, for temperature errors in other fiber optic-based measurements.

A. INITIAL DRIVE

The initial drive for the development of fiber optic sensors came from their potential use in military and aerospace applications where the cost factors of the introduction of new technology were less rigid and the working environment more hostile than is experienced with other areas of application. It is, however, disappointing to note that in spite of the wealth of information that has been published in the literature on fiber optic sensors, as yet the product range commercially available is small and few manufacturers are using fiber optic devices for actual industrial or process control. As a result, the industrial or medical sphere has not experienced the deployment of fiber optic sensors that was predicted up to 10 years ago in many of the earlier, somewhat overoptimistic, papers of the field.[4] Nonetheless, steady progress is being made and the advantages of the use of fiber optic sensors, as discussed in other chapters, particularly for temperature measurement, are evident. Contrasts and comparisons with the slow development of the use of the laser outside the laboratory can be made, as this technology has taken about 25 years to mature. Many commercial fiber optic products only now have evolved to product stage, following initial ideas and early research work of 10 years ago.[5]

It is, however, important to consider the major criteria for the acceptability of fiber optic techniques in modern industry. They must show equivalent accuracy, sensitivity, and physical robustness to the conventional sensor, at a cost which is allowable in the particular application, if not comparable. It is worth remembering that conventional sensing techniques are not a stationary art: indeed, the major developments in electronic sensing technology and electronic signal processing which are so advantageous for fiber optic sensor applications are also highly beneficial to the use of conventional techniques. However, the fundamental advantages of the use of an optical signal, unaffected by electromagnetic noise and potentially multiplexed with many other signals down a simple, thin, rugged, fiber optic cable, are uncontestable.

B. INTRINSIC AND EXTRINSIC TEMPERATURE-SENSING DEVICES

Most fiber optic chemical sensing systems are essentially point sensors, i.e., the fiber optic cable carries the light to and from the transducer where the interaction with the process occurs. As a result, the fiber optic is merely a light guide, making the constructing of this device simple and potentially compact for use in remote areas. In this chapter, such sensors are termed *extrinsic* sensors; a number of the fiber optic sensor techniques described are of the extrinsic point sensor type. These devices are likely to be simple and require comparatively unsophisticated signal processing. However, in addition there is the *intrinsic* fiber optic temperature sensor where the fiber is not merely used as a communication channel but actually acts as the sensing device itself. This results from some interaction, due to the temperature change of the characteristics of the fiber material and thus a simple sensor (usually requiring sophisticated signal processing to extract the temperature information) is available. This is a feature essentiallly unique to fiber optic physical sensors, i.e., temperature, pressure, etc., and has been exploited commercially most widely in temperature sensing.

C. SAFETY

The safety of the use of fiber optic sensors has been considered by a number of groups, and there is reason to believe that at the normal levels of optical power coupled into fiber optics, i.e., levels of up to several hundred milliwatts of optical power, there is almost no hazard with any accidental fracture of the cable and possible focusing of the optical radiation by lensing effects of the broken end. Particularly in the chemical industry, where highly explosive gases or gas mixtures may be used, this is an important consideration, but on the whole in normal use, fiber optic sensing systems can be considered intrinsically safe. A considerable discussion of this aspect has been made in reports to the U.K. Optical Sensors Collaborative Association by Tortishell[6] and others.

III. SENSOR DEVICES

A. CLASSIFICATION OF SENSOR SCHEMES

There are a number of possible ways of classifying fiber optic temperature sensor devices, including techniques which consider the modulation scheme which is used to encode the temperature information[7] or the nature of the device itself, i.e., interferometric sensors, frequency or resonant type, wavelength encoding, distributed, or even simply intensity-ratio-based devices. As a result it is difficult to make a consistent classification, but for the purposes of this work devices will be considered on the basis of whether they are intrinsic or extrinsic, and the nature of the interaction of the temperature measurand with the transducer will be considered. Details and discussions of optical fibers themselves, i.e., their modal propagation characteristics and their varied physical types, will not be considered except where this relates directly to the measurement itself.

B. EARLY WORK AND BEYOND ON THE DEVELOPMENT OF MULTIMODE FIBER OPTIC SENSORS

The early history of fiber optics for temperature sensors stretches back over 25 years, and it is interesting to note that this does coincide with earlier work in chemical sensing, such as that of Polanyi and Hehir who monitored the spectral reflectivity characteristics of oxygenated hemoglobin by optical techniques.[8] However, the earlier work on temperature sensors concentrated upon the conversion of conventional optical techniques to fiber optic methods. For example, the radiation thermometer has been well known and its principles discussed extensively in the literature.[2] In this approach the temperature of the target body is determined by the rate of energy emission from the surface and the relative spectral content of that emission. This is related to the Planck Radiation Law and thus the temperature can be determined. Problems of emissivity do exist, but as a well established technique it could readily be converted to fiber optic use.

In the early work of Dakin and Kahn[9] and others infrared pyrometry was extended using fiber optics to transmit the radiation to be examined. Attempts were made in this work to overcome the inaccuracies due to the unknown factors of emissivity of the surface, and although a claimed working region of 100 to 1500°C was reported, a more effective development of this technique has more recently been reported. Pyrometry is not the best technique to use for temperatures less than 500°C due to the difficulty in detecting either the very weak short wavelength emission or the long wavelength emission which is outside the sensitivity of most detectors. The high temperature of the ovens to be monitored using this technique make the use of optical fibers in direct contact somewhat difficult as many fibers have an upper temperature limit of 200°C, with certain special fibers useable up to 300 to 400°C. It is only highly specialist sapphire fibers which may be used up to the very high temperature region (1500 to 2000°C).

In addition, the variation of spectral radiance with wavelength and temperature is shown in Figure 1, and it can be seen that only at relatively high temperatures do the emissions occur in wavelength regions where simple solid state detectors such as silicon p-i-n photodiodes operate, e.g., at wavelengths less than 1.1 μm. Germanium diodes and other exotic devices can extend this range to about 2 μm. However, with the development of solid state detector devices such as PbSe (cooled) with sensitivities in the 1 to 10 μm range, low temperature pyrometry may further be developed. In recent work by Morden et al.[10] such a device operating in the range 60 to 150°C was described, using zirconium fluoride fibers.

The hybrid approach represents another aspect of the conversion of a conventional technique to fiber optic use. Hybrid devices may broadly be described as those which are optically powered but not using optical tchniques for the transduction. Electrical power does not have to be fed to the sensor directly from the source, and often the output of the sensor is interpreted optically or transmitted back via optical fibers. Thus, the data are free from data corruption problems by interference which could affect the electrically transmitted information. Some illustration of hybrid devices are included in this chapter to present an overall view of the subject. The resistance thermometer is used as the primary sensor, powered by photovoltaic conversion of light from a bright LED of several milliwatts output operating at 820 nm.[11] A bank of Si photodiodes converts the optical to a 5 V D.C. signal to drive the device. This is illustrated in Figure 2, and using microprocessor-based averaging, the temperature can be calculated to ±0.5°C. In this case the signal transmission is by fiber optics to and from the device, and so the signal is protected from interfering effects.

As temperature is a measurand which can readily be converted to a displacement, an effect that is seen through the unwanted expansion of materials or that is utilized in the metallic strip switch, a number of devices have been proposed which rely on this technique. For example, by fixing a mirror to a temperature-induced movement of an object, a simple fiber optic temperature sensor could be produced. However, as the device relied on the change of intensity as a function of temperature, it was necessary to be able to record the initial intensity level reaching the sensor head. Consequently, a referencing channel must be provided. This is an important limitation of any fiber optic intensity-based sensor, as it is often inadequate merely to monitor the intensity of light emitted into the fiber as bending, and other losses in

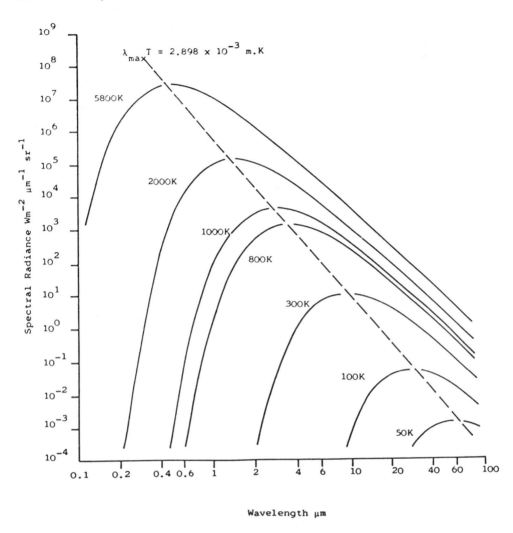

FIGURE 1.   Variation of spectral radiance with wavelength and temperature. (From Dixon, J., *Meas. Control,* 20, 11, 1987. With permission.)

the fiber may well reduce the light level significantly from the level of light launched, thereby providing significant errors.

Techniques to overcome this include the use of two different wavelengths[12] or the use of a fluorescent reference channel,[13] but by this means an effective intensity-based sensing device can be produced. The early "Fotonic" displacement sensor[14] used light reflection by the light emitted as a cone from the end of one fiber, reflected from the temperature displaced target so that light is incident on a second fiber. Although initially demonstrated for pressure sensing, the technique is equally applicable to temperature. Unfortunately, a highly nonlinear characteristic results. In further work, Bergstrom et al.[15] in 1979 proposed a reflector device which would provide a measurand and reference signal for use with submicron displacements. Other small displacement monitoring techniques proposed include those by Croft et al.[16] through the evanescent wave coupling from a critically cut fiber through to a second optical medium. By this means very small displacements of typically less than a micron could be measured. Early work using interferometric techniques have been discussed by Hocker[17] using hybrid fiber optic interferometers, demonstrating the very high sensitivity that can be achieved for temperature measurement through the use of a fiber optic analog of the optical interferometer.

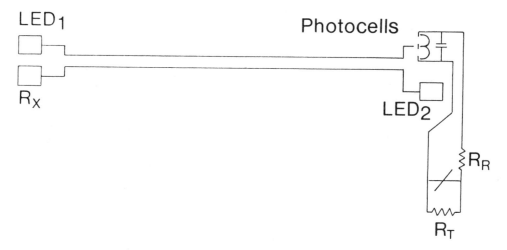

FIGURE 2. Diagram of the optically powered resistance thermometer. R_R, reference resistor which is in the head mounted transmitter; R_T, sensor resistor. (From Murray, R. T., *Int. J. Opt. Sens.*, 1, 27, 1986. With permission.)

The interferometer itself is a very basic tool of optical measurement techniques and it can offer a high dynamic range coupled with extremely high resolution. The measurand is coupled to a change in one of the optical paths of the device, and in recent years fiber optic versions of the familiar Michelson, Mach-Zehnder, and Fabry-Perot interferometers have been produced. Their use in early work was limited by the high component costs, but in recent years the reduction in the cost of such components has made these devices available at very reasonable prices. In the last decade considerable research has led to the use of monomode interferometric devices offering performance characteristics and stability which are comparable to those of open air path devices, as discussed later.

C. MORE RECENT WORK — EXTRINSIC SENSORS

It is only in more recent years, i.e., since about 1981 or so, that the wide variety of fiber optic temperature-sensing techniques now familiar have been reported. There has been something of an explosion of device proposals seen in the literature. As an example, the biological measurement range, i.e., 30 to 50°C, represents an important illustration, one ripe for development in terms of commercial exploitation. Accurate temperature sensors may be used in diagnosis and treatment, often being required where sophisticated, noise-generating medical equipment is present. The magnetic resonant image (MRI) scanning devices, for example, produce very large electrical and magnetic fields which interfere with signals carried by conventional transmission techniques. Fiber optic systems are ideal for monitoring uses in the environment of such instruments. The use of liquid crystals for measurement in this temperature region has been discussed, but early work showed that problems of stability exist.[18] More recently it has been proposed that liquid crystals are best suited to switching applications, under referenced conditions, where some of the stability problems can be overcome.[19]

Work is continuing by Ireland and Jones[20] to explore the use of these very interesting materials which show significant color changes with temperature. Screen printing techniques may be used to apply micron-thin films to give very rapid response for industrial or medical applications. Thermochromic materials have been explored and developed in a device by Brenci.[21] These sensors rely upon the monitoring of a significant color change, with temperature as shown in Figure 3. A cobalt chloride solution is encapsulated with an optical fiber to guide light from a white light source, and radiation at two different wavelengths which is reflected from the solution is monitored. One of these wavelengths provides a reference

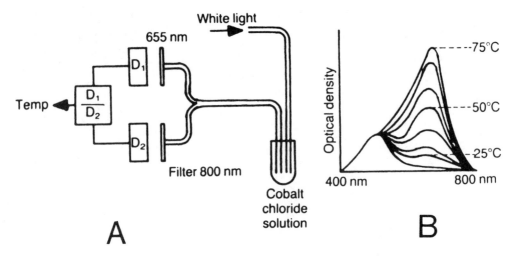

FIGURE 3. Thermochromic sensor. (A) Optical arrangement; (B) Optical characteristics as a function of temperature. (From Medlock, R. S., *Int. J. Opt. Sens.*, 1, 43, 1986. With permission.)

channel, the intensity of which is unchanged with temperature, and the second provides the temperature information as a result of the significant change in the absorption characteristic of the material with temperature. This can be seen in the figure over the 25 to 75°C range. The potential problem of such a system exists in that the thermochromic material may spill and could be a health hazard to the patient. This is obviously a major difficulty with the application of any technique for invasive biological use and careful encapsulation of the transducer must be ensured.

A simple device, particularly for use in the narrow biological range, has been described by Brenci et al.[22] The principle of this instrument is the change of dimensions of an air bubble with temperature allowing the expansion of a liquid thermosensitive cladding of a fiber at the probe, as shown in Figure 4. Thus, the numerical aperature and the attenuation of light carried by the fiber is modulated with temperature. A sensitivity of 0.1°C in an instrument of dimensions 1.5 × 10 mm is reported. Sensing for biological techniques is a particularly difficult area, due to this additional signal safety requirement, although by comparison with conventional techniques, the limitation of avoiding currents flowing in the body in a transducer can be overcome.

Some relatively early work on the subject of fiber optic sensing was done by Rogers,[23] mostly for current and voltage measurement. However, as a result of the need to monitor temperature for possible correction purposes in these sensors, a scheme was proposed using the birefringence of quartz which was also used in other sensing methods. This scheme[24] is somewhat complex, requiring a quartz block as the sensor and a number of expensive optical components. It is thus not well suited to simple sensor applications. Further, the cost of a gas laser as a source is significant and for the simple sensor regime can rarely be justified. It is well known that fiber bending induces losses due to the coupling of light from core to cladding as a result of the altered angles of incidence of light on the cladding-core interference. An early sensor was the microbend pressure sensor,[25] which could also be utilized to measure temperature-induced expansion effects. This approach has been used in a recent system by Arakawa and Yoshida[26] where the sensor is manufactured by slightly deforming a nylon jacket followed by covering it with a thermal shrinkable tube, to induce the temperature sensitive microbending loss. Although not strictly extrinsic sensors, they are "point" measurement devices and a transducer using these effects may be coupled (via fusion splicing) to a different type of fiber.

Optical filters which are often used as components in these devices may be birefringent, of semiconductor material, or of interference type, and they can often have temperature

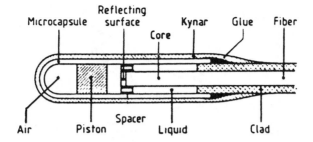

FIGURE 4. Temperature measuring device of Brenci et al. (From Brenci, M., Conforti, G., Falciai, R., Mignani, A. G., and Scheggi, A. M., *Proc. SPIE*, 701, 216, 1986. With permission.)

characteristics as a result of either their construction method or resulting from the fundamental physics of the materials themselves. This is often a disadvantage where they are used in actual devices, but it may mean that when they are used as the transducer element itself, they can act as temperature monitors. It is for this reason that the control and processing parts of many fiber optic sensor systems have temperature stabilization, but a sensor technique was described by Christiansen[27] over 100 years ago using a filter which becomes transmitting at a particular temperature due to refractive index matching conditions within the filter being met. As a result the temperature can be determined, in a non-intensity method, by a knowledge of the wavelength from a white light source which is transmitted by the filter. The well known temperature characteristic of the birefringence of the common electro-optic material, lithium niobate, has been studied extensively, where the difference for the ordinary and extraordinary ray passing through the material can be utilized in sensing. This has been exploited previously by Knox et al.,[28] but as a practical device there is, as with many, a requirement for significant expenditure on optical components. A monochromator is required to determine precisely the wavelength transmitted to yield the temperature details.

In many optical sensors the transducer itself is relatively simple, but optical processing can be complicated by the need for stable laser sources, expensive interference filters, etc. This, combined with the need for significant mechanical stability, may severely limit the practical usefulness of what is otherwise possibly an attractive sensor scheme. For this reason devices have been proposed which attempt to keep the optical aspects of a signal processing to a minimum. Wavelength division multiplexing, where the information is encoded as a function of the wavelength of light transmitted, reflected, or scattered by the material is very attractive (e.g., the device of Christiansen,[27] described earlier) but not if a sophisticated monochromator is required to determine that wavelength. A simple grating displacement sensor was reported by Jones and Spooncer[29] in 1984, described in pressure sensing mode but equally applicable to temperature measurement. This sensor incorporates a shutter modulator which consists of a pair of superposed optical gratings. The gratings are illuminated with normal incident light, and one grating traverses across the second in a direction perpendicular to the grating lines to that the transmitted light varies periodically. This is illustrated in Figure 5. As a sample device this shows excellent resolution and good stability. Referencing is achieved by measuring the transmitted intensity at two different wavelengths. The "reference" at the first wavelength passes through the shutter modulator unmodulated, while the signal at the second wavelength is modulated by the shutter. Both signals follow the same optical path, a feature highly desirable in wavelength-referenced systems. This device does however, require the use of a reflective grating and stable optical mounting.

Wavelength techniques are inherently less susceptible to instability, as they are not intensity-dependent. Wavelength modulation can be obtained using the displacement of a sensor element, and in temperature measurement this can readily be obtained. In contrast to the complex overlapping shutter system described, wavelength encoding can occur through the

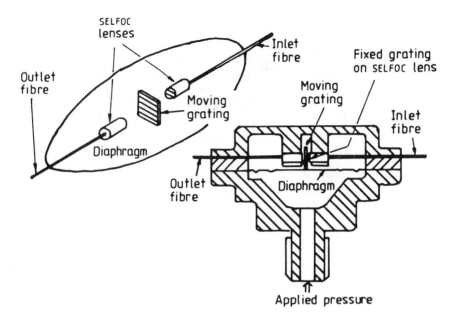

FIGURE 5. Shutter-modulator sensor device. (From Jones, B. E. and Spooncer, R. C., Proc. SPIE, 514, 223, 1984. With permission.)

temperature-induced rotation of a grating or prism, or the movement of the optical fiber receiving the information encoded. Alternatively, a simple-to-fabricate zone plate, undergoing movement, will provide this modulation in a displacement device discussed by Hutley.[30] Further, an achromatic lens with displacement of this element with respect to the receiving fiber may be used. In either case, basic simplicity is the essential element.

An important characteristic which has been exploited in temperature measurement is the change in the transmission (the "band edge shift") in semiconductors such as GaAs or doped glasses which contain semiconductors or metals. These effects have been known for some time, merely as difficulties in the use of these materials for other purposes, but they can readily be exploited for sensor measurement. Since 1979 work has been carried out by Saaski and Skangoet,[31] and early problems with the wavelength drift of light emitting diodes may be overcome through the use of the more stable high powered infrared LED sources now available. It is important to operate these systems below the maximum power to avoid problems in this area. The LED is an ideal optical source for fiber optic sensor purposes, being inexpensive and capable of being operated by low voltage, compact power supplies. However, it is well known that there are shifts in the output wavelength with temperature, although these shifts are less severe for the infrared emitting device than they are for the visible wavelength region. Recent work, however, has indicated that simple temperature stabilization can be provided for these very small devices through the use of Peltier coolers which will maintain the temperature of the source, where its optical characteristics are known. This does mean that the important wavelength stability that is required can be achieved.

There have been many proposals for systems which incorporate a referencing technique with these two wavelength devices. An example is the work of Kyuma et al.[12] using the optical absorption change of materials with temperature. A sensor was constructed from a small piece of GaAs where the shift in the absorption edge in a wavelength region around 0.9 μm was studied. A reference wavelength was provided at a wavelength of 1.3 μm, well outside the region of change in absorption characteristics of the transducer with temperature. Two solid-state LED sources were provided and as a result a simple sensor, ratioing the outputs at these two wavelengths could be built as the short wavelength is attenuated by the change in absorption and the long wavelength is not. One possible source of error is the fact that these two wavelengths are quite widely separated in wavelength terms, and it is preferable that they be closer together to avoid differential losses due to fiber bending, optical coupling, reflection losses, etc. Schoener et al.[32] have reported the similar scheme with two wavelengths much closer together and Theocharous[33] has developed a system using visible wavelengths which monitors the change in absorption of "ruby" glasses doped with metals. One of the disadvantages of this scheme is the use of a laser to generate the two wavelengths at 605 and 622 nm, although the two wavelengths being very close together eliminates the differential loss problem. A disadvantage is, however, that the range is limited due to the encroachment of the absorption edge on the reference wavelength at 622 nm.

These schemes are important and Kyuma et al.[12] have demonstrated them for temperature sensing in the practical application of the monitoring of temperature in a transformer on board a train in use in Japan. Very satisfactory results were obtained. Recent work by Grattan et al.[34] has demonstrated the use of absorbing glasses for temperature sensing using a second, longer wavelength for referencing generated by a fluorescent medium bonded to the glass. Such a "self-referenced" scheme has the advantage of both wavelengths having the same return light path. A number of other groups have described semiconductor sensitive elements, e.g., Andrew et al.[35] A "complete temperature measurement system" utilizes the semiconductor band-edge change in a four-probe miniature temperature sensor described by Christiansen et al.[36] Thus, this is still the subject of development and further research offers the promise for commercial exploitation.

D. EXTRINSIC SENSORS — LUMINESCENT DEVICES

One of the most important characteristics that has been utilized for fiber optic temperature sensors is the use of luminescent effects. When stimulated by incoming light, materials may emit radiation of a different wavelength, and the characteristics of that wavelength can often be indicative of the temperature of the source. Thus if the material is incorporated into a temperature sensor, i.e., by being placed at the end of an optical fiber, then temperature sensing is possible. This approach has been used in some of the few commercial devices which have been produced. One such early device, the Luxtron type 1000/2000 system,[37] is illustrated in Figure 6. Although involving a rather complex optical arrangement and the use of a conventional UV light source with dichroic mirrors for wavelength separation, this device relied upon the fact that the light from the lamp source induced fluorescence in a rare earth phosphor sample. A change in ratio of the emission at two different wavelengths generated by the sensor is monitored to give a measure of temperature. As two distinct wavelengths of emission are monitored and carried via approximately the same optical path, an internal intensity-referencing scheme is automatically provided.

During 1983, ASEA in Sweden[38] marketed a device using the change in the luminous spectrum of a semiconductor. This material was excited not by a UV lamp source but by light from a light emitting diode (LED) and an aluminum gallium arsenide sensor was used. The luminous spectra vary with temperature and, as shown in Figure 7, the wavelength of emission in two spectral regions can be used to determine the temperature. This simple approach makes the output largely independent of the installation. Temperatures in the region of 0 to 200°C with an accuracy claimed to be about 1°C can be obtained.

An important application of the use of luminescent phenomena in temperature measurement is the monitoring of the characteristic decay time of the fluorescence. It is well known that fluorescence emission will decrease in intensity as a function of time, and this frequently shows an exponential characteristic. Therefore, the intensity, I, at any time, t, can be given by the following equation: $I(t) = I_0 \exp(-t/\tau)$, where I_0 is the initial intensity (I at t = 0) and the characteristic decay time is represented by τ. As a result, an intensity independent method of the measurement of temperature can be envisaged. The temperature is thus characterized by

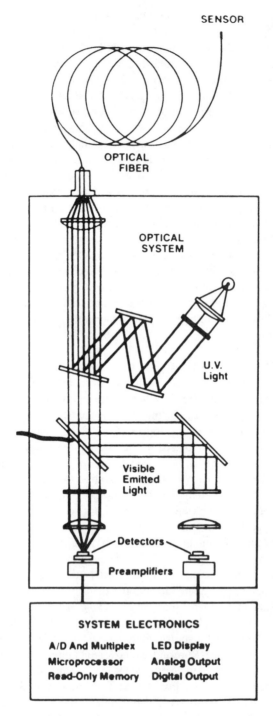

FIGURE 6. Luxtron type 1000/2000 fluorescent temperature sensor. (From Wickersheim, K. A. and Alves, R. V., Alan R. Liss, New York, 1982, 547. With permission.)

the decay time (the factor τ shown in the equation). This method has quite a long history in fiber optic sensor terms, although effective devices have only recently been discussed. A scheme using again UV light excitation of phosphors such as ZnCdS and ZnSe was discussed by James and Quick[39] in 1979. A paper was published by Scholes and Small[40] who discussed the application of the principle to the monitoring of temperatures in the biological range, i.e., 30 to 50°C. This work looked at the emission from a ruby crystal, but did not go on to develop a fiber optic probe system.

McCormack,[41] at GEC in the U.K., produced a working fiber optic sensor using the material BaClF:Sm^{2+}, which is a commonly used visible wavelength phosphor. This was shown to operate up to 150°C by monitoring the decay time of the material, again excited by a conventional lamp source which was mechanically modulated. A comparison of laboratory results on the characteristics of the phosphor with the results from the test of the fiber optic probe proved unsatisfactory, largely due to imperfect optical filtering techniques. However, the efficacy of the scheme was shown and recent work has developed it further. A publication by Bosselman et al.[42] showed the use of the characteristics of a chromium doped crystal, which was excited not from a lamp but from an LED emitting in the mid-visible range. As a result of the use of such a source, mechanical modulation techniques could be avoided and a device was produced to sense temperatures up to 200°C.

Other work by Grattan et al.[43,44] has considered the use of materials such as neodymium, alexandrite, and ruby for use in the fluorescent decay time thermometer. The attractive feature of the first of these materials is it has strong absorption bands in the near infrared at 750 to 810 nm. With the availability of LEDs and laser diodes with high power at low cost, these materials can be excited very readily with these sources. Thus, several milliwatts of optical power, which can be modulated directly and easily coupled into optical fibers, can be used to induce the luminescence at these longer wavelengths in neodymium, in the form Nd^{3+} doped into glasses or crystals. Unfortunately, the material has its strongest emission at the edge of the sensitivity of the silicon p-i-n diode at 1.06 μm, but this is still just within its sensitivity region. In addition, a major disadvantage is that the decay time of neodymium does not have a large change with temperature.

A study of the optical characteristics of the fluorescence of neodymium, when doped into the crystaline material YAG, has been carried out by Grattan et al.[45] This has revealed both a very small change with temperature and ambiguity for sensor purposes as the decay time initially increases and then decreases as a function of temperature. These characteristics can be explained in terms of the fundamental parameters of the material.[46] However, neodymium, when doped into glass at high concentrations (7 to 9%), does have the greatest change with temperature, and over a range of 0 to 150°C, a drop in the value of decay time of about 10% is seen. This is sufficient to produce a resolution on the order of 1.5°C under good operating conditions. However, it is advantageous that the decay time of the materials discussed are on the order of 10^{-4} to 10^{-3} sec (compared to the $\leq 10^{-6}$ sec decay times of many phosphors) so that these measurements can be made accurately, by modern electronic means, with respect to timing from inexpensive quartz oscillators operating at several megahertz. This again indicates that simple yet effective devices can be produced.

Alexandrite has been studied by Augousti et al.,[47] and this material is amenable to pumping by light from the convenient, inexpensive He–Ne laser. Unfortuately, in this measurement scheme, this laser has to be modulated, but electro-optic means can achieve this simply although at comparatively high cost. In addition, for any decay time measurement, the modulation can be in the form of a sinusoidal intensity variation of the light from the pump source, as this will lead to a fluorescent emission which shows the same sinuoidal variation at the same frequency, but *lags in phase* the pump light by a factor which is proportional to the decay time. The simple mathematics relating the decay time to this phase lag has been discussed by Augousti et al.[48] in a recent paper. A greater change in decay time with temperature is seen than for neodymium, and work to address this material with light from a light emitting diode is continuing.

However, there has been continued interest in the study of the characteristics of ruby, and when pumped by light from a visible light emitting diode, emission at 694 nm in the deep red

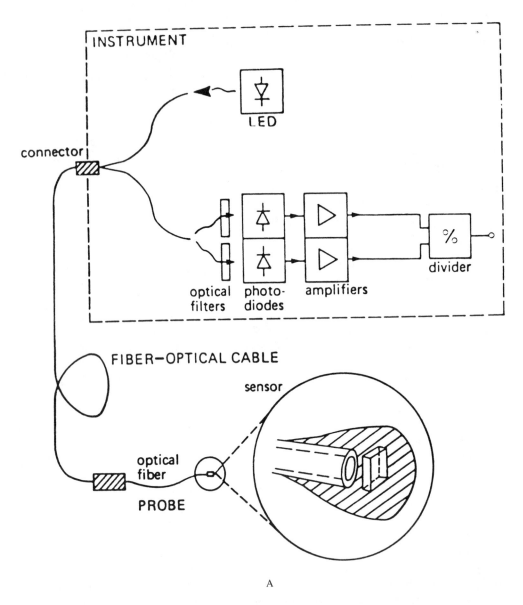

A

FIGURE 7. ASEA type 1010 temperature sensor. (A) Optical arrangement of the ASEA type 1010 temperature sensor. (B) Spectral characteristics. (From Grattan, K. T. V., *Meas. Control*, 20, 32, 1987. With permission.)

region of the spectrum shows a significant change in decay time with temperature, as shown in Figure 8. This material is stable, relatively inexpensive in its impure form, and inert. A detailed study of a ruby based fluorescence decay time sensor has been published recently,[49] and these studies suggest that it is a very suitable material for temperature sensor purposes. Resolutions up to 0.04°C, under laboratory conditions, have been demonstrated.[49]

One of the additional advantages of this decay time luminescent technique is that small changes in the emission wavelength of the source do not affect the information contained in the fluorescence emission and its characteristic decay time. LEDs do have a small change in their operating characteristics with temperature as the peak wavelength of emission increases as the device heats up. In systems where that emission wavelength is compared with the fluorescent wavelength, and that is used to determine the measurand, this can present difficul-

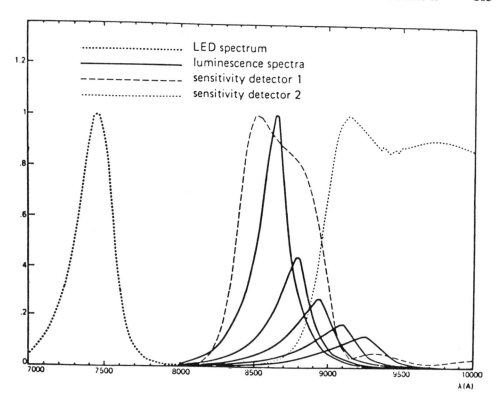

FIGURE 7B.

ties. In the decay time sensor, however, this merely results in a small change in the *intensity* of the fluorescence. As the device is intensity-independent, this should not affect the resulting reading of temperature.

Commercial devices based on these principles have been described by Luxtron Corporation in their model 750 fluoroptic sensor[50] which is illustrated in Figure 9. This device relies upon the change in the decay time characteristics of magnesium fluorogerminate to sense the measurand, but again uses a UV lamp source as the means of excitation of the material. This is stimulated at quite a slow frequency (1 to 10 Hz), which leads to a slow data acquisition rate, unless signal averaging is performed *more than once per pulse*. The significant change in decay time with temperature is shown in the figure. Recent work by Hirano[51] has resulted in the announcement of a new LED-based fiber optic fluorescence decay time sensor which can make measurements in the region from sub-room temperature to 200°C. A different type of phosphor transducer is used in this LED pumped device which again relies upon electronic signal averaging techniques, to achieve acceptable accuracies.

E. EXTRINSIC SENSORS — HIGH TEMPERATURE (>200°C) DEVICES

Standard optical fibers, i.e., those using plastic cladding, have an upper operating temperature region of the order of 200°C. There are many applications, such as in combustion diagnostics, where much higher temperatures are reached and where there is a great need for accurate temperature measurements. Intrusive sensors, such as thermocouples, can considerably modify the local temperature field and nullify the test being carried out. The use of fiber optic and phosphorescent techniques in such measurements has been demonstrated in recent work by Allison et al.[52] where the temperature-dependent properties of europium-doped phosphor is used to monitor engine temperature changes. Again, the characteristic decay time change with temperature is used, observing light emitted in the wavelength region of 500 nm.

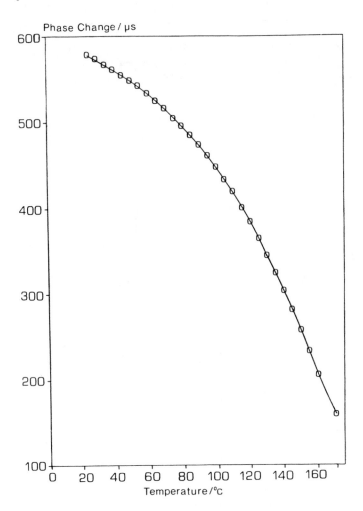

FIGURE 8. Plot of decay-time (recorded as phase difference in microseconds) as a function of temperature for a ruby-based probe. (From Grattan, K. T. V., Selli, R. K., and Palmer, A. W., *Rev. Sci. Instrum.*, 59, 1328, 1988. With permission.)

In a complex experimental configuration, using light from a neodymium YAG laser to excite the material, this phosphorescence which was emitted from a phosphor bonded to the part to be studied in the engine was observed. The use of this approach was tested up to 600°F (330°C) although in principle much higher temperatures could be reached.

Further work illustrating the use of this technique in a "fiber optic" approach has been demonstrated by Grattan et al.[53] in the sensing of temperatures up to 850°C through the use of fluorescence in neodymium: YAG. Being crystaline, this material is capable of responding to temperatures up to almost 2000°C without damage. In order to overcome the limitations of plastic clad fiber at the high temperatures required, an intermediate sapphire "fiber optic" was used. This ran from the hot region to a point outside the zone where the temperature was less than 200°C, where the light was then coupled to conventional optical fibers. Unfortunately, ambiguities result in this particular sensor, as shown in Figure 10, but work is continuing to investigate the use of other materials to produce an unambiguous high temperature sensor. This device has the advantage of running from below room temperature all the way througl to the very high temperatures discussed and in principle to even higher temperatures.

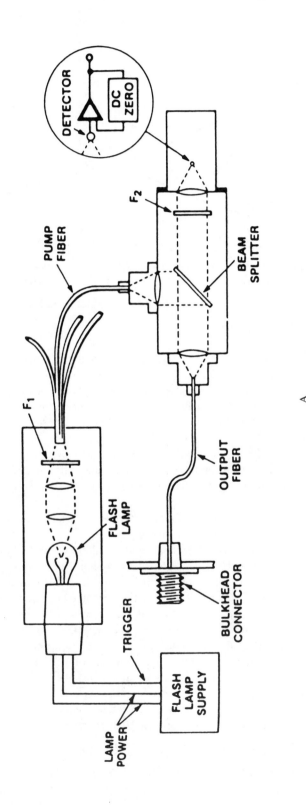

FIGURE 9. Luxtron type 750 fluoroptic temperature sensor. (A) Optical arrangement. (B) Decay time and fluorescent intensity changes with temperature. (From Wickersheim, K. A., Heinseman, S., Trans, H. N., and Sun, M. H., *Proc. ISA*, Paper 85-0072, 1985, 87. With permission.)

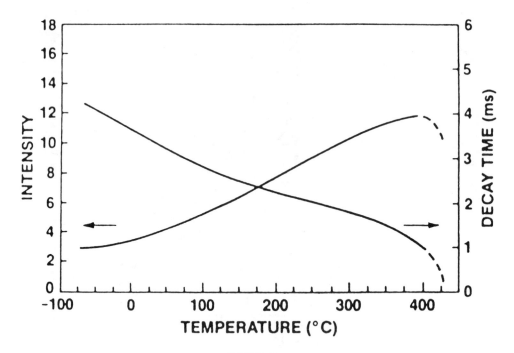

FIGURE 9B

A careful refinement of the optical pyrometry technique has been described in the "Accufiber" device[54] which is a conventional system with very high accuracy claimed — a resolution of approximately 1 to 0.5°C and an accuracy in the order of 0.0025%, illustrated in Figure 11. This pyrometric technique is capable of measuring temperatures using the black body radiation from a thin film deposited on a sapphire optical fiber. The temperature is determined from the emission on two narrow optical bands of the spectrum which is received by the detector. This device, being fiber optic based, is capable of making measurement in regions where optical observation of the black body radiation is either difficult or impossible, but is limited in its lower temperature range to those regions where the black body radiation is measurable by the sensitve detector system employed. A range of 500 to 2000 °C is reported by the manufacturers. This device is now used as a temperature standard in the 630 to 1060°C range by the National Bureau of Standards in the U.S.

For measurement of even higher temperatures (2000 to 6000°C) a multicolor pyrometry system has been developed by Foley et al.[55] for a nonintrusive measurement on solid and liquid surfaces. Again, applications in combustion diagnostics are particularly important. For these high temperature regions, optical fiber variations of conventional and newly developing techniques are possible, and indeed desirable. As an example, the use of CARS spectroscopy (Coherent anti-Stokes Raman Spectroscopy) is now well established for the measurement of combustion processes. This technique, which uses the fundamental Raman scattering processes which occur within the combustion region, relies for the measurement technique on the monitoring of light emitted. For regions where combustion is not readily visible, the use of sapphire high-temperature fibers can enable this measurement to be made remotely. As such a technique does not require the intrusion of either a probe or a measuring material into the combustion, it is truly noninvasive, and the fiber approach enables very small regions to be observed.

F. EXTRINSIC SENSORS — "FREQUENCY-OUT" DEVICES

An interesting new development in the field of fiber optic sensing, with applications to

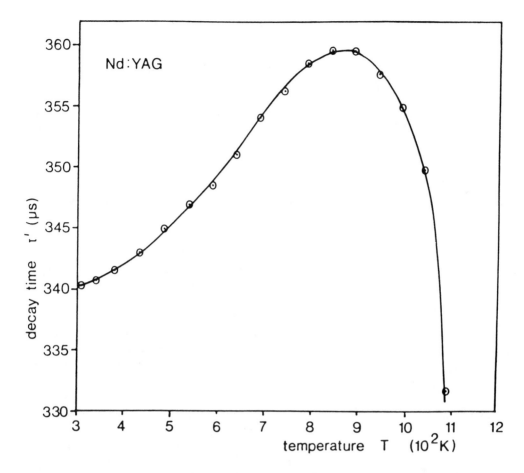

FIGURE 10. Illustration of the use of changes in the fluorescence decay time, τ, in Nd:YAG to measure high temperatures (to 1100 K, –850°C). (From Grattan, K. T. V., Manwell, J. D., Sim, S. M. L., and Willson, C. A., *IEE Proc.*, 134J, 291, 1987. With permission.)

temperature measurement, is the use of small or miniature resonator systems which yield the measurand in terms of frequency information. Resonating elements may be driven optically using the photothermal effect:[56] when certain materials absorb energy in the form of light, this optical energy can be converted to thermal energy. If such an absorbing material is deposited on a vibrating element, such as a miniature quartz structure, then the localized heat produces a bending force in the element due to the thermal stress. When light is removed, this is also removed and the material relaxes. As a result, if the light falling on or transmitted through the material is modulated, a flexural motion is seen; if this modulation occurs at a resonant frequency, then the maximum displacement of the element can be experienced. Thus, such an approach can be used where the change in temperature of a device is converted into the change in the stress on the optical crystal. This is a stress additional to that produced by the radiation from the modulating source.

Pressure sensors have been produced on this basis where the stress is induced by a pressure or weight change on a quartz sensor attached to a cantilever beam, but equally this stress could result from a change in temperature. The advantage of such a system is that being frequency based (and not relying upon intensity changes), the measurement of frequency is used and by electronic means this is both particularly simple and accurate. The readout of the frequency itself can be achieved by optical means, either by the techiques described in a paper by Lynch[56] or through evanescent wave coupling techniques (as used for pressure sensors) as illustrated

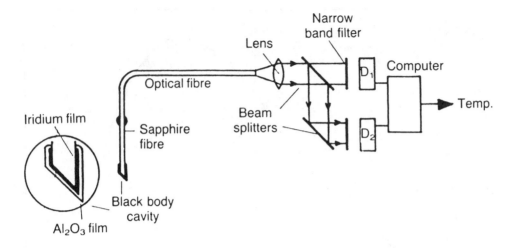

FIGURE 11. 'Accufibre' temperature measuring system. (From Medlock, R. S., *Int. J. Opt. Sens.*, 1, 43, 1986. With permission.)

by Grattan et al.[57] Further work in this area has discussed the use of micro-sized resonating elements which are produced from silicon, through the use of technology familiar in micro-electronic fabrication as shown in Figure 12. So far, this appears to have been used mainly for pressure sensing in the work of Venkatesh et al.,[58] but again there are possibilities for its use in temperature sensing.

For example, recent work has been reported by Culshaw[59] describing simple silicon-based vibrating bridges and diaphragms for the measurement of both temperature and pressure. Small power consumption only is required for these miniature silicon devices, and it is suggested that modification of the fabrication of these devices could permit the power threshold to come down to as low as nanowatts. In addition, this will simplify the multiplexing of a number of these sensors on a single fiber optic link, as the power consumption of each is very low and the interrogation system should be relatively simple. In semiconductor materials the process of optical absorption gives rise to a mechanical strain being set up in the material in a rather complex interaction. Therefore, careful design of the structure is required to optimize the use of this strain to give a measurable change in resonant frequency. Culshaw[59] suggests that a temperature transducer could be constructed having high resolution and accuracy (better than 0.01°C) with a rapid response time. Such a temperature sensor could be valuable in making corrections to measurements for pressure using a similar device, where the temperature-induced strain effects can be separated. Figure 12 shows the temperature response of the silicon diaphragm proposed for this use. Vibrating wire hybrid techniques have been used by Jones et al.[60] for pressure measurement, and by suitable transduction these could also be used for temperature sensing.

G. BIOLOGICAL APPLICATIONS

In the previous discussion there has been considerable mention of biological applications for fiber optic temperature sensors, and a number of groups have directed their work at this field. There has been definite progress in recent years in the development of sensors for the measurement of chemical parameters, e.g., pH, the monitoring of concentrations of certain chemicals such as glucose and gas sensing, as described in other chapters (see Chapter 19) and often it is required to correct for temperature effects when using these sensors. A recent paper by Gehrich et al.[61] showed work on a sophisticated *in vivo* sensor which incorporated a temperature monitoring thermocouple. It is envisaged that in the future, to be compatible with the other optical signals which will be carried by these devices, a fiber optic temperature measurement can be incorporated. The most likely approach to be a success from this field is the use of luminescent techniques, although these techniques themselves are often used for the monitoring of certain other materials, e.g., glucose.[62] It is thus important that fluorescent materials which are not susceptible to interference effects by the local environment are used. The success of this field will depend largely upon the quality of the development work by other groups, but it is safe to say that the fiber optic temperature sensor work which relates to these measurements is well advanced on that for purely chemical measurements.

IV. INTERFEROMETRIC TECHNIQUES

Interferometric techniques are well established for the high resolution measurement of length. As a result, many fiber optic techniques have been used where a measurand is converted directly into a change of displacement. In many cases temperature can be so converted and, in addition to the methods described, monitored using an interferometer. During the past decade considerable research and development has led to the introduction of monomode interferometric systems in addition to multimode devices. As fiber optic interferometers offer many advantages in terms of stability and convenience of use over open air devices, they are well suited to fiber optic sensing techniques. Optical sources usually required are gas lasers because of the high stability of the lasing wavelength and the fact that higher powers are available, e.g., from the Argon ion laser. This latter laser does increase cost quite considerably, but recent advances in solid state laser diodes has made it possible to produce laser-diode-based devices which will increase the applicability of inteferometric techniques outside the laboratory. It is beyond the scope of this chapter to discuss in detail the signal processing techniques used, but active homodyne and heterodyne techniques have been employed, and a detailed discussion of these is given in a recent paper by Jackson.[63]

A. OPERATION OF THE INTERFEROMETER

The operation of the basic conventional interferometer is discussed briefly, with reference to the diagrams of Figure 13. Taking, for example, the Michelson two-beam interferometer, light from the single frequency optical source is divided at the beam splitter and light is launched into two arms, one a "reference arm" and the other the "signal arm". The amplitudes of these are given by A_r and A_s, respectively, where these are as follows:[63] Reference — A_r $\exp[i(\omega_L t + 2kX_r)]$; signal — $A_s \exp[i(\omega_L t + 2kX_s)]$. X_r and X_s are the distances that the light traverses between the reference and the signal mirrors, k is the propagation constant given by $2\pi n/\lambda$ where n is the refractive index and ω_L is the angular frequency of the source of wavelength λ (in vacuo). The beams recombine after traversing the instrument and a current is detected at a photodetector, usually a simple photodiode. This current, I, is given by $I = \varepsilon$ $[1 + k'\cos \varphi (t)]$ where $\varphi (t)$ is a time-dependent phase difference in the interferometer, k' is a constant called the visibility, and the photodiode output as a function of phase is also shown in the figure.

The most ideal operating conditions can exist in the laboratory where the two amplitudes are the same and the polarization of the light beam in each arm is the same (as would be seen from a single source), the source is of narrow bandwidth of constant frequency, and the system is mechanically stable. Under such conditions very small path difference changes can be observed in this device, and this is the basis of the use of the interferometer as a displacement monitor. Periodic displacements of the order of 10^{-14} meters have been observed,[64] with the ultimate limit in detectability given by the noise conditions at the detector. Thus it is evident that there is a significant potential for the use of interferometric devices for optical sensing and in an optical fiber analog, for fiber optic sensor schemes.

However, there are several fundamental properties of interferometers, whether fiber optic or open-air, which can restrict their application for sensor purposes. As can be seen from

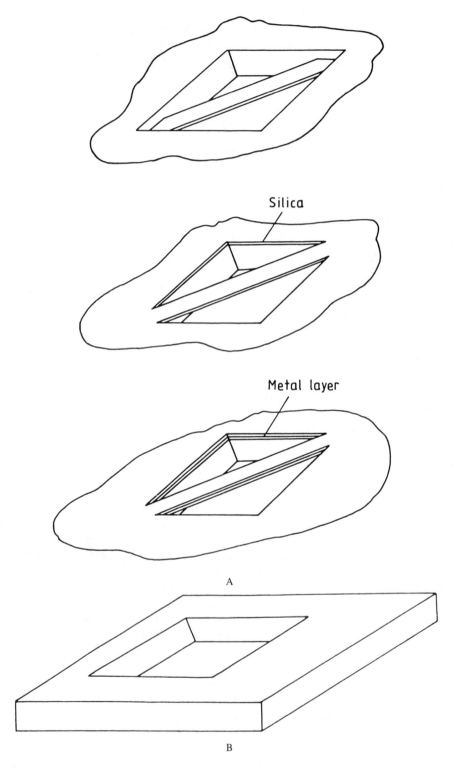

Silica

Metal layer

A

B

FIGURE 12. Silicon-etched structures for resonant sensors. (A) Bridge structure. (B) Diaphragm structure. (C) Temperature response of the diaphragm. (From Culshaw, B., *Short Meeting Series,* 7, Institute of Physics, London, 1987, 33. With permission.)

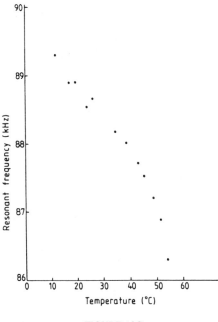

FIGURE 12C.

Figure 14, the output of the interferometer varies periodically, equivalent to a change in relative mirror separation equal to one half wavelength of the light used. In sensor applications where there is a requirement for the output to be uniquely a function of zero displacement, then the sensor must be used in the so called "zero path length" mode and the operational range reduces to one quarter wavelength, typically 2×10^{-7} meters, for wavelengths in the red or near infrared region. This would be very difficult in practice, and as a consequence most interferometers are operated with an arbitrary imbalance between the two arms; displacement measurements are therefore made relative to an initial condition which is not specified in the instrument itself.

It is this arbitrary starting point which can present problems when the instrument is switched off. The point at which the instrument is at "zero" is lost and this can be a problem. Techniques are under development to overcome this. In particular, if the sensor is used for slowly varying measurands, such as a small temperature change, this lack of an intial "zero" point is a major difficulty. Further, the output for the device is nonlinear varying periodically, and at a maximum where the phase difference, φ, equals $\pm m\pi \pm \pi/2$ and zero when φ equals $m\pi$ where m is an integer. There is a so-called quadrature position where maximum sensitivity is reached corresponding to $(2m + 1)\ \lambda/4$, and similar processing arrangements can be made to enable the device to operate in this mode.

B. FIBER OPTIC INTERFEROMETRIC DEVICES

In order to produce the fiber optic equivalent interferometer, a monomode fiber optic device is required. As a result, the same optical components, including mirrors, polarizers, and beam splitters, are needed. Further optical phase and frequency monitors, compatible with fiber optic devices, may also be required. Much work has been expended in recent years in the development of such devices and up until very recently their high cost made optical fiber interferometric devices considerably more expensive than the types of extrinsic point sensors discussed earlier. However, with developments in the technology, the prices of optical couplers in particular are decreasing and devices such as the fused directional coupler (which

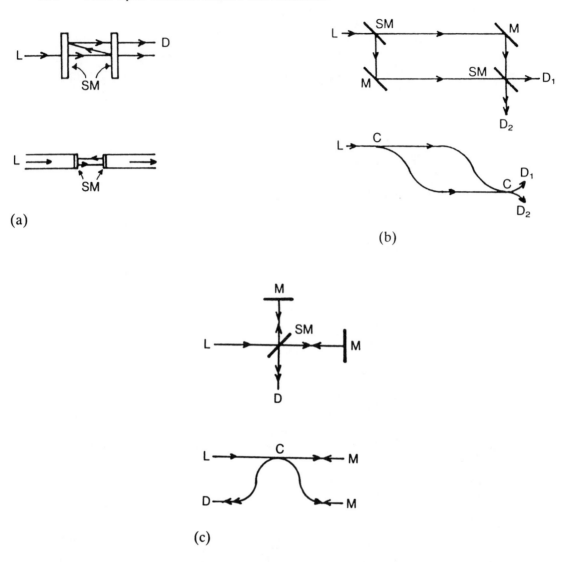

FIGURE 13.   Illustration of open-air and fiber interferometers. (a) Fabry-Perot. (b) Mach Zehnder. (c) Michelson. L — laser, D — detector, M — mirror, SM — semi-mirror, C — coupler. (From Medlock, R. S., *Int. J. Opt. Sens.*, 1, 43, 1986. With permission.)

has a fixed splitting ratio of light into the arms of the device) has been successfully used in the field trials of a fiber optic interferometric sensor. This was the hydrophone developed at the Naval Research Laboratories in the U.S., the details of which are reported by Giallorenzi et al.[65] Other components such as high quality mirrors can easily be fabricated on optical fibers using conventional mirror technology as can polarizers. In addition, other techniques to make fibers, themselves polarization dependent have been developed and such fibers are commercially available.

Fibers can easily be joined for use in interferometers using fusion techniques developed for communications purposes, and fiber optic phase modulators were first introduced by Davies and Kingsley[66] for multimode fiber optic sensors. Here the fiber is tightly wound onto a large piezoelectric cylinder. When a voltage is applied to this cylinder, the guiding of the light is redistributed among the modes of the multimode fiber. A similar approach has been developed by Jackson et al.[67] for monomode fibers where the optical pathlength of the fiber is changed

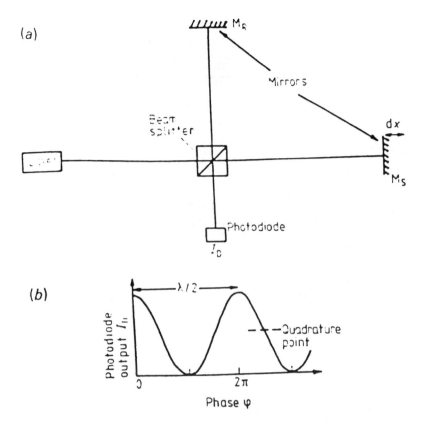

FIGURE 14. (a) Michelson interferometer. (b) Transfer function, photodiode current I_d as a function of the relative phase ϕ. (From Jackson, D. A., *J. Phys. E*, 18, 987, 1985. With permission.)

and "moderate" voltages on the device can produce pathlength changes greater than 100 μm. The absolute frquency of the electromagnetic wave in the device can be changed through the use of a Bragg cell, in an analog of open-air optical devices.

The two most widely used configurations for optical fiber interferometers are those of the Mach-Zehnder and the Michelson, as illustrated in Figure 13. These devices can be constructed relatively easily from the optical components described. Although the Michelson configuration uses fewer components, severe problems with the stability of the source may be encountered due to the high levels of optical feedback which arise in this arrangement. The Mach-Zehnder arrangement overcomes these problems and is perhaps the most commonly used interferometric type in optical fiber sensors. The Sagnac device is, however, the most widely discussed for optical sensor applications, but is essentially limited in its application to the very important optical fiber gyroscope which has very significant potential application in the military and aerospace markets.

The fiber optic analog of the Fabry-Perot interferometer is one of the simplest arrangements that can be envisaged, and in recent work a number of potential sensing devices using this arrangement have been discussed. This follows it being largely ignored in early work, although it does show promise as a small remote point sensor with ready applications to optical multiplexing. In addition, a relatively simple differential interferometer has been discussed by Jackson[63] where for temperature measurements fibers with very high levels of built-in linear birefringence are used, rather than conventional communication-type fibers. This gives additional sensitivity, although as yet the ideal fiber for this arrangement with high circular birefringence is not commercially available.

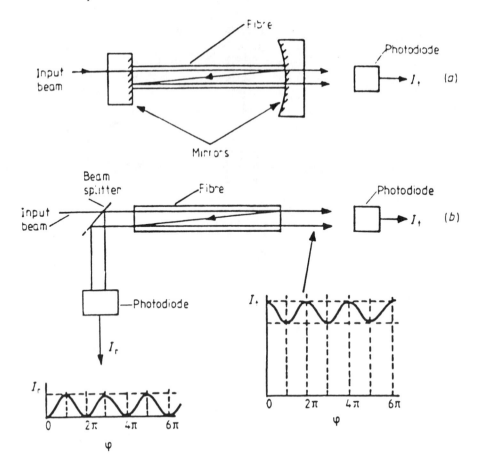

FIGURE 15. Fiber Fabry-Perot configurations. (a) Mirrors attached to the fiber to enhance the reflectivity of the mirror cavity. (b) Accurate cleaving of fiber ends. (Inserts show transfer functions for reflection I_r v Q and I_t v Q, for transmission.) (From Jackson, D. A., *J. Phys. E*, 18, 987, 1985. With permission.)

The most familiar optical source for the interferometer is the commercial Helium-Neon gas laser operating on a single frequency. This can be controlled very accurately to ±10 MHz. Unfortunately, gas lasers are fragile and somewhat expensive by comparison to the solid state diode laser which has recently been made available with high powers at low cost. Such a device would be ideal if it could be stabilized in terms of the laser center frequency. Although much development work has continued, these devices are not available at the moment.

However, recent work has been continuing with the use of low coherence sources and diode lasers which provide synthetic multiple wavelengths. These low coherence sources can be used very effectively in multiplexing systems where coherence tuning between the sensor and the measuring interferometer is exploited.[68]

C. SIGNAL PROCESSING

As discussed earlier, the sensitivity of an interferometer to an induced optical phase change is not constant but relies upon the periodic nature of the interferometer function. This causes problems with fading of the signal, and the device is best operated in a linear range. There are a number of approaches which can achieve linearization on the output, broadly classified as active homodyne, passive homodyne, or heterodyne. These have been developed for both classical interferometric and other applications in optical processing, and their discussion in this chapter will not be detailed. Considerable information on these techniques of signal processing are to be found in a recent publication by Jackson.[69]

D. OPTICAL FIBER INTERFEROMETERS APPLIED TO TEMPERATURE SENSING

The application of the interferometric technique to temperature sensing is hardly surprising in view of the ready conversion of temperature to displacement of the mirror in one arm of an interferometer. In principle, high resolution devices can be realized, although as yet surprisingly few practical schemes have been described. The problems with initialization and absolute calibration may well be the reason for this. Work by Hocker[70] first demonstrated the principle that temperature (and strain) could be measured in this way with fiber optics. The first high resolution monomode temperature sensor with signal processing was described by Musha et al.[71] Using a Mach-Zehnder approach, a sensitivity of 1200 rad K^{-1} was demonstrated, but this required an expensive stabilized laser source. Subsequent experiments based on a simple unbalanced optical fiber interferometer in the Michelson arrangement with this type of solid state laser with pseudo-heterodyne signal processing and fringe counting have demonstrated[72] practical thermometers with working ranges in the useful room temperature to 250°C range with a resolution of 10^{-3} K.

Polarimetric techniques have been used that have significantly reduced sensitivities when compared with the conventional configurations of the fiber optic interferometer. Such devices have a small dynamic range, and in practice it is difficult to use the device because the *total* length of the fiber is used for sensing. Early work in this field was undertaken by Eickhoff.[73] Using highly birefringent monomode fiber, this scheme has also been described by Kersey et al.[74] It has a large dynamic range and pseudo-heterodyne signal recovery. The basic optical arrangement is one with a short length of birefringent fiber fused and spliced with a similar long length of fiber such that the relative orientation of their axis is at 45°. The actual sensing element is the short length of fiber. This overcomes the problem of the entire length of this arm of the interferometer being used for the sensing purpose. Light is launched from a linearly polarized source to excite one mode of the fiber. After passing into the sensing fiber, the orthogonally polarized beams are reflected back and they recombine and produce light of polarization with an arbitrary ellipticity. The displacement sensitivity is very much lower for this sensor, and it is possible that this device could replace conventional sensing elements with the added advantage of enhanced range and resolution.

An accuracy better than 0.2 μm has been reported using the approach discussed which is sufficient to enable the current fringe number of the sensing interferometer to be determined. A variation of this technique is the use of both the "white light" *and* a laser source where the system is initialized using the "white light" approach. Then, the high precision mode using the laser source is employed to obtain both a wide range and high accuracy from these devices.

The Fabry-Perot interferometer, illustrated by Figure 16, has recently been employed by Corke et al.[75] in a device which extends the measurement range by combining polarimetric and interferometric configurations in a single. Recently, this scheme has been the subject of work by various groups. A temperature-sensing device was reported by Schultheis et al.[76] in which a miniature Fabry-Perot cavity was used for temperature sensing, the characteristics of the device being mechanically displaced by temperature resulting in a change in the optical characteristics of the system. This arrangement was both compact and able to fit at the end of a fiber itself, making this a very useful potential optical fiber point sensor yet one using interferometric techniques.

The Fabry-Perot (F-P) interferometer has been the subject of work by Seki and Noda[77] using pseudo-heterodyne techniques, to achieve high sensitivity and stable operation. Electronic circuitry is proposed to overcome the problem of recognizing whether the temperature is rising or falling. The cavity is formed from a length of single mode fiber, of 1 m, wound around a piezoelectric cylinder to vary periodically the fiber length. Its averaged length in the modulation period varies due to temperature variation. Therefore, the instantaneous phase of the reflected and transmitted light through the F-P interferometer was determined by both the average length as a function of temperature and the length variation of the modulator. Initially, a phase difference at the specified temperature is measured, then the phase difference change based on this measurement is proportional to the temperature change from the specified value. The resolution of the system is high at 7.6×10^6 m^{-1}. This depends on the phase-meter resolution used and assuming a typical value gives a temperature resolution of 0.01 K.

E. PROBLEMS WITH INTERFEROMETRIC FIBER OPTIC SENSORS

The major disadvantage of fiber optic interferometers is as discussed. The device is not an absolute device, and therefore it requires calibration and the absolute relative phase or the fringe number is lost when the device is switched off. This may well account for the fact that, although fiber optic sensors using interferometers appear very attractive, as yet few commercial fiber optic interferometer-based sensors are available, except for the related Laser Doppler Velocimeter.[77] However, progress is being made in the developing of techniques to extend range, one of which is the use of dual wavelength processing.[78] In this technique, two wavelengths are used to illuminate the interferometer ($\lambda 1$ and $\lambda 2$). This scheme relies upon the generation of a synthetic wave whose wavelength is given by the product $\lambda 1 \lambda 2/(\lambda 1 - \lambda 2)$, and this is achieved by the use of "ramping" of the drive current to the laser diode that is used as the source. With typical wavelengths of laser diodes in the near infrared, an extension of a factor of 100 or so in the useful range of the device could result by comparing the operation of the synthetic wave interferometer with that of the original wave using only $\lambda 1$. In addition, a technique such as white light interferometry[79] has the additional advantage that it allows the possibility of multiplexing several sensors, allows for sensor initialization at switch on, and the use of an inexpensive white light source for the entire sensor system.

The operation of the technique is as follows: the broadband source illuminates the interferometer, and the coherence length of that source is shorter than the optical path imbalance of the device.[80] Therefore, the interference effects upon which the device depends will only be observed at the output of the receiving interferometer when its path imbalance is zero and identically equal to the path imbalance in the sensing interferometer. This is discussed in detail by Jackson[69] and illustrated in Figure 17. As a result of their extreme sensitivity and the possibility of a number of additional sophisticated techniques (some borrowed from previous studies on radar techniques), it is expected that fiber optic interferometric sensors will, in time, become established. Significant work is advancing in the field of the fiber optic gyroscope, for which there are tremendous rewards in the aerospace industry, and much of the work that will be done on signal processing, etc. has applicability to fiber optic temperature sensors.

V. DISTRIBUTED OPTICAL FIBER SENSORS

Distributed optical fiber sensors utilize the intrinsic sensing capabilities of optical fibers themselves. Optical fibers are essentially a one-dimensional medium and they can allow certain manipulation of the sensed data to be performed in the fiber itself. The line-integration property afforded by optical fibers allows for high sensitivity, as is seen with the optical fiber gyroscope. In addition, the line-differential characteristics allows the determination of the spatial distribution of the measurand. This is a particularly important characteristic as it means that not only can information on the *magnitude* of the measurand be obtained but also the *locality* of the measurand. There are, therefore, a number of potential applications of these characteristics in fiber optic sensing of all parameters, and in particular in the sensing of temperature, where intrinsic sensor schemes have been demonstrated. As a result, *any* optical fiber which is installed for *any* sensor or communication purpose may be utilized to make these measurements. This is particularly important in situations where the exact locality of an area where an unexpected temperature excursion could occur is not known, but there are a number of suspect sites.

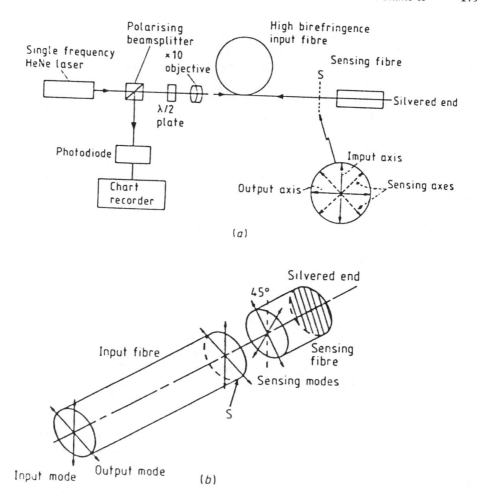

FIGURE 16. Remote polarimetric sensor (after Corke et al.) (From Jackson, D. A., *J. Phys. E*, 18, 987, 1985. With permission.)

The use of extrinsic point sensors would require one individual sensor at each location, whereas the intrinsic distributed sensor technique simply allows the connection of those sites by a single fiber which is then connected to an optical source and detector. The spatial information on a location of the unexpected temperature elevation can be obtained from the signal processing of the received data. As fiber optics are small, flexible, and insulating dielectric media, they are ideally suited for this application in industrial plants. These techniques are discussed in detail in a number of papers by Rogers.[81,82]

Conventional industrial instrumentation cannot provide this distributed facility that is available from optical fibers, and thus a unique sensing opportunity exists. In conventioanl techniques there would be considerable problems of multiplexing and calibration, with the consequent expense. In addition, the small size of the optical fibers means that they can be installed retrospectively in many applications where it is desired subsequently to make these temperature measurements.

A. PRINCIPLES AND CLASSIFICATION OF DISTRIBUTED FIBER OPTIC SENSORS

In the literature in general, there are discussed two main types of distributed fiber optic sensor are discussed. The first, the so-called "quasi-distributed system" or "multiplex system" is one where a series of individual sensors are linked together on a single continuous fiber optic. Therefore, measurements can only be made at discrete, predetermined points or along specific, limited lengths of fiber. This is contrasted with the fully distributed system where the measurement can be made at *any* point along the fiber and the measurand is determined as a function of position. Clearly this offers considerably greater flexibility for sensing applications than the first, but a cost penalty from the greater complexity of the signal processing.

B. QUASI-DISTRIBUTED SYSTEMS

Early work on the development of quasi-distributed systems is due to Nelson et al.,[83] where a narrow pulse of light is launched into the fiber and subsequent measurements of the differential delay are made. A small fraction of the initial light pulse is split by a coupler at the measurement point, modulated by the sensor, and then reflected back to the launch end. There, accurate "time of flight" measurements are made and from the temporal information and a knowledge of the propagation velocity of the fiber (determined from the fiber refractive index), the positional information can be obtained. Optical-time-domain-reflectometry (OTDR) is a technique that has been used extensively in communications applications to determine faults in either conventional or optical fibers. Essentially, "time of flight" information is related to spatial information. In these distributed sensing applications, this OTDR approach may be used in a number of different ways.

1. Differential Absorption Distributed Thermometry

This was one of the first systems to be described and is illustrated schematically in Figure 18. It is an extension of the use of differentially absorbing glass elements in simple extrinsic point sensors to their development in a distributed sensor. A number of absorbing doped glass plates are positioned at the point where the measurement is required and they are connected together by a length of conventional optical fiber. Light from the laser source provides the two wavelengths required, one for referencing and the second to indicate the degree of absorption using the technique discussed earlier. The position individual glass plates can be identified from the backscatter resulting from simple Rayleigh scattering in the fiber. At any one particular sensor the light backscattered from the front end of the fiber is compared with that backscattered after passing through the absorber in order to determine the degree of absorption. Signal averaging is used to eliminate some of the noise problems with this type of arrangement. With a response time of the order of the second, a very satisfactory thermometer can be obtained. However, the main disadvantage is that a high power laser light source is required due to the attenuation of the signal at each absorbing stage. As a result, there is a limit to the number of measuring sensors that can be so connected together, and in the work of Theocharous[33] a maximum of about ten plates seems to be the effective limit.

C. DISTRIBUTED TEMPERATURE SENSORS

In considering possible schemes for truly distributed temperature sensors, it is important to be aware of those interactions which will yield temperature information and give adequate signal levels for that information to be extracted. Thus, there are some temperature-dependent interactions which may occur within the fiber which would require highly sophisticated signal processing to extract the temperature-dependent data and therefore are not practical as sensor schemes. For example, the use of Brillouin scattering, a nonlinear optical phenomenon which depends upon temperature, may in principle be used but the very small wavelength shifts which result from this scattering are not easy to measure. Rayleigh backscattering is a wavelength-dependent phenomenon which is small and in communication fibers is the process which results in the gradual attenuation of signal upon which "chemical" losses are superimposed.

One of the conflicts in optical fiber sensing is that often fibers are designed primarily for

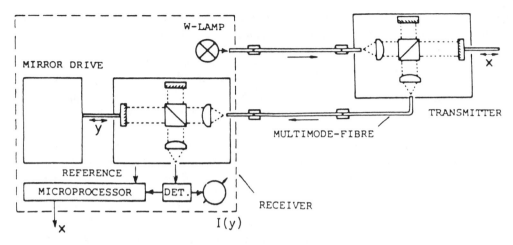

FIGURE 17. 'White-light' fiber coupled interferometer. (From Bosselmann, Th. and Ulrich, T., Conference on Optical Fiber Sensors (OFS '84), VDE-Berlin, 1985, 361. With permission.)

the much larger communications market and not for the sensor market. Thus, fibers are designed to have very low Rayleigh scattering coefficients. For sensor applications, however, a higher scattering coefficient, where this is used as a means to determine the measurand, is often desirable. In the work of Hartog,[84] the Rayleigh backscattering in an optical fiber is used as a means of temperature sensing. In order to obtain a high Rayleigh scattering coefficient, a nonsilica fiber was used. In this case, the fiber was one with a liquid core, the liquid itself being a transparent organic solvent. The arrangement for this scheme is shown in Figure 19 where the temperature of the liquid-cored fiber is determined with respect to the temperature of the adjoining fiber. Light from a laser generating a pulse on a nanosecond timescale is launched into the fiber, following optical processing of the signal to limit the propagation to low-order modes to reduce the effect of temperature on numerical aperture. Measurements were made in the region from below room temperature up to 70 or 80°C.

The upper limit on the use of this technique is set by the fact that the refractive index of the core material chosen becomes equal to that of the cladding at a temperature of about 160°C, the guiding action stops, and the device ceases to function. Higher temperature regimes could be explored with the use of solid cored fibers, but there the disadvantage of poor sensitivity is seen. Although interesting results have been demonstrated, a major problem is the cost and unsuitability for many purposes of the liquid core fiber. The possibilities of fracture and spillage of the liquid exist, and in small quantities this fiber would be expensive to manufacture.

1. Use of Doped Fibers in Distributed Sensing

In order to enhance the backscattered signal and ease problems of identifying the information carrying optical signal in the presence of noise, schemes have been proposed which use fibers doped with other materials. Again, this represents a problem in that the specialist fibers are not available for communication purposes and therefore they are more expensive to produce than conventional fibers. Early work using such systems was reported from United Technologies by Moray et al.[85] in 1983 who described a sensor which relied upon the differential spectral absorption characteristics of the material when doped with neodymium. This material has been discussed for use in fluorescent sensors, but the particular application here is one where an absorption band centered at 840 nm decreases with a corresponding increase in the band centered at 860 nm. This ratio is measured at the temperature-dependent function.

Recent work has been carried out by Farries et al.,[86] again using the same dopant in an optical fiber. In this case, measurements were made at 904 nm which show a dependence of an attenuation on temperature. Initial work indicated a range from −50 to about 100°C with a ±2°C accuracy and a spatial resolution of 15 meters over a length of 140 meters. More recent work[87] indicated the potential of this system with specialist fibers to make measurements up to much higher temperatures, approaching 1000°C, again using sophisticated signal processing. Difficulties arise from nonuniformity of the dopant but better fabrication techniques could overcome such problems.

An advantage of this approach is that these fibers are now being manufactured commercially, with rare earth materials like neodymium and erbium, for use as optical fiber lasers.[88] Such devices are very convenient and easily pumped by solid state diode laser sources. If this market develops, there is every expectation that these fibers will be available at lower cost.

2. Distributed Anti-Stokes Raman Thermometry

This technique is one relying upon a nonlinear optical effect, the familiar Raman effect. Light is scattered inelastically by atoms in an optical fiber, and as a result emerges at an optical wavelength which is different from that of the source, in contrast to Rayleigh scattering which occurs at the same optical wavelength. Additionally, it is independent of the fiber material and hence can be used on already existing optical fiber installations. More fundamentally, it does not require calibration as the temperature information is fundamental to the scattering process itself. The technique operates through a measurement of the ratio of the Stokes and anti-Stokes backscattered radiation levels where the Stokes line frequency, v_s is given by $v_s = v_i - v$, and the anti-Stokes line is given by $v_a = v_i + v$ where v_i represents the frequency of the incident radiation and v is the frequency of the Raman scatter, corresponding to an energy level separation of the scattering material. As a result it can be shown that the ratio, K of the anti-Stokes to Stokes power levels is given by $K = (v_a/v_s)^4 \exp(-hv/kT)$. Thus, it can be seen that the absolute temperature can be determined, in principle, from the other parameters without further calibration.

The Raman effect is not discussed in detail here except to say it is in effect where light propagating in the medium is modulated by the vibrations and rotations at a molecular level. Unfortunately, silica does not have a large Raman scattering coefficient, but its convenience as a fiber optic material does help to overcome this difficulty. An experimental system has been described by Dakin et al.[89] using light from a 514.5 nm argon ion laser which produces short (15 ns) pulses. The high power from this laser is required due to the weak nature of the scattering, and the Raman spectra are seen at a wavelength separation of about 10 nm from the Rayleigh lines as shown in Figure 20. Results from this work indicate that a resolution of about 5°C in temperature and 5 meters in length can be obtained. The main difficulties lie in the low level of the scatter signal and interference from Rayleigh backscatter. In order to obtain the information on temperature, long integration times have been required to extract the signal from the noise.

In order to overcome the difficulties of the use of an expensive and bulky argon ion laser, recent work by Dakin et al.[90] and Hartog et al.[91] has concentrated upon the use of semiconductor lasers. These latter experiments are reporting accuracies of 1°C over 1 km with a spatial resolution of 7.5 meters. More recently, a commercial system has been marketed by York Technology Ltd.[92] using the ratio of the Raman to Rayleigh scattering in a silica fiber. As the anti-Stokes line has a lower intensity than the Stokes line, this avoids the problem that arises from the comparative weakness of this line using only Raman scattering, although similar signal processing schemes are used.

The further development of these techniques will rely to a large extent on the developments in sophisticated signal processing and upon the manufacture of suitable fibers for sensor applications. Some work has already been seen with the development of specially doped fibers

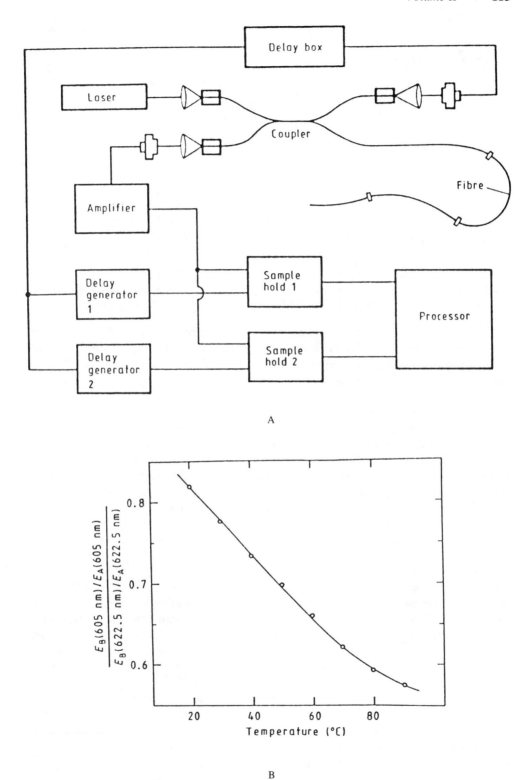

A

B

FIGURE 18. Illustration of differential absorption distributed thermometry. (A) Optical arrangement. (B) Temperature dependence. (From Theocharous, E., *Conference on Optical Fiber Sensors (OFS '84),* VDE-Berlin, 1985, 199. With permission.)

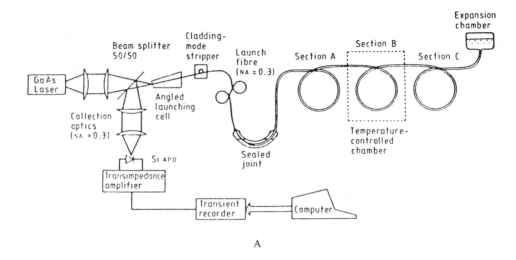

A

FIGURE 19.   Distributed optical fiber thermometer using liquid core fiber. (A) Optical arrangement. (B) Temperature dependence. (From Hartog, A. H., J. Lightwave Technol., LT1, 489, 1983. With permission.)

and liquid cored fibers. However, the small market for these fibers does make their price rather higher than is desirable. In addition, other workers have considered the possibilities of the use of other physical techniques for this measurement. Detailed comments on the mathematical analysis of these sensor approaches have been made by Rogers.[82]

3. Temperature Sensor — Cryogenic Leak Switch

A temperature sensor which is not designed to require accurate calibration has been devised by Pilkington Security Systems[93] in order to monitor the severe temperature change that arises when cryogenic fluids leak onto an optical fiber. The scheme is one where a continuous length of optical fiber, encased in a stainless steel jacket, is run back and forth underneath a cryogenic storage tank. If there is any leakage or spillage from the tank, the liquid falls upon the fiber and reduces its temperature very severely to below −60°C. At this stage the refractive index change of the core and cladding is such that the guiding action of the fiber is severely interrupted, increased attenuation of light propagating in the fiber is seen, and thus the alarm is sounded due to the drop in the optical signal.

The system is a distributed sensor but in the particular application used, it is essentially unnecessary to have a spatial resolution better than the total length of the fiber required for one tank. This eases considerably the signal processing aspects of the sensor and an effective device results. The system is "fail-safe" in that if the fiber is accidentally tampered with or fractured, the alarm sounds and the fiber can be inspected due to the optical loss experienced. A prototype has been successfully installed by British Gas and the response of the device, which is on a timescale of several seconds, is shown in Figure 21. Again, the slow response is not a problem as this is primarily used as an alarm system. Similar techniques have been employed on fiber optic security systems[94] where the loss is pressure induced.

4. Other Intrinsic Devices

Recent work by Ibe et al.[95] reported the use of a sensor utilizing optical polarization changes due to the alteration of the temperature. The transducer itself was composed of polarization maintaining fibers and no other components, such as prisms or lenses, are used. However, it does consist of four polarization-maintaining fibers arc-fusion spliced to each other with their optical axes at 45° to each other. The optical source required was a wavelength scanning distributed feedback laser, the wavelength shift of which was obtained by heating. The scheme

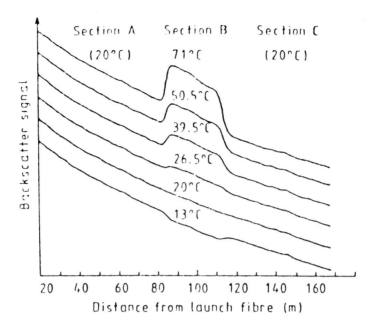

Section A   Section B   Section C

(20°C)   71°C   (20°C)

FIGURE 19B.

could not be considered simple, and the analysis was quite complex, but for a measurement range of 100°C, an accuracy of 1% full scale was reported.

D. DISTRIBUTED SENSORS — POSSIBLE FUTURE DEVELOPMENTS

The importance of the technique described cannot be overemphasized and the advantage of an optical approach, combined with spatial and temporal information, is indeed valuable for applications to industry and in research.

The work described so far has demonstrated the feasibility of this measurement technique and future developments lie in two directions. One is the use of faster and more sophisticated signal processing to extract the encoded measurand more quickly and more accurately. On the other hand, developments on the optical technology are continuing and other physical effects, coupled with the use of specially designed fibers, may be employed. The range of nonlinear optical effects familiar since the invention of the laser and before is large; it may be possible to exploit these for sensor applications. Further, with higher powers available from laser diodes, the problems with the use of gas laser sources may be overcome. The economic rewards of success in these areas could be significant.

VI. ASSESSMENT OF FIBER OPTIC TEMPERATURE SENSORS

As yet, many of these fiber optic devices described have been used only under laboratory conditions, and little or no testing under operational conditions has been done. This is partly due to the fact that many devices have not reached commercial production. For those few that have, a recent assessment was carried out by Harmer[96] to address the possible use of fiber optic instrumentation for process control in offshore oil rig installations. The report was concerned with investigating whether devices met the maufacturers' design specifications, and tests were performed to consider such characteristics as accuracy, reproducibility, sensitivity, linearity, effects of supply voltage, handling considerations, and aging. In addition to a number of other devices, three prototype commercial temperature sensors were tested. The results are shown in Table 1. Sensor A was a guide fiber bundle with a bimetallic strip and reflector varying the

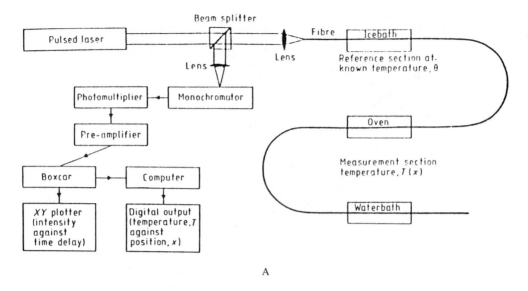

FIGURE 20. Distributed sensor using Raman thermometry. (A) Optical arrangement. (B) Received data intensity vs. distance. (C) Temperature dependence. (From Dakin, J. P., Pratt, D. J., Bibby, G. W., and Ross, J. N., *Electron. Lett.*, 21, 569, 1985. With permission.)

intensity of the backreflected light. Sensor B used the temperature-dependent fluorescent intensity of a semiconductor crystal, and sensor C depended on the pulse decay of fluorescence from a fluorophor. The comments may be summarized in that with sensor A, problems associated with intensity modulation were seen, due to connector and cable losses which could be misconstrued to be a temperature error, but sensor B could achieve accuracies of better than ±0.4°C with individual calibration of the probe. Consistent reliable readings were seen over a period of testing of 2 weeks. For the pulse decay type device, errors greater than specifications were seen, although the device was at an early stage of production.

It is important that fiber optic temperature sensing devices be subjected to this intensive scrutiny, in order to show their capabilities by comparison to conventional sensing techniques. In the test reported, sensor B did appear to perform very well when individually calibrated. As yet the market is still quite young, and many of the devices are at an early stage of development. With greater production numbers, many of the early problems may be eliminated through further development.

VII. FIBER OPTIC TEMPERATURE SENSORS — REVIEW

A. THE MARKET

At present the market for fiber optic sensors is quite small but is forcast to grow rapidly over the next few years.[97] Work is being done in the field by a large number of manufaturers, and the number of products which are beginning to appear on the market is increasing. It is estimated that in the 1990's a global market of the size of $300M to $400M for fiber optic sensors is expected, with the major application being in the military/aerospace region. Of this, fiber optic gyroscopes will represent a major area of expenditure, but there is a considerable application for fiber optic temperature sensors in the midst of that. By comparison with the important Japanese market in 1985, fiber optic sensors represented approximately 0.4% of the optoelectronics market of $4 billion, i.e., about $10 million in value. This market is approximately 25% of the world market, with major applications also being seen in the western European nations, especially the U.K., France, and Germany, and the U.S. However, it is significant that more fiber optic sensor products are available from Japanese manufacturers

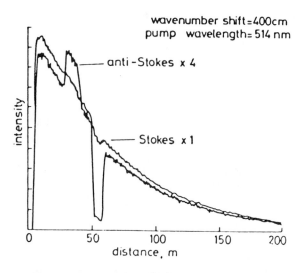

wavenumber shift = 400 cm
pump wavelength = 514 nm

— anti-Stokes x 4

— Stokes x 1

FIGURE 20B.

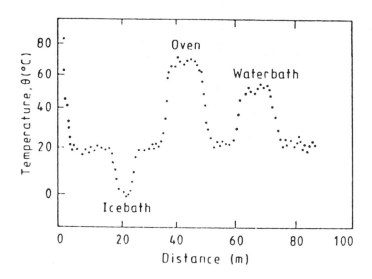

FIGURE 20C.

than from those of many other countries. Given the dominance of that country in the optoelectronics market in general, Japanese manufacturers are not expected to be idle in the development of fiber optic sensors for process control and for the aerospace and biomedical industries. Following closely behind the military and aerospace applications are the medical applications of fiber optic sensors and here the particular benefits of the fiber optic approach to sensors may be seen. Temperature sensors may not have such a ready application in this market, as conventional techniques may well be sufficient, with the bulk of this market looking toward fiber optic chemical and gas sensing for physiological use.

B. CONCLUSIONS

It is difficult to draw conclusions in the field of fiber optic temperature sensors beyond saying that there exists a vast wealth of ideas and scientifically interesting schemes, in addition to one of the most significant market penetrations with commercial devices. These products have been shown to hold up well under rigorous testing. The optoelectronics (and with it the

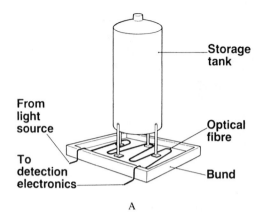

A

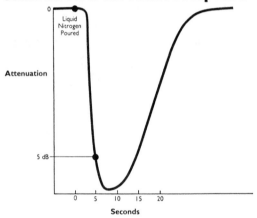

B

FIGURE 21.   Schematic of cryogenic temperature sensor. (A) Installation arrangement. (B) Response to sudden cooling with liquid nitrogen. (From Murray, R. T., *Int. J. Opt. Sens.,* 1, 27, 1986. With permission.)

fiber optic sensor) markets are predicted to grow significantly over the next few years. Some major laboratories are reporting a "lead time" of at least 10 years for their products.[5] As a result, it is not surprising that many of these ideas described have not ben yet brought to the commercial market place.

However, this situation is expected to change over the next few years and a healthy future for fiber optic sensors is seen. It is envisaged, however, that this will still remain a specialist part of the overall temperature monitoring market, as competition with a device as basic as the thermocouple still exists. The distinctive advantages of fiber optic sensors in their own fields will remain, along with their attractiveness to potential users.

REFERENCES

1. **Rusby, R. L.,** The basis of temperature measurement, *Meas. Control,* 20, 7, 1987.
2. **Dixon, J.,** Industrial radiation thermometry, *Meas. Control,* 20, 11, 1987.
3. **Midwinter, J.,** Current status of optical communications technology, *J. Lightwave Tecnol.,* LT3, 927, 1985.
4. Institute of Measurement and Control, UK, Optical Sensors and Optical Techniques in Instrumentation, Symp. London, UK, November, 1981, collected papers, Pub: Institute of Measurement and Control, London, 1982.
5. **Harmer, A. L.,** Fibre Optic Sensors for Sale? Proc. Optical Fiber Sensors Conf. OFS '88 New Orleans, January, 1988, Pub: OSA, 1, 2, 1988.
6. **Tortishell, G.,** Reports to UK Optical Sensors Collaborative Assoc., unpublished, 1987, 1988.
7. **Medlock, R. S.,** Review of modulating techniques for fibre optics sensors, *Int. J. Opt. Sens.,* 1, 43, 1986.
8. **Polanyi, M. L. and Hehir, R. M.,** New reflection oximeter, *Rev. Sci. Instrum.,* 31, 401, 1960.
9. **Dakin, J. P. and Kahn, D. A.,** A novel fibre optic temperature probe, *Opt. Quant. Electron.,* 9, 540, 1977.
10. **Mordon, S., Zonde, S., and Brienetaud, J. M.,** Non contact temperature measurement with a zirconium glass fiber, in Proc. Optical Fiber Sensors Conf. OFS '88 New Orleans, January, 1988. Pub: OSA (Optical Society of America) 2, 502, 1988.
11. **Murray, R. T.,** Fibre optic sensors for the chemical industry, *Int. J. Opt. Sens.,* 1, 27, 1986.
12. **Kymura, K., Tai, S., Sawada, T., and Nunonshita, M.,** Fiber Optic Instrument for Temperature Measurement, *IEEE J. Quant. Electron.,* QE18, 676, 1982.
13. **Grattan, K. T. V., Selli, R. K., and Palmer, A. W.,** Fluorescence referencing for fiber optic thermometers using visible wavelengths, *Rev. Sci. Instrum.,* 59, 256, 1988.
14. Fotonic sensor, Mechanical Technology, Inc., Latham, NY, 1967.
15. **Bergstrom, J., Betibon, H., Brogardh, T., and Perrson, A.,** U.K. Patent Appl. GB2010476A, 1979.
16. **Croft, J. R., Palmer, A. W., and Valsler, R.,** Evanescent wave fiber optic sensor, *Proc. SPIE,* 375, 206, 1983.
17. **Hocker, G. B.,** Fiber optic sensing of pressure and temperature, *Appl. Opt.,* 18, 1445, 1979.
18. **De Rossi, D.,** A new fibre-optic crystal catheter for oxygen saturation and blood flow measurements in the coronary sinus, *J. Biomed. Eng.,* 2, 256, 1980.
19. **Augousti, A. T., Grattan, K. T. V., and Palmer, A. W.,** A liquid crystal fibre optic temperature switch, *J. Phys. E.,* 21, 817, 1988.
20. **Ireland, P. T. and Jones, T. V.,** The response time of a surface thermometer employing encapsulated thermochromic liquid crystals, *J. Phys. E,* 20, 1195, 1987.
21. **Brenci, M.,** Thermochromic transducer optical fiber temperature sensor, *Proc. SPIE,* 514, 155, 1984.
22. **Brenci, M., Conforti, G., Falciai, R., Mignani, A. G., and Scheggi, A. M.,** Optical fiber temperature measuring instrument, *Proc. SPIE,* 701, 216, 1986.
23. **Rogers, A. J.,** Optical measurement of current and voltage on power systems, *IEE J. Electric Power Appl.,* 2, 120, 1979.
24. **Rogers, A. J.,** Optical temperature sensor for high voltage applications, *App. Opt.,* 21, 882, 1982.
25. **Fields, J. N.,** Fiber optic pressure sensor, *J. Acoust. Soc. Am.,* 67, 816, 1980.
26. **Arakawa, K. and Yoshida, K.,** Compact fiber-optic temperature sensor utilizing thermally induced bending loss, *Trans. Inst. Electron. Inf. Comm. Eng. C,* J70C, 559, 1987.
27. **Christiansen, C.,** Untersuchungen uber die optischen Eigenschaften von fein vertheilten Korpern 1, Wiedemann, X., Ed., *Ann. Phys. Chem.,* 23, 298, 1884.
28. **Knox, J. M., Marshall, P. M., and Murray, R. T.,** Birefringent filter temperature sensor, Proc. 1st Conf. Opt. Fiber Sens. (OFS 83) 1983, Pub: IEE, London, p.1.
29. **Jones, B. E. and Spooncer, R. C.,** An optical fibre pressure sensor using a holographic shutter modulator with two wavelength intensity referencing, *Proc. SPIE,* 514, 223, 1984.
30. **Hutley, M. C.,** Wavelength encoded optical fibre sensors, *Proc. SPIE,* 514, 111, 1984.
31. **Saaski, E. W. and Skangoet, R. I.,** 7th IEEE/PES Trans. and Dist. Conf., 208, 1979.
32. **Shoener, G., Bechtel, J. H., and Salour, M. H.,** Novel fiber coupler for optical fiber temperature sensor, *Proc. SPIE,* 514, 203, 1984.
33. **Theocharous, E.,** Displacement and temperature multimeasurant transducer, Proc. 2nd Int. Conf. on Opt. Fiber Sensors, Stuttgart, Pub: VDE-Verlag Berlin, 1985, 199.
34. **Grattan, K. T. V., Selli, R. K., and Palmer, A. W.,** A prism configuration internally referenced temperature sensor, *Int. J. Opt. Sens.,* 1, 507, 1986.
35. **Andrew, A. T., Angelov, A. K., and Zafirova, B. S.,** Fibre-optic sensor with semiconductive sensitive element (for temperature measurement), *C.R. Acad. Bulg. Sci.,* 40, 51, 1987.

36. **Christiansen, D. A. and Ives, J. T.,** Fiber optic temperature probe utilizing a semiconductor sensor, in *Optical Fiber Sensors. Proc. NATO. Adv. Study Institute,* Trieste, Italy, Reidel, Dordrecht, 1987, 361.

37. **Wickersheim, K. A. and Alves, R. V.,** Fluoroptic Thermometry: A New RF-Immune Technology: Biomed. Thermol. Pub: Alan Liss, New York, 1982, 547.

38. **Ovren, C., Adoleson, M., and Hök, B.,** Fiber optic systems for temperature and vibration measurements in industrial applications, in Proc. Int. Conf. Optical Techniques in Process Control, The Hague, 1983, Pub: BHRA - British Hydromechanical Research Association, Cranfield, Bedford, MK43 OAJ, U.K., 67.

39. **James, K. A. and Quick, W. H.,** Fiber Optics — The way to True Digital Sensors, Control Eng. (February 1979), 30.

40. **Sholes, R. R. and Small, J. G.,** Ruby fluorescent decay-time temperature sensor, *Rev. Sci. Instrum.,* 51, 882, 1980.

41. **McCormack, J. S.,** Remote optical measurement of temperature using luminescent materials, *Electron. Letts.,* 17, 630, 1981.

42. **Bosselman, T., Ruele, A., and Schröder, J.,** Fiber optic temperature using fluorescent decay time, in *Proc. 2nd Opt. Fiber Sensors Conf. (OFS 84)* Stuttgart, Pub: VDE - Verlag, GmbH, Berlin, 1985, 151.

43. **Grattan, K. T. V. and Palmer, A. W.,** A fibre optic temperature sensor using fluorescence decay, *Proc. SPIE,* 492, 535, 1984.

44. **Grattan, K. T. V., Selli, R. K., and Palmer, A. W.,** Infra-red fluorescence 'decay-time' temperature sensor, *Rev. Sci. Instrum.,* 56, 1784, 1985.

45. **Grattan, K. T. V., Manwell, J. D., Sim, S. M. L., and Willson, C. A.,** Fibre optic temperature sensor with wide temperature range characteristics, *IEE Proc.,* 134J, 291, 1987.

46. **Grattan, K. T. V., Manwell, J. D., Sim, S. M. L., and Willson, C. A.,** Lifetime investigation of fluorescence from neodymium:yttrium aluminium garnet at elevated temperatures, *Opt. Commun.,* 62, 104, 1987.

47. **Augousti, A. T., Grattan, K. T. V., and Palmer, A. W.,** A laser-pumped temperature sensor using the fluorescent decay time of alexandrite, *J. Lightwave, Technol.,* LT5, 759, 1987.

48. **Augousti, A. T., Grattan, K. T. V., and Palmer, A. W.,** Visible LED pumped fiber optic temperature sensor, *IEEE Trans. Instrum. Meas.,* 37, 470, 1988.

49. **Grattan, K. T. V., Selli, R. K., and Palmer, A. W.,** Ruby decay-time fluorescence thermometer in a fiber-optic configuration, *Rev. Sci. Instrum.,* 59, 1328, 1988.

50. **Wickersheim, K. A., Heinseman, S., Trans, H. N., and Sun, M. H.,** Advances in fluoroptic thermometry, in Proc. Digtech. '85 Boston, U.S.A. Proc., ISA Instrument Society of America paper 85-0072, 1985, 87.

51. **Hirano, M.,** Characteristics and Applications for some types of fiber-optical temperature sensor, in Tech. Digest 4th Int. Conf. on Opt. Fiber Sensors, (OFS 86) — Informal Workshop, Tsukuba Science City, Japan, October 1986, VII1-VII8, pub: Institute of Electronics and Communications Engineers of Japan, Tokyo, 1986.

52. **Allison, S. W., Coates, M. R., Scudiere, M. B., Bentley, H. T., III, Borella, H., and Marshall, B.,** Remote thermometry in combustion environment using the phosphor technique, *Proc. SPIE,* 788, 90, 1987.

53. **Grattan, K. T. V., Manwell, J. D., Sim, S. M. L., and Willson, C. A.,** A wide range fiber optic sensor for temperature measurement to 1100 K, in Proc. Conference on Lasers and Electro Optics, Baltimore, 1987, Digest of Tech. Papers Pub: Optical Society of America, 1987, 306.

54. **Accufibre Corp.,** Vancouver, Canada, publicity data.

55. **Foley, G. M., Morse, M. S., and Cezrairligan, A.,** Temperature measurement and control in science and industry, 5, 447, 1982, Pub: Amer. Inst. of Phys.

56. **Lynch, B.,** Photo-acoustic Resonators Fibre Sensors, Short Meetings Series, 7, 109, 1987; Pub: Institute of Physics, London, 1987.

57. **Grattan, K. T. V., Palmer, A. W., and Saini, D. P. S.,** Optical vibrating quartz crystal pressure sensor using frustrated-total-internal-reflection readout technique, *J. Lightwave Technol.,* LT5, 972, 1987.

58. **Ventakesh, S. and Culshaw, B.,** An optically activated vibration of micromachined silicon structure, *Electron. Lett.,* 21, 315, 1985.

59. **Culshaw, B.,** Optically Excited Resonant Sensors Fibre Optic Sensors, Short Meetings Series, 7, 33, 1987. Pub: Institute of Physics, London, 1987.

60. **Jones, B. E. and Philp, G. S.,** A vibrating wire sensor with optical fibre links for force measurement, in Proc. 1st Conf. Sensors and their Applications, Manchester, 1983, Pub: Institute of Physics, London, 1983, 86.

61. **Gehrich, J. L., Lübbers, D. W., Optiz, N., Hansmann, D. R., Miller, W. W., Tusa, J. K., and Yafuso, M.,** Optical fluorescence and its application to an intravascular blood gas monitoring system, *IEEE J. Biomed. Eng.,* BME33, 117, 1986.

62. **Schultz, J. S., Mansouri, S., and Goldstein, I. J.,** Affinity sensor: a new technique for developing implantable sensors for glucose and other metabolites, *Diabetes Care,* 5, 245, 1982.

63. **Jackson, D. A.,** Overview of Fibre Optic Interferometric Sensors, Fibre Optic Sensors, Short Meeting Series 7, 1, 1987, Pub: Institute of Physics, London, 1987.

64. **Moss, G. E., Miller, L. R., and Forward, R. L.,** Photon-noise — limited laser transducer for graitational antennae, *Appl. Opt.*, 10, 2495, 1971.

65. **Giallorenzi, T. G., Bucaro, J. A., Dandridge, A., Singel, G. H., Cole, J. H., Rashley, L., S.C., and Priest, R. G.,** Optical fiber sensor technology, *IEEE J. Quant. Electron.*, QE18, 626, 1982.

66. **Davies, D. E. N. and Kingsley, S. A.,** Method of phase modulating signals in optical fibres: application to optical telemetry systems, *Electron. Letts.*, 10, 21, 1974.

67. **Jackson, D. A., Priest, R., Dandridge, A., and Tvefen, A. B.,** Elimination of drift in a single-mode optical fibre interferometer using a piezo-electrically stretched coiled fibre, *Appl. Opt.*, 29, 16, 1980.

68. **Brooks, J. L., Wentworth, R. C., Youngquist, T. T., Kim, B. Y., and Shaw, H. J.,** Coherence multiplexing of fiber-optic interferometric sensors, *J. Lightwave Technol.*, 3, 1062, 1985.

69. **Jackson, D. A.,** Monomode optical fibre interferometers, *J. Phys. E*, 18, 987, 1985.

70. **Hocker, G. B.,** Fibre optic sensing of pressure and temperature, *Appl. Opt.*, 18, 3679, 1979.

71. **Musha, T., Kanimura, J., and Kakazawa, M.,** Optical phase fluctuations thermally induced in a single mode optic fiber, *Appl. Opt.*, 21, 694, 1982.

72. **Corke, M., Kersey, A. D., Jackson, D. A., and Jones, J. D. C.,** All fibre 'Michelson' thermometer, *Electron. Letts.*, 19, 471, 1983.

73. **Eickhoff, W.,** Temperature sensing by mode-mode interference in birefrigent optical fibres, *Opt. Lett.*, 6, 204, 1981.

74. **Kersey, A. D., Corke, M., and Jackson, D. A.,** Linearized polarimetric optical sensor using a 'heterodyne-type' signal recovery scheme, *Electron. Lett.*, 20, 209, 1984.

75. **Corke, M., Jones, J. D. C., Kersey, A. D., and Jackson, D. A.,** Dual Fabry-Perot interferometer implemented in parallel on a single mode optical fibre, in Tech. Digest 3rd Int. Conf. on Opt. Fibre Sensors (OFS '85) San Diego, (Pub: Optical Society of America) 1985, 128.

76. **Schultheis, L.,** A simple fiber-optic Fabry Perot temperature sensor, in Proc. Conf. on Opt. Fiber Sensors, OFS 88, Pub: OSA (Optical Society of America), 2, 506, 1988.

77. **Seki, Y. and Noda, K. I.,** Pseudo-heterodyne, optical-fiber thermometer using Fabry-Perot interferometer, in Proc. Conf. on Opt. Fiber Sensors (OFS '86) Tokyo, Japan, p. 227/30, 1986. Pub: Institute of Electronics and Communications Engineers of Japan, Tokyo, 1986.

78. **Meggitt, B. T., Palmer, A. W., and Grattan, K. T. V.,** Fibre optic sensors using coherence properties-signal processing aspects, *Int. J. Optelectron.*, - to be published 1989.

79. **Bosselmann, Th. and Ulrich, T.,** High accuracy position sensing with fiber coupled white light interferometers, in Proc. Conf. on Opt. Fiber Sensors (OFS '84) Stuttgart, 1984, Pub: VDE - Verlag, Berlin, 1985, 361.

80. **Al-Chalabi, S. A., Culshaw, B., and Davies, D. E. N.,** Partially coherent sources in interferometric sensors, in Proc. 1st Int. Conf. on Opt. Fibre Sensors, (OFS '83), London Pub: IEE London, 1983, 132.

81. **Rogers, A. J.,** Distributed optical-fibre sensors, *J. Phys. D*, 19, 2237, 1986.

82. **Rogers, A. J.,** Distributed and multiplexed sensor systems, Fibre Optic Sensors, Short Meetings Series 7, 55, 1987, Pub: Institute of Physics, London, 1987.

83. **Nelson, A. R., McMaohon, D. H., and Gravel, R. L.,** Passive multiplexing system for fiber optic sensors, *Appl. Opt.*, 19, 2917, 1980.

84. **Hartog, A. H.,** A distributed temperature sensor based on liquid core optical fibres, *J. Lightwave Technol.*, LT1, 489, 1983.

85. **Morey, W. W., Glenn, W. H., and Snitzer, E.,** Fiber optic temperature sensors, in Proc. 29th Int. Instrumentation Symp., Albuguerque, NM, May 1983, Proc. Pub by ISA (Instrument Society of America), 1983, 261.

86. **Farries, M. C., Fermann, M. E., Lamming, R. I., Poole, S. B., Payne, D. N., and Leach, A. P.,** Distributed temperature sensor using Nd^{3+}-doped optical fiber, *Electron. Lett.*, 22, 418, 1986.

87. **Farries, M. C. and Ferman, M. E.,** Thermal sensing by thermally induced absorption in a neodymium doped optical fiber, *Proc. SPIE*, 798, 115, 1987.

88. **Payne, D. N.,** Progress in rare earth doped fiber lasers, Proc. CLEO, Conference on Lasers & Electro Optics, Baltimore, 1987, 206, pub: OSA Optical Society of America, Washington, DC, 1987.

89. **Dakin, J. P., Pratt, D. J., Bibby, G. W., and Ross, J. N.,** Distributed anti-Stokes ratio thermometry, in Proc. 3rd Int. Conf. Opt. Fiber Sensors (OFS '85), Post Deadline Paper Pub: OSA (Optical Society of America), 1985.

90. **Dakin, J. P., Pratt, D. J., Bibby, G. W., and Ross, J. N.,** Distributed optical fibre Roman temperature sensor using a semiconductor light source and detector, *Electron. Lett.*, 21, 569, 1985.

91. **Hartog, A. H., Leach, A. P., and Gold, M. P.,** Distributed temperature sensing in solid core fibres, *Electron. Lett.*, 21, 1061, 1985.

92. **York, VSOP Ltd.,** Chandlers Ford, Hampshire, S05, 3DG, England, manufacturer's publicity data.

93. **Pilkington Security Systems,** St. Asaph, Clywyd, Wales, U.K., manufacturer's data.

94. **Leung, C. Y., Chang, I - Fan, Hsu, S.,** Fiber optic line sensing system for perimeter protection against intrusion, in Proc. 2nd Conf. on Opt. Fiber Sensors (OFS '84), Stuttgart, Pub: VDE - Verlag, Berlin, 1985, 113.

95. **Ibe, H., Furuyama, H., and Shibagaki, T.,** Optical temperature sensor based on wavelength scanning technology, in Proc. Conf. on Optical Fiber Sensors (OFS '86), Tokyo, Japan, 231, 1986. Pub: Institute of Electronics and Communications Engineers of Japan, Tokyo, Japan, 1986.

96. **Harmer, A. L.,** Fiber optic sensors for offshore process control instrumentation, in Proc. Optical Fiber Sensors Conf. (OFS '86), Informal Workshop at Tusukuba Science City, 1986, VI1, Pub: Institute of Electronics and Comunication Engineers of Japan, Tokyo, 1986.

97. **Long, T.,** Future markets for fiber optic sensors in Europe, Fiber Optic Sensors, Short Meetings Ser., 7, 129, 1987. Pub: Institute of Physics, London, 1987.

98. **Jones, B. E.,** Introductory technology overview of intensity and wavelength-based sensors, Fibre Optics Sensors, Short Meeting Series, 7, 45, 1987, Pub: Institute of Physics, London, 1987.

99. **Grattan, K. T. V.,** The use of fibre optic techniques for temperature measurement, *Meas. Control*, 20, 32, 1987.

100. **Theocharous, E.,** Differential Absorption Distributed thermometry, in Proc. 1st Opt. Fibre Sensors Conf. (OFS '83), Pub: IEE London, 1983, 10.

Chapter 16

TRANSDUCER-BASED AND INTRINSIC BIOSENSORS

Mark A. Arnold and Julie Wangsa

TABLE OF CONTENTS

I. INTRODUCTION

The term biosensor generally refers to an analytical device which incorporates a biologically active material in intimate contact with an appropriate transduction device. Ideally, biosensors can be used to determine reversibly and selectively the concentration or activity of a chemical species. Isolated enzymes, intact bacterial cells, whole sections of mammalian or plant tissue, antibodies, and bioreceptor proteins are examples of biological agents that have been used for the preparation of biosensors. A variety of transduction devices, such as amperometric and potentiometric electrodes, optical fibers, field effect transistors (FETs), piezo-electric crystals, and surface acoustic wave (SAW) devices, have been used for sensor fabrication.

This chapter reviews biosensors in which the biologically active material is a biocatalyst and the transduction element is a fiber-optic sensing device. Fiber-optic biosensors based on immunoglobulins, receptor proteins, and other selective complexing biomolecules are covered in Chapter 17 of this text. Recent reviews and monographs for the general topic of biosensors, and especially for the specific topic of electrode-based biosensors, are available.[1-10]

The field of fiber-optic biocatalytic biosensors is currently in a stage of rapid growth and development. Sensors for new analytes and innovative sensor designs are being reported with great frequency. Although we have tried to include as many of the fiber-optic biocatalytic biosensors reported to date, we fear that this review will be somewhat outdated by the time the text is released. This review includes reports that have appeared through August 1988.

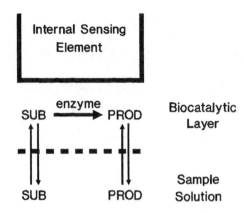

FIGURE 1. General schematic of a biocatalytic biosensor.

II. FIBER-OPTIC BIOCATALYTIC BIOSENSORS IN GENERAL

Biocatalytic biosensors are kinetic devices based on the establishment of a steady-state concentration of a detectable species at the sensing surface of the transduction element. Figure 1 shows a general response scheme for such a sensor. An analyte consuming reaction is catalyzed by the immobilized biocatalyst and typically a product of this reaction is monitored. The purpose of the biocatalyzed reaction is to mediate between the analyte of interest and the transducer by converting this analyte to a detectable species.

Sensor operation involves the diffusion of the analyte to the biocatalytic layer where the analyte is converted to a detectable species. A steady-state concentration of the detected product is established when the rate of product generation is exactly counterbalanced by the rate of product removal. A steady-state condition is likewise established when the consumption of a cosubstrate is monitored. In both cases, a steady-state signal is attained and the magnitude of this signal is related to the concentration of analyte in the sample through a previously prepared calibration curve.

Fiber-optic biocatalytic biosensors are based on the generation or consumption of an optically detectable species. Here, the biocatalyst is immobilized at the surface of either a single fiber or a bundle of fibers and the detected species is measured by either absorbance, fluorescence, bioluminescence, or chemiluminescence (see Chapter 20). The optical fibers serve as conduits through which light travels to and from the sensor tip. The transduction element in this case is an optoelectronic transducer (i.e., photomultiplier tube [PMT] or photodiode) which converts the resulting optical signal into an electrical signal. The transduction element is connected to the biocatalytic system through the optical fiber device.

Figure 2 is a block diagram that shows the instrumental components needed for a fluorescence-based fiber-optic biosensor. Radiation sources such as tungsten-halogen lamps, mercury arc lamps, and argon ion lasers have been used to supply the incident radiation. Interference filters and monochromators are generally used as the wavelength selection device, although the use of GRIN lenses has been proposed.[11] For the example of a fluorescence-based sensor, the excitation radiation is selected and sent to the sensor tip through an optical fiber. This excitation radiation interacts with a fluorescence species at the sensor tip, and a fraction of the resulting luminescence is collected by the optical fiber device. This emitted radiation is then guided to a second wavelength selection device and then to a detector. The second wavelength selector is needed to remove undesired excitation radiation and to reduce the amount of stray or ambient radiation. The resulting electrical signal is recorded on a strip chart recorder and the steady-state signal is identified.

Fiber-optic biosensors can be built with single or multiple optical fibers. Figure 3 shows

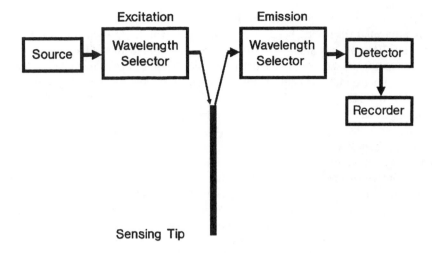

FIGURE 2. Block diagram for a fluorometric detection scheme for a fiber-optic biosensor.

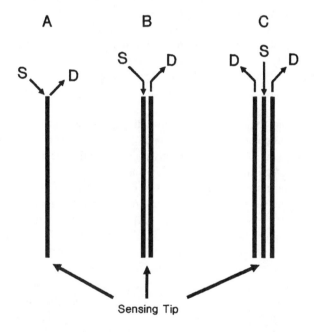

FIGURE 3. Schematic diagrams for a single (A), dual (B), and multiple (C) fiber arrangements.

several popular sensor designs based on single (A), dual (B), and multiple (C) fiber arrangements. A single optical fiber can be used to transport excitation radiation from the source optics to the sensor tip and emitted radiation from the sensor tip to the detection optics. A laser is used to supply a high intensity, narrow beam of incident radiation. After passing through a hole in a mirror, the incident radiation enters the optical fiber and travels to the sensor tip. Returning light exits the fiber and is reflected by the mirror to the detection optics. This design was originally reported by Block and Hirschfeld[12] in their early work with fiber-optic chemical sensors (FOCSs). The design has been subsequently used by many research groups for fiber-optic biosensors.

Alternatively, two separate fibers can be used with one fiber transmitting the source radiation to the sensor tip and the other fiber collecting light from the sensor tip and directing this light to the detection optics. Figure 3B shows a sensor arrangement based on this dual fiber design. A bifurcated fiber optic bundle can be used in a similar fashion where the common end of the bundle combines multiple fibers from the two bundle arms. One arm is connected to the source optics and the other to the detector optics. Custom-made fiber bundles have also been reported. For example, a bundle composed of a central fiber surrounded by a ring of outer fibers (Figure 3C) has been used for fluorescence-based sensors.[13] The central fiber provides the excitation radiation and the outer fibers collect the emitted radiation. This final probe design is similar to that reported by Schwab and McCreey in their work on fiber-optic Raman spectro-electrochemistry.[14]

Single and multiple fiber arrangements have relative advantages and disadvantages. The primary advantage of the single fiber configuration is the small sensing tip which allows measurements in small sample volumes. Also, faster response times are generally observed for sensors with small sensing tips. The single fiber arrangement is also more efficient optically because the overlap between the incident and collected radiation is maximized. On the other hand, the multiple fiber arrangements are generally easier to construct, and when optical bundles are employed, higher optical throughput can be achieved. The higher throughput advantage of fiber bundles allows the use of nonlaser sources and relatively inexpensive detection systems. Typically, sensor size vs. optical throughput is the key consideration when selecting between single and multiple fiber designs. A reasonable strategy seems to be the initial use of multiple fibers to establish and characterize the chemistry required for a given sensor. Subsequently, the task of miniaturizing the sensor can be undertaken.

As with all biocatalytic-based biosensors, enzyme loading and solution conditions are primary factors that must be considered when developing a fiber-optic biosensor. To a large extent, these factors control the sensor's sensitivity, dynamic range, response time, selectivity, and operational lifetime.

Enzyme loading refers to the amount of active enzyme at the sensor tip. The rate of product generation at the sensor tip is governed by the rate of the enzymatic reaction and/or by the rate at which the substrate approaches the biocatalytic layer. Under high enzyme loading conditions, the rate limiting step is the rate at which the substrate enters the biocatalytic layer. Here, the rate of substrate consumption is mass transport limited and the sensor's response is first order with respect to the substrate concentration. As long as the amount of enzyme activity is sufficiently high, the sensor's response is independent of the actual amount of activity present. Biosensor lifetimes can be extended significantly by loading the sensor tip with an excess of enzyme. As the enzyme activity decreases due to natural degradation processes, the sensor will still give the same response until the amount of active enzyme drops below the critical level and substrate mass transport is no longer solely responsible for the kinetics at the sensor tip. For this same reason, sensors with high enzyme loading are *less* susceptible to the effects of enzyme inhibitors and activators.

The kinetics of the enzyme reaction are rate limiting when enzyme loading is low. Enzyme reaction kinetics are typically described by the Michaelis-Menten equation. This equation is derived based on the following reaction scheme:

$$ATP + luciferin + O_2 \longrightarrow AMP + oxiluciferin + PP_i + CO_2 + light$$

where Sub, Prod, Enz, and Sub·Enz are the enzyme substrate, reaction product, enzyme, and substrate-enzyme complex, respectively, and k_1, k_2, and k_3 are the rate constants for the indicated reactions. The following rate expression relates the initial velocity of the catalyzed reaction (v_i) to the substrate concentration ([Sub]):

$$V_i = \frac{k_3[\text{Enz}][\text{Sub}]}{K_M + [\text{Sub}]} \tag{1}$$

where K_M is termed the Michaelis constant. When the substrate concentration is much lower than the K_M value, the rate of product generation, and consequently the sensor response, is first order with respect to the substrate concentration. When the substrate concentration is much greater than the K_M value, the sensor response is independent of the substrate concentration. Under conditions of low enzyme loading, the K_M value sets the upper limit of detection. A decrease in the dynamic range of response for a sensor is typically an indication that the enzyme activity has decreased significantly and that the rate limiting process has shifted from being mass transport limited to one which is limited by enzyme reaction kinetics. With low enzyme loading, the sensor response is directly proportional to the amount of active enzyme at the sensor tip; hence, the sensor is susceptible to activity modulators such as pH, heavy metals, etc. Overall, high enzyme loading is preferred.

High enzyme loading requires efficient immobilization of the biocatalyst. It is important to attach or hold the enzyme at the sensor tip in a fashion that retains as much of the original enzyme activity as possible. Numerous enzyme immobilization procedures are available. The three most commonly used procedures are covalent attachment, protein crosslinking, and physical entrapment. No single immobilization technique is ideal for all enzymes. A suitable method must be identified for each particular enzyme/surface combination. Numerous monographs and reviews are available for a complete discussion of enzyme immobilization techniques.[15,16]

Enzyme loading is of particular interest when attempting to build a biosensor with a single fiber. Here, the small surface area available for enzyme immobilization drastically restricts the amount of enzyme that can be retained by the sensor. Figure 4A shows a schematic of a sensor tip where the enzyme is covalently attached to the glass core of a single fiber. One enzyme is immobilized at each point of attachment to the glass. Walt and co-workers[17,18] have developed an alternative immobilization scheme where multiple enzymes are immobilized at each point of attachment to the glass tip (Figure 4B). Higher enzyme loading and consequently superior biosensors are obtained with this approach.

Solution conditions must also be considered. Solution parameters of interest include pH, temperature, and ionic strength. Stabilizers, protectors (inhibitor scavangers), activators, and prosthetic groups are commonly added to enhance the sensor response characteristics. Again, the overall goal is to adjust conditions so that a maximum amount of enzyme activity is present. Depending on the particular application, it might not be possible to adjust certain solution parameters. In such cases, the application dictates the conditions under which the sensor must operate.

Sensor selectivity refers to the ability of the sensor to detect a particular analyte in the presence of other substances. The selectivity of a fiber-optic biosensor is a combination of the chemistry and spectroscopy at the sensor tip. Many enzymes are quite selective in the reactions that they catalyze. Others, however, are general in nature and catalyze reactions for entire classes of substrates. The mode of detection used at the sensor tip also influences the sensor selectivity. For example, endogenous species that absorb the incident radiation or that fluoresce under the influence of incident radiation are potential interferences. In this respect, chemiluminescence reactions are more selective than fluorescence measurements, and both of these detection modes are more selective than absorbance measurements.

Fiber-optic biosensors can be divided into two main classes. One class includes sensors where the optically detected species is an intrinsic part of the biocatalyzed reaction. A second class consists of systems where the biocatalyzed reaction does not include an optically detectable species. In the latter case, an opto-chemical transduction reaction is required to convert either a cosubstrate or product of the biocatalyzed reaction into an optically detectable species. Transducer-based fiber-optic biosensors (i.e., those involving an opto-chemical transduction reaction) use a combination of a biocatalyzed reaction and a subsequent chemical equilibrium to generate an optical signal that can be related to the concentration of the analyte

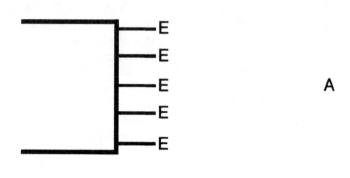

A

FIGURE 4. Schematic representation of single (A) and multiple (B) enzyme immobilization strategies.

of interest. Intrinsic biosensors, on the other hand, use only a biocatalyzed reaction. The remaining sections of this chapter present details of these two classes of fiber-optic biosensors.

III. TRANSDUCER-BASED BIOSENSORS

A. GENERAL CONCEPTS

An opto-chemical transduction reaction is required for fiber-optic biosensors when an optically detectable species is not part of the biocatalyzed reaction. Here, the opto-chemical reaction mediates between the enzymatic reaction and the optical sensing device. Opto-chemical reactions for the detection of oxygen, ammonia, and pH have been used. For oxygen, the opto-chemical transduction reaction is based on dynamic quenching of an immobilized fluorophore. The ammonia sensor is based on an immobilized pH indicator dye in combination with a gas-permeable membrane. An immobilized pH indicator dye is also used for the pH sensing system, but there is no gas-permeable membrane. These fiber-optic detection systems for oxygen, ammonia, and pH represent independently functioning FOCSs.

Biosensors are fabricated by coupling these FOCSs with suitable enzymes. In this configuration, the FOCS is the internal sensing element for the biosensor and the production or consumption of oxygen, ammonia, or protons is detected. Specifically, the oxygen FOCS can be used with oxidase enzymes and oxygen consumption is detected, the ammonia FOCS can be used with deaminating enzymes and the production of ammonia is detected, and the pH FOCS can be used with any enzyme that catalyzes a reaction during which a proton is either consumed or produced. In many ways, transducer-based fiber-optic biosensors are similar to the corresponding electrode-based systems in which electrochemical sensors for oxygen, ammonia, and pH are employed as the internal sensing elements.[1,7] Details for the oxygen (Chapter 10), ammonia (Chapter 11), and pH (Chapter 8) FOCSs are presented elsewhere in this text. Table 1 lists the various transducer-based biosensors that have been reported.

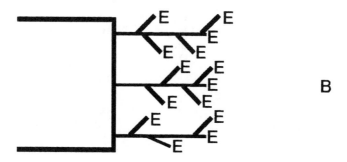

FIGURE 4B.

B. OXYGEN-BASED BIOSENSORS

Fiber-optic biosensors based on oxygen detection have been reported for glucose, lactate, xanthine, ethanol, and hydrogen peroxide. In each case, a selective oxidase enzyme is immobilized at the tip of a fiber-optic oxygen sensor. The consumption of oxygen is detected and related to the analyte concentration in the sample solution.

Fiber-optic oxygen sensors are based on dynamic quenching of fluorescent dyes such as pyrenebutryic acid[19] and perylenedibutyrate.[20] Sensors are constructed by placing the oxygen-sensitive fluorescent dye in a hydrophobic reagent layer. Oxygen from the sample permeates the reagent layer and quenches the fluorescence of the dye. The resulting luminescence is measured through a fiber-optic system, and the fluorescence intensity is related to the oxygen partial pressure in the sample through the Stern-Volmer equation:

$$\frac{I_0}{I} = 1 + K\, p_{O_2} \tag{2}$$

where I and I_0 are the fluorescence intensities in the presence and absence of oxygen, respectively, K is the quenching constant, and P_{O_2} is the oxygen partial pressure in the sample. Excellent selectivity is possible with such a design because only volatile components in the sample can enter the indicator solution, and of those, only volatile components that influence the fluorescence of the indicator are detected.

Glucose oxidase selectively catalyzes the following reaction:

$$\beta\text{-}D\text{-}glucose + O_2 \longrightarrow gluconic\ acid + H_2O_2$$

Many researchers have used this reaction in conjunction with an oxygen sensor (both fiber-optic and electrochemical) to prepare glucose biosensors.

Figure 5 shows a schematic diagram of a fiber-optic glucose biosensor. Glucose and oxygen diffuse into the enzyme layer where glucose is oxidized to gluconic acid. The consumption of oxygen is detected as an increase in the fluorescence intensity owing to a

TABLE 1
Tranducer-Based Fiber-Optic Biosensors

Analyte	Detected species	Primary Enzyme
Ethanol	O_2	Alcohol oxidase
Glucose	O_2	Glucose oxidase
Glucose	pH	Glucose oxidase
Hydrogen peroxide	O_2	Catalase
Lactate	O_2	Lactate oxidase
Penicillin	pH	Penicillinase
Urea	NH_3	Urease
Urea	pH	Urease
Xanthine	O_2	Xanthine oxidase

decrease in the extent of quenching. A steady-state fluorescence intensity is established and related to the glucose concentration in the sample solution. A linear calibration curve for glucose is obtained based on a modified Stern-Volmer equation as follows:

$$\frac{I_0}{I} = 1 + Kp_{O_2} - K'[\text{glucose}] \tag{3}$$

where K' is an empirical parameter which accounts for various diffusional processes and enzyme kinetics. The dynamic range of the glucose response is governed by the initial amount of oxygen in the sample, the extent of enzyme loading, and the permeability of membranes.

Several different methods for construction of fiber-optic glucose sensors have been published. Lübbers and co-workers crosslinked glucose oxidase with bovine serum albumin (BSA) and glutaraldehyde.[21-23] The resulting enzyme layer was about 50 μm thick. The chosen indicator was pyrenebutyric acid dissolved in bis(2-ethylhexyl)phthalate and stabilized with 1% ethyl cellulose. A layer of this highly viscous hydrophobic indicator solution was placed along a wall of a flow-through cell, and the enzyme layer was then formed on top of this indicator layer. The aqueous sample flowed past the enzyme/indicator solution layers and the fluorescence intensity was continuously monitored. Sensor response times (i.e., the time to reach 90% of the steady-state signal) were approximately 1 min. These researchers have also utilized this strategy to prepare biosensors for lactate,[24,25] xanthine,[25] and ethanol.[22,25,26] Lactate oxidase, xanthine oxidase, and alcohol oxidase, respectively, have been used as the biocatalyst.

A similar approach has been reported by Kroneis and Marsoner where glucose oxidase and catalase have been co-immobilized by the BSA/glutaraldehyde crosslinking method.[27] Catalase was added to the sensor tip to remove the hydrogen peroxide formed during the glucose oxidase reaction. The reaction catalyzed by catalase is as follows:

$$2H_2O_2 \longrightarrow O_2 + H_2O$$

The response range for this glucose sensor was from 20 to 500 mg/dl and response times were on the order of minutes. A flow-through cell was also used by these researchers.

Wolfbeis and co-workers have dissolved decacyclene in a thin silicone gas-permeable membrane and used this indicator layer as the internal sensing element for a glucose biosensor.[28] Glucose oxidase was covalently attached to a porous nylon membrane. This attachment was achieved through the active carboxyl groups on the commercially available immudyne immunoaffinity membrane.[29] The resulting biosensor possessed a detection limit of 0.1 mM glucose with an analytical range up to 20 mM. The time required to reach 90% of the steady-

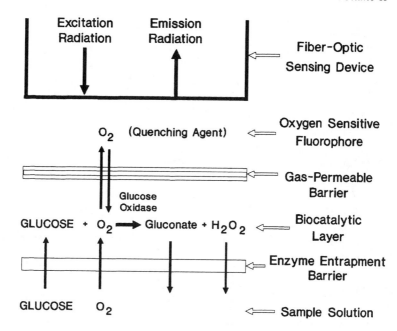

FIGURE 5. Membrane phases and reactions for a glucose-selective fiber-optic biosensor.

state response was from 1 to 6 min and sensor reproducibility was about 2.4%. A flow-through cell was employed for sensor characterization.

A prototype fiber-optic oxygen sensor has been reported and the feasibility of a glucose biosensor based on this oxygen sensor has been evaluated.[30] This oxygen sensor is based on the entrapment of 9,10-diphenylantracene in a poly(2-hydroxyethylmethacrylate) hydrogel layer. Response times for oxygen with 50 μm thick hydrogel layers were on the order of 6.5 to 9 min. Additionally, glucose oxidase was co-entrapped in this polymer. Polymer particles that contained both the enzyme and the indicator were characterized from a sensor point of view. Results indicate that such an enzyme/indicator system is suitable for the preparation of a fiber-optic glucose biosensor. As in the previously mentioned glucose biosensor studies, a flow-through cell and optical bundles were used.

The first transducer-based fiber-optic biosensor was that for the measurement of ethanol. This system, published by Völkl et al.,[26] was based on a thin layer of alcohol oxidase held at the surface of a Teflon membrane. The enzyme was entrapped in an agar gel and the gel was held adjacent to the Teflon membrane with a cellophane membrane. The Teflon membrane served to separate the sample solution from an internal solution composed of the fluorescent indicator (pyrenebutyric acid). A flow-through cell arrangement was employed.

A similar fiber-optic ethanol biosensor has been described by Wolfbeis and Posch.[31,32] In their system, alcohol oxidase, catalase, and the oxygen sensitive fluorescent dye (tris-(2,2′-dipyridyl)-ruthenium(II) dichloride) were co-immobilized in a gas-permeable silicon rubber membrane along one side of a flow-through cell. Initially, these reagents were dissolved in an aqueous buffer solution. Porous silica particles were then added to the buffer. The reagent solution filled the pores of these particles and the resulting reagent-loaded silica particles were embedded in a thin layer of silicone rubber. Sensor operation required the diffusion of oxygen and ethanol through the silicone rubber membrane into the entrapped reagent phase where the biocatalytic reaction occurs. As before, oxygen consumption was measured as a decrease in fluorescence quenching (increase in fluorescence intensity).

Response characteristics for the resulting biosensor include a dynamic range from 50 to 500 mM, a detection limit of 10 mM, response times of approximately 2 min, and a useful lifetime of 2 weeks. Sensor selectivity is somewhat limited in this case because alcohol oxidase catalyzes the oxidation of several volatile alcohols. Hence, these other alcohols are potential interferences. In addition, variations in the oxygen level in the sample will interfere with ethanol determinations (see Equation 3). Wolfbeis and Posch propose a dual sensor approach to account for such variations. The dual sensor approach includes one sensor with the enzymes and another without. A differential measurement is then performed. This approach is similar to that initially proposed for glucose biosensors based on the amperometric oxygen electrode.[33]

The novel feature of the Wolfbeis ethanol biosensor is the separation of the biocatalyst from the sample solution. By having the enzyme located in the silicon rubber membrane, there is no direct interaction between the enzyme and the sample. The only substances in the sample that can pose a problem are those which can rapidly permeate the silicon rubber membrane. Problems from nonpermeable species, such as common enzyme inhibitors and activators, are eliminated. A major drawback of this type of sensor design is the requirement that the analyte must be able to permeate the membrane under the operating conditions. Depending on the reactions involved, shorter operational lifetimes might also be encountered for this type of sensor owing to the small volume of solution that contains the enzyme. The composition of this solution cannot be easily controlled or altered after sensor construction. Changes in pH during the biocatalyzed reaction can rapidly overcome the buffer capacity of the small quantity of solution, and these pH changes can adversely affect the enzymatic activity and the sensor response.

Posch and Wolfbeis[34] have also evaluated three unique arrangements for hydrogen peroxide biosensors. The first uses catalase and tris(2,2'-dipyridyl)-ruthenium(II) dichloride co-immobilized in silica particles and embedded in a silicone rubber membrane. The second also uses catalase and tris(2,2'-dipyridyl)-ruthenium(II) dichloride, but these reagents are immobilized in separate batches of silica particles and these reagent-containing particles are then mixed and suspended in a silicone rubber membrane. The last system uses a layer of dispersed silver in silicone rubber on an inner silicone rubber layer with immobilized tris(2,2'-dipyridyl)-ruthenium(II) dichloride. All three systems are based on the catalytic decomposition of hydrogen peroxide to form oxygen and water. The increase in oxygen causes an increase in fluorescence quenching of the tris(2,2'-dipyridyl)-ruthenium(II) dichloride. Although the reported system does not employ fiber-optic, adaptation to a fiber-optic design should be straight forward. Response characteristics for this system include a working range from 0.1 to 10 mM with response times from 2.5 to 5.0 min.

An important aspect concerning the operation mechanism of this hydrogen peroxide sensor is the permeability of the hydrogen peroxide into and throughout the silicon rubber layer. Posch and Wolfbeis suggest that hydrogen peroxide is able to diffuse into the silicon rubber membrane and interact with the embedded catalyst. It is not clear, however, if the required catalyst/hydrogen peroxide interaction occurs in the silicon rubber membrane or only at the membrane/solution interface. No evidence is provided to rule out this latter possibility. The selectivity for this sensor is strongly influenced by the response mechanism.

C. AMMONIA-BASED BIOSENSORS

Many biosensors have been prepared by immobilizing a selective deaminating enzyme at the surface of the ammonia gas-sensing electrode.[1,7,8,35] Biosensors based on the fiber-optic ammonia sensor[36,37] should also be possible. A recent report demonstrates this point with a fiber-optic biosensor for urea.[38]

Figure 6 shows the various membrane phases involved in the response of the fiber-optic ammonia gas sensor. The transducer layer consists of a pH indicator dye and ammonium chloride.[37] This indicator solution is separated from the sample solution by an ammonia-permeable microporous Teflon membrane. At equilibrium, the ammonia partial pressure is the same on both sides of the gas-permeable membrane. A change in the sample ammonia concentration changes the pH of the indicator solution. This pH change is measured optically by detecting the relative concentrations of the protonated and nonprotonated forms of the indicator by either absorbance or fluorescence detection.[36-38] The acid dissociation constant of the indicator is the most important parameter when designing such a sensor. Combinations of indicators can effectively extend the dynamic range for fiber-optic gas sensors.

Urease is the enzyme used for the fiber-optic urea biosensor. This enzyme catalyzes the following reaction:

$$NH_2 - CO - NH_2 + H_2O \longrightarrow 2NH_3 + CO_2$$

The urea biosensor is constructed by immobilizing urease at the sensing tip of the fiber-optic ammonia probe with the common BSA-glutaraldehyde crosslinking method. Sensors based on both absorbance and fluorescence detection have been prepared and their response characteristics have been directly compared.[38] Nitrazine yellow (pK_a = 6.4) and bromothymol blue (pK_a = 7.1) have been used as the indicators for the absorbance-based sensor and 5-(and 6)-carboxyfluorescein (pK_a = 6.5) and 2',7'-bis-(carboxyethyl)-5-(and 6)-carboxyfluorescein (pK_a = 7.0) have been used for the fluorescence-based sensor. Figure 7 shows the response curves obtained for urea with these two sensors. Response functions to describe the steady-state response for each of these sensors have been developed and are as follows:

$$A_{total} = \sum \frac{\left(\Sigma\varepsilon_{ij}\right) b[HIn]i\, (Ka)_{HIni}\, \alpha[Urea]}{(Ka)_{amm}\, C_{amm} - (Ka)_{amm}\, \alpha[Urea] + (Ka)_{HIni}\, \alpha[Urea]} \qquad (4)$$

$$\frac{I_F}{I_R} = \sum \frac{2 \cdot 3\, b\, \left(\phi_i \varepsilon_i\right)[HIn]_i\, (Ka)_{HIni}\, \alpha[Urea]}{\chi\left((Ka)_{amm}\, C_{amm} - (Ka)_{amm}\, \alpha[Urea] + (Ka)_{HIni}\, \alpha[Urea]\right)} \qquad (5)$$

where A_{total} is the total absorbance of the indicator solution, ε_{ij} is the molar absorptivity of indicator i at wavelength j, b is the effective optical path length, $[HIn]_i$ is the concentration of indicator i, $(Ka)_{HIni}$ is the acid dissociation constant for indicator i, α is the fraction of the total ammonia nitrogen in the form of ammonia at the specified pH, [Urea] is the concentration of urea in the sample, $(Ka)_{amm}$ is the acid dissociation constant for ammonium ions, C_{amm} is the total ammonium ion concentration in the indicator solution, I_F is the fluorescence intensity, I_R is the intensity at a reference wavelength, and ϕ_i is the quantum efficiency of indicator i. The indicators are assumed to respond independently and the total signal is the summation of signals from the individual indicators. The solid lines in Figure 7 represent the calculated response curves based on the appropriate response function.

The analytical utility of these sensors has been judged by comparing their steady-state and dynamic response properties. Steady-state response characteristics (i.e., limit of detection and dynamic range) are essentially identical for the two sensors. This similarity is expected because the steady-state response is principally controlled by the acid dissociation constants of the indicators and the indicators used have similar values. Response times for the fluorescence-based system are significantly faster than those for the absorbance-based sensor. Response times for the absorbance sensor range from 12 to 23 min, whereas those for the fluorescence sensor range from 2 to 8 min for the same urea concentrations. As is typical with gas sensors, faster response is attained at higher analyte concentrations. Enhancements in the response times with fluorescent indicators are possible because lower indicator concentrations and smaller volumes of the indicator solution can be used while maintaining the steady-state response.

The fluorescence-based fiber-optic urea sensor has been applied to the determination of

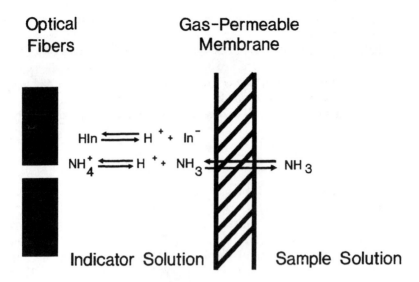

FIGURE 6. Membrane phases and chemical equilibria for the fiber-optic ammonia gas sensing probe.

urea in human serum samples. Results from the sensor have been directly compared to those from a conventional method for the determination of urea in serum. The conventional method is based on the quantitative conversion of urea to ammonia with subsequent determination of the generated ammonia by measuring NADH formation with the glutamate dehydrogenase catalyzed reaction.[39] A correlation plot of the data from the two methods, over a concentration range from 0.5 to 25 mM, has a slope of 1.01 ($\pm$0.02), a y-intercept of 0.1 ($\pm$0.2) mM, and a correlation coefficient of 0.999. Excellent agreement between the two methods is indicated.

The strategy of combining enzymes with fiber-optic gas-sensing probes can be easily extended to other deaminating systems. In addition, similar biosensors can be developed by coupling decarboxylating enzymes with a fiber-optic carbon dioxide sensor.[40,41]

D. BIOSENSORS BASED ON pH PROBES

The third class of transducer-based fiber-optic biosensors is based on the detection of a pH change at the surface of a fiber-optic pH probe where this pH change is a result of the biocatalyzed reaction. Sensors are prepared by co-immobilizing the enzyme and an appropriate pH indicator dye at the distal tip of a fiber-optic probe. Examples of this type of biosensor include sensors for glucose,[42,43] urea,[42] and penicillin.[18,42,44,45]

The first pH-based fiber-optic biosensors were reported by Goldfinch and Lowe.[42] These researchers developed flow-through devices with penicillinase, urease, and glucose oxidase individually immobilized with a chromophoric pH indicator dye. These reagents were co-valently attached to a transparent film, and this reagent film was sandwiched between a red-LED and a solid-state photodiode detector. A small volume of sample flowed past the reagent film. The presence of the enzyme substrate caused a pH change in the microenvironment of the immobilized indicator, and a fraction of the LED incident radiation was altered by a change in the relative concentrations of the protonated and nonprotonated forms of the indicator.

These penicillin and glucose biosensors are based on the production of an acid during the biocatalyzed reaction. Penicillinase catalyzes the cleavage of the β-lactam ring of penicillin to produce penicillinoic acid. Glucose oxidase catalyzes the oxidation of glucose to produce gluconic acid. In both cases, the resulting decrease in pH is detected with bromocresol green ($pK_a = 5.1$), and the rate of detector voltage change is measured. This rate increases with an increase in the substrate concentration. Linear responses have been reported from 0.5 to 5 mM

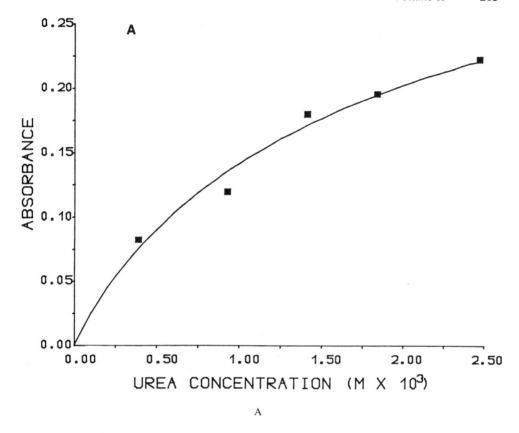

FIGURE 7. Response curves for the fiber-optic urea biosensor based on absorbance (A) and fluorescence (B) detection modes.

for penicillin and from 0 to 70 mM for glucose. Total assay times of 5 min have been demonstrated with excellent reproducibility (relative standard deviations less than 2%).

The pH-based fiber-optic urea biosensor is based on the consumption of a proton at the sensor tip. Urease and bromothymol blue (pK_a = 7.1) are co-immobilized, and the rate of change of the detector output is related to the sample urea concentration. A dynamic range from 0.1 to 40 mM is reported.

A fiber optic biosensor for glucose has also been reported by Trettnak et al.[43] An enzyme/ fluorescent pH indicator dye layer in a hydrogel matrix is placed on a polyester film. An outer layer of hydrogel with a suspension of charcoal is used to optically isolate the sensor from both ambient radiation and endogenous fluorophores in the sample. Glucose oxidase is the enzyme and hydroxypyrene trisulfonate (HPTS) is the indicator. The sensor uses a bifurcated fiber-optic bundle in a flow-through arrangement. Glucose calibration curves from 0.1 to 2 mM are reported with response times from 8 to 12 min.

Fiber-optic biosensors for penicillin have been reported by Walt and co-workers[18] and by Christian and co-workers.[44] In both cases, penicillinase has been co-immobilized with a fluorescent pH indicator dye at the distal tip of a single optical fiber. A steady-state fluorescence intensity is measured, and this signal is related to the solution penicillin concentration.

Two immobilization strategies are compared in the report by Walt et al.[18] Immobilization of penicillinase by entrapment in a polyacrylamide gel and by crosslinking with BSA and glutaraldehyde are compared. In the first case, fluorescein serves as the indicator dye, and in the second, hydroxypyrene trisulfonate (HPTS) is used. For both systems, the enzyme/ indicator reagents are covalently attached to the tip of a single fiber. The resulting sensor has a tip size of 200 μm. An argon ion laser is used as the excitation source and a lock-in amplifier-

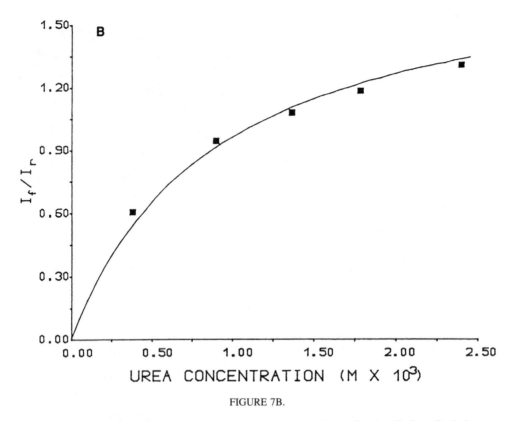

FIGURE 7B.

based photon counting detection scheme is used to detect the emitted radiation. Both immobilization schemes provide suitable detection with little difference in sensor response characteristics. A linear response for penicillin G from 0.25 to 10 mM is reported with sensor response times from 20 to 60 s. Such response times are significantly shorter than corresponding biosensors based on glass pH electrodes. The small surface area of the active sensor tip is primarily responsible for these response time improvements.

Penicillinase and fluorescein have been co-immobilized on a porous glass bead which is attached to the distal tip of a single optical fiber in the system reported by Christian et al.[45] Again, an argon laser is used to excite the fluorescent indicator and the luminescence is collected by the same fiber and detected with a photon counting apparatus. The resulting sensor has a detection limit of 0.1 mM penicillin and response times range from 20 to 45 s (95% of the total signal).

The most significant aspect of the work by the Walt and Christian research groups is the demonstration that fiber-optic biosensors based on single fibers are possible. The question of enzyme loading has always been an issue because of the small surface area that is available with single fibers sensors. Innovative immobilization schemes, such as that reported by Walt et al.,[17,18] are necessary to load sufficient quantities of enzyme at the fiber tip. It remains to be seen if single fiber sensors with relatively low-cost optics are possible.

Fiber-optic biosensors based on pH measurements suffer from the same drawbacks as do biosensors based on the pH electrode.[1] The major problem is the pH dependency of the biocatalytic activity. Because the ability of the enzyme to catalyze the reaction is pH dependent, the amount of effective enzyme at the sensor tip can change during sensor operation. The overall sensor response may or may not be repeatable depending on the pH stability of the biocatalyst. The buffer capacity of the sample matrix strongly influences the sensor response. A high buffer capacity compensates for the formation or consumption of an acid and no pH change is detected. A low buffer capacity, on the other hand, can result in large pH change which can inactivate the enzyme and limit the dynamic range of response. A compromise is generally required between the small signals over a wide concentration range with high buffer capacities and large signals over a narrow concentration range with low buffer capacities. A difference in buffer capacity between the standards and the sample cannot be tolerated. In addition, fiber-optic pH sensors are subject to interference by quenching agents and by ionic strength and temperature changes. Overall, biosensors of this type will find limited use in special circumstances where these important parameters can be controlled.

IV. INTRINSIC BIOSENSORS

A. GENERAL CONCEPTS

Intrinsic or nonmediated fiber-optic biosensors are constructed by immobilizing a biocatalyst at the distal tip of a fiber-optic detection device. The biocatalyzed reaction, in this case, is monitored directly by following the production or consumption of an optically detectable product or co-substrate. This type of biosensor is limited to systems where the biocatalyzed reaction induces a change in an optical property at the sensor tip. Intrinsic biosensors based on absorbance, fluorescence, chemiluminescence, and bioluminescence have been developed. Table 2 lists the various intrinsic fiber-optic biosensors that have been reported to date.

B. ABSORBANCE-BASED BIOSENSORS

A biosensor responsive to p-nitrophenylphosphate illustrates the concept of fiber-optic biosensors based on the detection of a chromophoric reaction product.[46] In this system, the enzyme alkaline phosphatase is immobilized at the common end of a bifurcated fiber-optic bundle. The catalyzed reaction is as follows:

$$p\text{-nitrophenylphosphate} \xrightarrow[\text{phosphatase}]{\text{alkaline}} p\text{-nitrophenol} + \text{phosphate}$$

The reaction product, p-nitrophenol, strongly absorbs 404 nm radiation in basic solutions. The molar extinction coefficient for p-nitrophenol at this wavelength is 19,050 (cm^{-1} M^{-1}) at pH 10.4. The configuration of the p-nitrophenylphosphate biosensor is shown in Figure 8. The sensor is composed of two sheets of nylon mesh held at the common end of the fiber bundle. Alkaline phosphatase is covalently attached to the inner nylon membrane by the procedure of Mascini and co-workers.[47] Incident radiation is guided to the sensor tip by one arm of the bifurcated bundle. The nylon mesh scatters this radiation, and a fraction of the back-scattered light is collected by the second half of the bifurcated bundle and transported to a PMT detector. A 404.7 nm narrow band pass filter is positioned before the detector. The intensity of light at 404.7 ± 10 nm is monitored and recorded on a strip chart recorder. Production of p-nitrophenol decreases the returning light intensity corresponding to an increase in the amount of incident radiation absorbed by the reaction product. A steady-state absorbance value is measured when a steady-state concentration of the chromophore is established at the sensor tip. A linear relationship has been found between the measured steady-state absorbance and the concentration of p-nitrophenylphosphate in the sample solution.[46]

C. FLUORESCENCE-BASED BIOSENSORS

Biosensors based on the fluorometric detection of reduced nicotinamide adenine dinucleotide (NADH) have been reported recently by several research groups.[48-51] Figure 9 shows a schematic diagram of the basic sensor design and the processes which occur at the sensor tip. A dehydrogenase enzyme is immobilized at the end of a fiber-optic device. This enzyme catalyzes a reaction that involves the analyte of interest and NADH. As the analyte diffuses to the biocatalytic layer, NADH is either consumed or produced and a corresponding steady-state concentration of NADH is established. The NADH that results at the sensor tip is

TABLE 2
Intrinsic Fiber-Optic Biosensors

Analyte	Detected species	Enzyme
ATP	Photon	Firefly luciferase
Bile acids	NADH	3α-Hydroxysteroid dehydrogenase
Ethanol	NADH	Alcohol dehydrogenase
Glucose	NADH	Glucose dehydrogenase
Glucose	Photon	Glucose oxidase/ peroxidase
H_2O_2	Photon	Peroxidase
Lactate	NADH	Lactate dehydrogenase
NADH	Photon	Bacterial luciferase
p-Nitrophenylphosphate	p-Nitrophenol	Alkaline phosphatase
Pyruvate	NADH	Lactate dehydrogenase

detected by a fluorescence measurement where the wavelength of excitation is 350 nm and the wavelength of emission is 450 nm. The measured steady-state fluorescence intensity is related to the analyte concentration in the bulk solution.

The first report of NADH-based biosensor was by Wangsa and Arnold[48,49] for the detection of lactate and pyruvate. Lactate dehydrogenase (LDH) was the enzyme used for both analytes. This enzyme catalyzed the following reaction:

$$\text{lactate} + \text{NAD}^+ \xleftrightarrow{\text{LDH}} \text{pyruvate} + \text{NADH} + \text{H}^+$$

The thermodynamically favored direction for this reaction can be set by adjusting the solution pH. Pyruvate is the favored product at high pH and lactate is favored at low pH. For the measurement of lactate, the solution pH should be high to favor the production of pyruvate. In addition, NAD$^+$ must be added to the sample as the reaction co-substrate. NADH is produced at the sensor tip and an increase in the measured fluorescence signal is detected. For the measurement of pyruvate, the solution pH should be low and NADH must be added to the sample. Now, the consumption of NADH is detected and a decrease in fluorescence intensity is measured for an increase in the pyruvate level.

Figure 10 shows a typical calibration curve for lactate. As expected, an increase in lactate concentration corresponds to higher fluorescence signals owing to the production of NADH at the sensor tip. The best response for lactate is obtained at pH 9.8. The dynamic range of sensor response extends from the detection limit (S/N = 3) of 2 μM to an upper limit of approximately 50 μM. This detection limit is based on the use of a relatively weak excitation source (100 W tungsten-halogen lamp) in conjunction with a common PMT detector operated at 500 V. Improvements in the detection limit are possible with greater intensity sources and more sensitive detectors. Response times range from 5 to 12 min, with faster response times at higher lactate concentrations. Relative standard deviations for lactate detection range from 5 to 9%, with higher values at lower lactate concentrations. The lifetime of the lactate sensor is 3 d.

The fiber-optic lactate biosensor has been evaluated for use in the determination of lactate in serum samples.[52] Fourteen human serum samples were obtained from the University of Iowa Hospitals and Clinics. These samples, which had been previously assayed for lactate levels by the Clinical Chemistry Laboratory at the hospital, were analyzed for lactate with the NADH-based lactate biosensor. A correlation line between the biosensor results and those from the laboratory has a slope of 0.98, a y-intercept of 0.006 mM, and a correlation

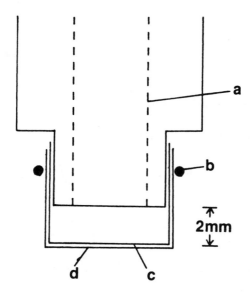

FIGURE 8. Configuration for the p-nitrophenylphosphate fiber-optic biosensor where (a) is the common end of a bifurcated fiber-optic bundle, (b) is an o-ring, (c) is the inner nylon mesh onto which the enzyme is immobilized, and (d) is the outer nylon membrane.

coefficient of 0.9953. This excellent correlation between the fiber-optic biosensor method and the conventional method demonstrates the potential utility of the fiber-optic biosensor.

Figure 11 shows a typical calibration curve for pyruvate. Here, a decrease in the fluorescence signal is measured with an increase in the pyruvate concentration because NADH is consumed during the biocatalyzed reaction. The best pH for pyruvate detection is 7.4. The detection limit in the NADH consumption mode depends on the concentration of NADH in the bulk solution. High NADH concentrations generate high background fluorescence intensities on top of which the sensor response must be measured. Detection of a small concentration of analyte, which generates only a small decrease in the NADH fluorescence, is difficult with a high background signal. On the other hand, low NADH levels limit the upper region of response. When the analyte concentration equals the concentration of NADH in the bulk solution, all the NADH in the region of the biocatalytic layer is consumed and no response is possible for higher analyte concentrations. The NADH concentration must be adjusted to allow the needed lower and upper limits of detection for a particular application. A detection limit (S/N = 3) of 1 μM for pyruvate has been measured with a sample NADH concentration of 0.05 mM. Relative standard deviations for pyruvate measurements range from 1 to 8%, with higher values at higher concentrations of pyruvate (lower fluorescence intensities). Sensor lifetimes of 7 d have been measured.

Narayanaswamy and Sevilla have independently developed a NADH-based biosensor for the measurement of glucose.[50] Their biosensor uses glucose dehydrogenase which catalyzes the following reaction:

$$\text{glucose} + \text{NAD}^+ \longrightarrow \text{gluconolactone} + \text{NADH}$$

This sensor is constructed by covalently attaching the enzyme to a nylon mesh and holding this mesh at the tip of a single optical fiber. The resulting tip diameter is 3 mm. Excitation radiation at 350 nm is supplied by the source of a Perkin-Elmer LS-5 luminescence spectrometer and the emitted radiation at 450 nm is detected by the detector of the same spectrometer. Connection between the fibers and the spectrometer have been made in the sample compartment.

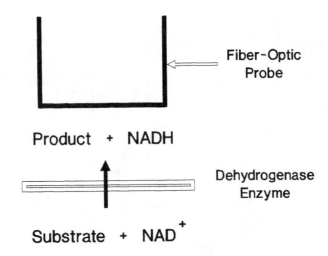

FIGURE 9. Schematic diagram showing the general configuration of an NADH-based fiber-optic biosensor.

Response characteristics for this fiber-optic glucose biosensor have been measured by operating the sensor at the optimum pH of 7.0 and the optimum NAD^+ concentration of 2.2 mM. A linear calibration from 1.1 to 11 mM and a detection limit of 0.6 mM have been measured. For a steady-state signal to be established, 10 to 20 min are required. Because of these long response times, the analytical signal has been taken as the change in fluorescence intensity after the sensor has been exposed to a fresh sample for 5 min.

Klainer and Harris[51] have also reported NADH-based biosensors. They are developing biosensors for the determination of bile acids. 3-Hydroxysteroid dehydrogenase is used to convert the 3-hydroxyl group of the bile acid to a ketone with the concomitant conversion of NAD^+ to NADH. Their sensor is not selective for an individual bile acid, but responds to the total bile acid concentration in solution. By using a selective enzyme, however, response to an individual acid is possible. Their system has the enzyme covalently attached to the tip of a single 200 μm silica fiber. A laser is used to supply the excitation radiation, and a photon counting apparatus is used for detecting the emitted radiation. The rate of fluorescence intensity change is measured and related to the sample concentration of the bile acid. Bile acid concentrations from 2 to 200 μM can be detected with this system.

Two important points must be made concerning NADH-based biosensors. First, many enzymes have been identified and are readily available to catalyze $NADH/NAD^+$ and $NADPH/NADP^+$ dependent reactions. Hence, a wide variety of biosensors based on these reactions can be developed. Second, the consequences of the need to add either NAD^+ or NADH to the sample must be considered. The degree to which this requirement affects the utility of the sensor depends on the application of interest. There are many applications where the addition of a reagent is not a problem. For instance, the addition of reagents is no problem when the biosensor is used to monitor continuously the release of a compound from a tissue material when the tissue is perfused with an externally supplied buffer solution.[53] The required co-substrate can be conveniently added to the perfusion buffer. On the other hand, for *in vivo* applications, where the sensor is positioned directly in a system to be studied, the addition of reagents to the site of the measurement is extremely difficult, if not impossible. Clearly, the versatility of a sensor is limited when a reagent must be added to the sample. Previous attempts to develop reagent-less probe configurations for NADH-based sensors have had only moderate success, with the major problem being the depletion of the immobilized reagent. Although regeneration strategies have been attempted with cycling reactions, a satisfactory system has not yet been developed. A tremendous contribution to the field of biosensors will be made when a general strategy for reagent-less NADH-based biosensors is developed.

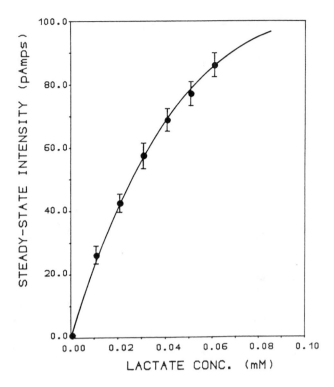

FIGURE 10.   Lactate response curve from the NADH-based fiber-optic biosensor with lactate dehydrogenase.

A unique type of biosensor, termed an internal enzyme biosensor, has been introduced. Instead of placing the enzyme in contact with the sample solution, as is the case with conventional biosensors, the enzyme is separated from the sample solution by a perm-selective membrane. The analyte must cross this membrane before interacting with the enzyme. Figure 12 shows a schematic diagram of the sensor design. The internal enzyme solution consists of the enzyme and all other reagents required for the analytical reaction (co-substrate, buffer, etc.). As the analyte enters the enzyme solution, the enzyme catalyzed reaction takes place and the rate of this reaction is detected by following either the generation of a reaction product or the consumption of a co-substrate. A variety of detection devices, such as fiber-optic, can be used to monitor the analytical reaction. The measured reaction rate (d[P]/dt) is related to the concentration of the analyte in the sample.

The first example of a fiber-optic internal enzyme biosensor is that for ethanol.[48,54] The internal solution is composed of alcohol dehydrogenase and NAD^+ in a pH 8.7 buffer solution. The following reaction takes place during sensor operation:

$$\text{ethanol} + NAD^+ \xrightarrow[\text{dehydrogenase}]{\text{alcohol}} \text{acetaldehyde} + NADH + H^+$$

and the rate of NADH production is measured through a bifurcated fiber-optic bundle. The perm-selective membrane in this case is microporous Teflon. The ethanol biosensor reported by Wolfbeis and Posch[31] is quite similar in design except oxygen depletion is measured by fluorescence quenching (see above).

A linear calibration curve for ethanol is obtained from 0.9 to 9 mM. The clinically important range for ethanol in serum is approximately one order of magnitude higher than this. Hence, the internal enzyme sensor should be applicable for serum ethanol measurements after the sample is diluted. The average standard deviation for between-sensor reproducibility is 6.4%. Response times for this sensor are short because the reaction rate is measured, as

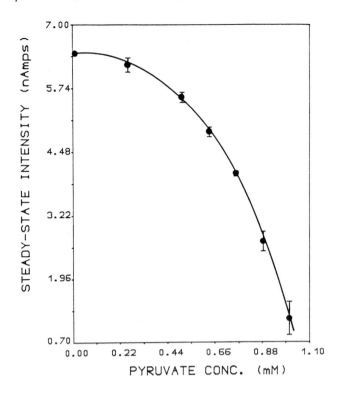

FIGURE 11.   Pyruvate response curve from the NADH-based fiber-optic biosensor with lactate dehydrogenase.

opposed to a steady-state concentration, and this rate is readily measured shortly after the reaction begins.

The selectivity of this internal enzyme biosensor is based on the combined selectivities of the the gas-permeable membrane and the enzymatic reaction. The gas-permeable membrane prevents nonvolatile interferences from entering the internal solution. For instance, this sensor is usable in turbid solutions because the particulate material cannot pass through the gas-permeable membrane and interfere with the optical measurement. At the same time, however, this membrane restricts the sensor to volatile substrates.

D. BIOSENSORS BASED ON CHEMI- AND BIO-LUMINESCENCE

The first example of a fiber-optic biosensor was reported by Freeman and Seitz.[55] In their system, peroxidase was immobilized in a polyacrylamide gel at the distal tip of a fiber-optic bundle. This enzyme catalyzes the chemiluminescence reaction between luminol and hydrogen peroxide. The resulting photon is guided through the fiber-optic bundle to a PMT detector. An increase in the concentration of hydrogen peroxide results in the generation of more photons and a larger signal. A detection limit of 1 μM hydrogen peroxide and response times of 4 s are reported for this sensor.

Aizawa et al. have reported a similar sensor design where the peroxidase enzyme is immobilized directly on the face of a photodiode detector.[56] These researchers demonstrate that their hydrogen peroxide biosensor can be expanded to monitor other analytes by co-immobilizing a second enzyme that catalyzes the production of hydrogen peroxide. An example is a glucose sensor with glucose oxidase. Detection limits for hydrogen peroxide and glucose are 2 and 50 mM, respectively, with this system.

A preliminary report of a biosensor based on bacterial luficerase has appeared.[57] Here, the following reactions are used to generate light at the distal tip of a fiber-optic bundle:

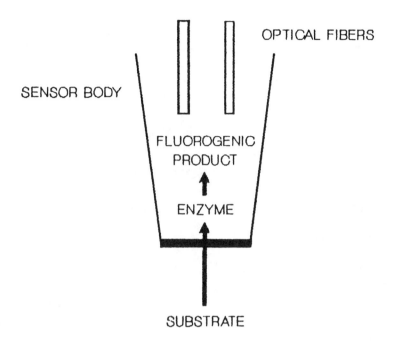

FIGURE 12. Schematic diagram of the internal enzyme biosensor.

$$NADH + H^+ + FMN \longrightarrow NAD^+ + FMNH_2$$

$$FMNH_2 + RCHO + O_2 \longrightarrow FMN + RCOOH + H_2O + light$$

where FMN and $FMNH_2$ represent the oxidized and reduced forms of flavin mononucleotide, respectively, RCHO represents a long chain aldehyde, and RCOOH represents the corresponding long chain acid. The first reaction is catalyzed by NADH:FMN oxidoreductase enzyme, and the second reaction is catalyzed by bacterial luciferase. An increase in the steady-state detector signal is obtained with an increase in the solution NADH concentration. Response times range from less than 1 to 2 min, and responses in the micromolar concentration region are easily obtained.

Coulet and co-workers have recently reported three fiber-optic biosensors based on bioluminescence or chemiluminescence reactions.[58] Their first sensor is also based on the combined activities of oxidoreductase and the bacterial luciferase. They provide more detail concerning the proper operating conditions for this bioluminescence-based biosensor.

Their second sensor uses firefly luciferase for the determination of adenosine triphosphate (ATP). This enzyme catalyzes the following reaction:

$$ATP + luciferin + O_2 \longrightarrow AMP + oxiluciferin + PP_i + CO_2 + light$$

where AMP represents adenosine monophosphate and PP_i represents pyrophosphate. Luciferin is a required co-substrate and must be added to the sample solution.

The final system uses peroxidase to catalyze the chemiluminescence reaction between hydrogen peroxide and luminol. This is the same reaction system used by Freeman and Seitz[55] and by Aiazawa and co-workers.[56] The following reaction scheme is used:

$$2H_2O_2 + luminol + OH^- \longrightarrow 3 - aminophtalate + H_2 + 3H_2O + light$$

In all three systems, the enzymes are covalently immobilized onto nylon membranes. The enzyme membranes are positioned at the distal end of a fiber bundle and the light generated at the sensor tip is detected by a PMT detector. Detection limits (S/N = 2) for these sensors are 2.8 ¥ 10^{-10} for ATP, 3 ¥ 10^{-10} for NADH, and 1 ¥ 10^{-7} for hydrogen peroxide. The low detection limits for ATP and NADH illustrate the potential of bioluminescence-based fiber-optic biosensors for measuring trace levels of biochemical species in solution. The principles of these biosensors can be extended by coupling these biosensors with ATP or NAD(P)H producing enzymes.

V. MISCELLANEOUS SENSORS

The concept of a biosensor has been turned around and used for the determination of the activity of an enzyme in solution by immobilizing the reagent phase at the distal tip of the optical fiber. The first example of such an enzyme-sensing fiber-optic biosensor has been reported by Wolfbeis.[59] In this system, fatty acid esters of 1-hydroxypyrene trisulfonate (HPTS) are electrostatically immobilized on an anion exchange membrane and held at the distal tip of an optical fiber. An esterase enzyme in solution catalyzes the hydrolysis of the ester. The resulting HPTS product is highly fluorescent, and the generated fluorescence is measured through the optical fiber. Higher concentrations of esterase give faster rates of hydrolysis. The rate of fluorescence increase is measured and related to the sample esterase concentration through a previously prepared calibration curve. Sensor arrangements with the enzymatic substrate immobilized along the walls of an optical fiber core and at the tip of a single fiber with a dialysis membrane have also been reported.[60] In any configuration, this type of sensor suffers from the problem that the reagent is consumed during sensor operation. For this reason, sensor calibration is difficult. Potential applications are likely limited to "spot test" type applications.

Finally, Seifert et al.[61] have reported an optical biosensing system based on the determination of the refractive index of very small volumes of solution (i.e., 10^{-4} µl). They show that the adsorption of a biomolecule can be detected as a change in index of refraction. This type of detection scheme might prove useful for the development of fiber-optic immunosensors. In addition, these researchers have demonstrated the ability of this detection scheme to detect the change in the composition of a solution after an enzyme catalyzed reaction. Here, urease is immobilized in a column reactor. As a urea-containing sample passes through the column, the urea is hydrolyzed to ammonia and bicarbonate. The change in the refractive index of the solution before and after passing through the enzyme reactor is measured optically. These researchers show that millimolar concentration levels can be monitored with such a device. Although this system is not a biosensor because the biocatalyst and detection element are separated, this concept might be suitable for the development of fiber-optic biosensors by immobilizing the enzyme along the sides of a fiber core. In the vicinity of the core, the biocatalyzed reaction can alter the index of refraction thereby altering the amount of light that effectively propagates through the fiber.

REFERENCES

1. **Arnold, M. A. and Meyerhoff, M. E.,** Recent advances in the development and analytical applications of biosensing probes, *CRC Critical Reviews in Analytical Chemistry*, 20, 149, 1988.
2. **Rechnitz, G. A.,** Biosensors, *Chem. Eng. News*, 66(36), 24, 1988.
3. **Rechnitz, G. A.,** Bioselective membrane electrodes using tissue materials as biocatalysts, *Methods Enzymol.*, 137, 138, 1988.
4. **Guilbault, G. G.,** Enzyme electrode probes, *Methods Enzymol.*, 137, 14, 1988.
5. **Rechnitz, G. A.,** Bioselective membrane electrode probes, *Science (London)*, 214, 287, 1981.
6. **Turner, A. P. F., Karube, I., and Wilson, G. S.,** *Biosensors: Fundamentals and Application*, Oxford University Press, New York, 1987.
7. **Carr, P. W. and Bowers, L. D.,** *Immobilized Enzymes in Analytical and Clinical Chemistry*, John Wiley & Sons, New York, 1980.
8. **Arnold, M. A.,** Biocatalytic membrane electrodes, *Am. Lab.*, 15, 34, 1983.
9. **Rechnitz, G. A.,** Biosensors: an overview, *J. Clin. Lab. Anal.*, 1, 308, 1987.
10. **Guilbault, G. G.,** *Analytical Uses of Immobilized Enzymes*, Marcel Dekker, New York, 1984.
11. **Fuh, M.-R. S. and Burgess, L. W.,** Wavelength division multiplexer for fiber optic readout, *Anal. Chem.*, 59, 1780, 1987.
12. **Block, M. J. and Hirschfeld, T. B.,** Apparatus including optical fiber for fluorescence immunoassay, U.S. Patent 4,582,809, 1986.
13. **Wangsa, J. and Arnold, M. A.,** NADH-based fiber optic biosensors, presented at the 3rd Chemical Congr. North America and 195th Nat. Meet. American Chemical Society, Toronto, June 1988.
14. **Schwab, S. D. and McCreery, R. L.,** Versatile, efficient raman sampling with fiber optics, *Anal. Chem.*, 56, 2199, 1986.
15. **Barker, S. A.,** Immobilization of the biological component of biosensors, in *Biosensors: Fundamentals and Applications*, Turner, A. P. F., Karube, I., and Wilson, G. S., Eds., Oxford University Press, New York, 1987, chap. 6.
16. **Weetall, H. H.,** *Immobilized Enzymes, Antigens, Antibodies, and Peptides*, Marcel Dekker, New York, 1975.
17. **Munkholm, C., Walt, D. R., Milanovich, F. P., and Klainer, S. M.,** Polymer modification of fiber optic chemical sensors as a method of enhancing fluorescence signal for pH measurement, *Anal. Chem.*, 58, 1427, 1986.
18. **Kulp, T. J., Camins, I., Angel, S. M., Munkholm, C., and Walt, D. R.,** Polymer immobilized enzyme optrodes for the detection of penicillin, *Anal. Chem.*, 59, 2849, 1987.
19. **Peterson, J. I., Fitzgerald, P. V., and Buckhold, D. K.,** Fiber-optic probe for in vivo measurement of oxygen partial pressure, *Anal. Chem.*, 56, 62, 1984.
20. **Lübbers, D. W. and Opitz, N.,** Optical fluorescence sensors for continuous measurements of chemical concentrations in biological systems, *Sensors Actuators*, 4, 641, 1983.
21. **Uwira, N., Opitz, N., and Lübbers, D. W.,** Influence of enzyme concentration and thickness of the enzyme layer on the calibration curve of the continuously measuring glucose optode, *Adv. Exp. Med. Biol.*, 169, 913, 1984.
22. **Opitz, N. and Lübbers, D. W.,** Electrochromic dyes, enzyme reactions and hormone-protein interactions in fluorescence optic sensor (optode) technology, *Talanta*, 35, 123, 1988.
23. **Opitz, N. and Lübbers, D. W.,** Fluorescence-based optochemical sensors (optodes) and related biosensors using enzyme-catalyzed biochemical and antibody-linked immunological reactions, *GBF Monogr.*, 10, 207, 1987.
24. **Lübbers, D. W., Voelkl, K. P., Grossman, U., and Opitz, N.,** Lactate measurements with an enzyme optode that uses two oxygen fluorescence indicators to measure the PO2 gradient directly, in *Progress in Enzyme and Ion-Selective Electrodes*, Lubbers, D. W., Acker, H., Buck, R. P., Eisenman, G., Kessler, M., and Simon, W., Eds., Springer-Verlag, New York, 1981.
25. **Voelkl, K.-P., Grossman, U., Opitz, N., and Lübbers, D. W.,** The use of an oxygen-optode for measuring substances as glucose by using oxidative enzymes for biological applications, *Adv. Physiol. Sci. Proc. Int. Cong.*, 25, 99, 1981.
26. **Voelkl, K.-P., Opitz, N., and Lübbers, D. W.,** Continuous measurement of concentrations of alcohol using a fluorescence-photometric enzymatic method, *Fresenius Z. Anal. Chem.*, 301, 162, 1980.
27. **Kroneis, H. W. and Marsoner, H. J.,** Enzyme sensors using fluorescence based oxygen detection, *GBF Monogr.*, 10, 303, 1987.
28. **Trettnak, W., Leiner, M. J. P., and Wolfbeis, O. S.,** Optical sensors, part 34. Fibre optic glucose biosensor with an oxygen optrode as the transducer, *Analyst*, 113, 1619, 1988.
29. **Pall, D. B.,** Process for preparing hydrophilic polyamide membrane filter media and product, U.S. Patent 4,340,479, 1982.

30. **Shah, R., Margerum, S. C., and Gold, M.,** Grafted hydrophilic polymers as optical sensor substrates, *Proc. SPIE*, 906, 65, 1988.

31. **Wolfbeis, O. S. and Posch, H. E.,** Optical sensors, part 20. A fibre optic ethanol biosensor, *Frensenius Z. Anal. Chem.*, 301, 162, 1980.

32. **Wolfbeis, O. S.,** Fibre-optic sensors for chemical parameters of interest in biotechnology, *GBF Monogr.*, 10, 197, 1987.

33. **Updike, S. J. and Hicks, G. P.,** The enzyme electrode, *Nature (London)*, 214, 986, 1967.

34. **Posch, H. E. and Wolfbeis, O. S.,** Optical sensor for hydrogen peroxide, *Mikrochim. Acta*, Part I, 41, 1989.

35. **Arnold, M. A.,** Potentiometric sensors using whole tissue sections, *Ion-Selective Electrode Rev.*, 8, 85, 1986.

36. **Arnold, M. A. and Ostler, T. J.,** Fiber optic ammonia gas sensing probe, *Anal. Chem.*, 58, 1137, 1986.

37. **Rhines, T. D. and Arnold, M. A.,** Simplex optimization of a fiber-optic ammonia sensor based on multiple indicators, *Anal. Chem.*, 60, 76, 1988.

38. **Rhines, T. D. and Arnold, M. A.,** Fiber-optic biosensor for urea based on sensing of ammonia gas, *Anal. Chim. Acta,* 227, 387, 1989.

39. **Kaplan, L. A.,** Renal disease: urea, in *Methods in Clinical Chemistry*, Pesce, A. J. and Kaplan, L. A., Eds., C. V. Mosby, St. Louis, 1987, chap. 4.

40. **Vurek, G. G., Feustel, P. J., and Severinghaus, J. W.,** A fiber optic PCO2 sensor, *Ann. Biomed. Eng.*, 11, 499, 1983.

41. **Wolfbeis, O. S., Weis, L. J., Leiner, M. J. P., and Ziegler, W. E.,** Fiber-optic fluorosensor for oxygen and carbon dioxide, *Anal. Chem.*, 60, 2028, 1988.

42. **Goldfinch, M. J. and Lowe, C. R.,** Solid-phase optoelectronic sensors for biochemical analysis, *Anal. Biochem.*, 138, 430, 1984.

43. **Trettnak, W., Leiner, M. J. P., and Wolfbeis, O. S.,** Fibre-optic glucose sensor with a pH optrode as the transducer, *Biosensors*, 4, 15, 1988.

44. **Fuh, M. S., Burgess, L. W., and Christian, G. D.,** Single fiber-optic fluorescence enzyme-based sensor, *Anal. Chem.*, 60, 433, 1988.

45. **Yerian, T. D., Christian, G. D., and Ruzicka, J.,** Flow injection analysis as a diagnostic tool for development and testing of a penicillin sensor, *Anal. Chem.*, 60, 1250, 1988.

46. **Arnold, M. A.,** Enzyme-based fiber optic sensor, *Anal. Chem.*, 57, 565, 1985.

47. **Mascini, M., Iannello, M., and Palleschi, G.,** Enzyme electrode with improved mechanical and analytical characteristics obtained by binding enzymes to nylon nets, *Anal. Chim. Acta*, 146, 135, 1983.

48. **Arnold, M. A.,** Development of biosensors using optical fibers, *GBF Monogr.*, 10, 223, 1987.

49. **Wangsa, J. and Arnold, M. A.,** Fiber-optic biosensors based on the fluorometric detection of reduced nicotinamide adenine dinucleotide, *Anal. Chem.*, 60, 1080, 1988.

50. **Narayanaswamy, R. and Sevilla, F., III,** An optical fibre probe for the determination of glucose based on immobilized glucose dehydrogenase, *Anal. Lett.*, 21, 1165, 1988.

51. **Klainer, S. M. and Harris, J. M.,** The use of fiber optic chemical sensors (FOCS) in medical applications: enzyme-based systems, *Proc. SPIE*, 906 (Optical Fibers in Medicine, III), 65, 1988.

52. **Wangsa, J. and Arnold, M. A.,** unpublished results.

53. **Diaz, G. V., El-Issa, L. H., Arnold, M. A., and Miller, R. F.,** Feasibility of continuous glutamate monitoring in perfused retinal tissue with a potentiometric biosensing probe, *J. Neuro. Sci. Methods*, 23, 63, 1988.

54. **Walters, B. S., Nielsen, T. J., and Arnold, M. A.,** Fiber-optic biosensor for ethanol, based on an internal enzyme concept, *Talanta*, 35, 151, 1988.

55. **Freeman, T. M. and Seitz, W. R.,** Chemiluminescence fiber optic probe for hydrogen peroxide based on the luminol reaction, *Anal. Chem.*, 50-9, 1242, 1978.

56. **Aizawa, M., Yoshihito, I., and Kuno, H.,** Photo-voltaic determination of hydrogen peroxide with a biophotodiode, *Anal. Lett.*, 17, 555, 1984.

57. **Arnold, M. A.,** Fiber optic biosensing probes for biomedically important compounds, *Proc. SPIE*, 906, (Optical Fibers in Medicine, III), 128, 1988.

58. **Blum, L. J., Gautier, S. M., and Coulet, P. R.,** Luminescence fiber-optic biosensor, *Anal. Lett.*, 21, 717, 1988.

59. **Wolfbeis, O. S.,** Fiber-optic probe for kinetic determination of enzyme activities, *Anal. Chem.*, 58, 2874, 1986.

60. **Wolfbeis, O. S.,** The development of fibre-optic sensors by immobilization of fluorescent probes, in *Analytical Uses of Immobilized Biological Compounds for Detection, Medical and Industrial Uses*, Guilbault, G. G. and Mascini, M., Eds, D. Reidel, 1988, 219.

61. **Seifert, M., Tiefenthaler, K., Heuberger, K., Lukosz, W., and Mosbach, K.,** An integrated optical biosensor (IOBS), *Anal. Lett.*, 19-1&2, 205, 1986.

Chapter 17

FIBEROPTICS IMMUNOSENSORS*

T. Vo-Dinh, G. D. Griffin, and M. J. Sepaniak

TABLE OF CONTENTS

* This work was jointly sponsored by the National Institutes of Health under contract number GM 34730 and the Office of Health and Environmental Research, U.S. Department of Energy, under contract number DE-AC05-84OR21400 with Martin Marietta Energy Systems, Inc.
"The submitted manuscript has been authored by a contractor of the U.S. Government under contract No. DE-AC05-84OR21400. Accordingly, the U.S. Government retains a nonexclusive, royalty-free license to publish or reproduce the published form of this contribution, or allow others to do so, for U.S. Government purposes."

I. INTRODUCTION

The recent development of sensors using immunological techniques, which offer excellent selectivity through the process of antibody-antigen recognition, has revolutionized many aspects of chemical and biological sensor technologies. The exquisite specificity and high sensitivity of immunoassays permit the measurement of many important compounds at trace levels in complex biological samples. Radioimmunoassay (RIA) utilizes radioactive labels and has been the most widely used immunoassay method. Radioimmunoassays have been applied to a number of fields including pharmacology, clinical chemistry, forensic science, environmental monitoring, molecular epidemiology, and agricultural science. The usefulness of RIA, however, is limited by several shortcomings, including the cost of instrumentation, the limited shelf life of radioisotopes, and the potential deleterious biological effects inherent to radioactive materials. For these reasons, there are extensive research efforts aimed at developing simpler, more practical immunochemical techniques and instrumentation which offer comparable sensitivity and selectivity to RIA.

Recent advances in spectrochemical instrumentation, laser miniaturization, biotechnology, and fiberoptics research have provided opportunities for novel approaches to the development of sensors for the detection of environmental and human exposure to toxic chemicals and biological materials. The operating principles and applications of fiberoptic immunosensors (FIS) are presented in this chapter. The basic principles of antibody-antigen interaction will be reviewed. The development of different types of FIS systems based on various spectrochemical schemes such as absorption, total internal reflection, fluorescence, evavescent-field emission, and surface plasmon resonance will be discussed. Examples of measurements will illustrate the application of several FIS devices developed for the detection of important environmental chemicals and biological compounds.

II. PRINCIPLE OF IMMUNOSENSORS

A. GENERAL CONSIDERATIONS ON ANTIBODY ANTIGEN OF IMMUNOASSAY

The basis for the specificity of immunoassays is the antigen-antibody (Ag-Ab) binding reaction, which is a key mechanism by which the immune system detects and eliminates foreign matter.[1] The enormous range of potential applications of immunosensors is due, at least in part, to the astonishing diversity possible in one of their key components, antibody molecules. Antibodies are complex biomolecules, made up of hundreds of individual amino acids arranged in a highly ordered sequence. The antibodies are actually produced by immune system cells (B cells) when such cells are exposed to substances or molecules, which are called antigens. The antibodies called forth following antigen exposure have recognition/binding sites for specific molecular structures (or substructures) of the antigen. The way in which antigen and antigen-specific antibody interact may perhaps be understood as analogous to a lock and key fit, in which specific configurations of a unique key enable it to open a lock. In the same way, an antigen-specific antibody "fits" its unique antigen in a highly specific manner, so that hollows, protrusions, planes, ridges, etc. (in a word, the total three-dimensional structure) of antigen and antibody are complementary. Further details of how such complementarity is achieved will be discussed later in this chapter. It is sufficient at this point simply to indicate that, due to this three-dimensional shape fitting and the diversity inherent in individual antibody makeup, it is possible to find antibodies which can recognize and bind to a huge variety of molecular shapes. This unique property of antibodies is the key to their usefulness in immunosensors: this ability to recognize molecular structures allows one to develop antibodies that bind specifically to chemicals, biomolecules, microorganism components, etc. One can then use such antibodies as specific "detectors" to identify and find an analyte of interest that is present, even in extremely small amounts, in a myriad of other chemical substances. The other property of antibodies of paramount importance to their analytical role in immunosensors is the strength or avidity/affinity of the antigen-antibody interaction. Because of the variety of interactions which can take place as the antigen-antibody surfaces lie in close proximity one to another, the overall strength of the interaction can be considerable, with correspondingly favorable association and equilibrium constants. What this means in practical terms is that the antigen-antibody interactions can take place very rapidly (for small antigen molecules, almost as rapidly as diffusion processes can bring antigen and antibody together), and that once formed, the antigen-antibody complex has a reasonable lifetime. The following text describes various ways in which these biomolecules may be used in immunosensors.

B. LABELS FOR IMMUNOASSAY

Until the mid-1980s, the most commonly used type of immunoassay has been RIA. Small molecules such as drugs and steroid hormones can often be radiolabeled to high specific activity using ^3H or ^{14}C. Another radioisotope, ^{125}I, is also commonly used since counting techniques for ^{125}I are considerably easier and cheaper than scintillation counting. There are, however, major disadvantages associated with the use of ^{125}I. Due to the short half-life of ^{125}I (60 d) labeled antigens have to be prepared frequently. Radioiodination techniques involve the use of relatively large amounts of radioiodine and thus present a potential health hazard. For these reasons, several other types of nonradioactive labels have been developed for biomedical applications. Table 1 provides a list of several types of labels that can be used in immunosensors: radioactive labels, enzyme-based labels, particles, fluorescent labels, chemi- or bioluminescent tags, and other types of labels. Labels that can induce a change of an optical signal are amenable for use in fiberoptics immunosensors. A variety of fluorescent labels are utilized in fluorescence immunoassays, with the most common one being fluorescein isothiocyanate (FITC).

C. GENERAL CONSIDERATIONS ON FIBEROPTIC SENSORS

Fiberoptics sensors may be divided into two general classes. The first class of sensors uses the waveguide, usually an optical fiber, as a simple lightpipe taking light to and from a sensing device, such as a microcell containing a dye that exhibits changes in absorbance or luminescence with varying concentrations of a substance or a physicochemical parameter, for example, CO_2, O_2, or pH. For this class of sensors, one of the most important factors involves the light transmission efficiency through the fiber. Such optical chemical sensors can be used as the host transducer of an immunoassay detector. The second category of sensors involves an intrinsic change in the properties of the optical fiber itself, which serves as the sensing element. For example, this type of sensor may exploit the evanescent field associated with light propagating in a waveguide and has already led to several immunosensors that are at the

TABLE 1
Several Types of Labels for Immunosensors

Types of labeles	Examples of label systems
Radioactive	^{3}H, ^{14}C
	^{125}I, ^{75}Se, ^{57}Co
Enzyme-based	Alkaline phosphatase
	β-D-Galactosidase
	Peroxidase
Fluorescent	Fluorescein isothiocyanate (FITC)
	Rhodamine B isothiocyanate (RBITC)
	Dansyl chloride (DANS)
	Umbelliferones
	Rare earth metal chelates
Chemi- or bioluminescent	Luciferase/luciferins
	Luminol (and derivatives)
	Acridinium esters
	Peroxidase
Particles	Latex particles
	Red blood cells
Other types of labels	Coenzymes
	Proteins
	Viruses
	Free radicals
	Substrates

prototype development stage within industrial research laboratories. This second category may be further subdivided into two subcategories: evanescent field immunosensors and surface plasmon resonance (SPR) immunosensors.

Fiberoptics immunosensors may use either bifurcated or single-strand fibers (Figure 1). In a device based on bifurcated design, separated fibers carry the excitation and emission radiation. In the single-fiber device, a dichroic filter or mirror-pinhole assembly is generally used to separate the excitation and emission radiation. When compared to multiple (bifurcated) fiber designs, sensors that utilize a single optical fiber to transmit excitation radiation to the sample, and emission radiation to the sample detector, have the advantages of good signal collection efficiency and small size. The small-size attribute is important for eventual *in vivo* applications. Single-fiber sensors, however, exhibit high optical background levels because a significant amount of excitation radiation, which is reflected at the fiber tip, is not efficiently rejected by spatial filtering. Attempts to minimize this backscatter radiation problem include the use of high rejection-ratio double monochromators, and grinding fiber faces at angles that minimize the reflected optical background. Disadvantages of these correction techniques include the low throughput of double monochromators, and the difficulty of producing small quartz surfaces with precise angles.

The properties of the optical fibers are important factors determining FIS device capabilities. Etching the surface of the fiber may increase the surface area and may allow binding of increased amounts of antibodies onto the sensing tip. Quartz tends not to be as effectively etched as glass. The material of the optical fibers also determines the usable spectral range. Fused-silica optical fibers permit measurements in the ultraviolet down to 220 nm, but are relatively expensive. Glass is less costly but is only appropriate for measurement in the visible. Plastic fibers are less expensive, but are restricted to spectral ranges above 450 nm.

D. VARIOUS TYPES OF ASSAYS FOR FIBEROPTICS IMMUNOSENSORS

Immunosensors can be designed to use different types of immunoassays for specific

a)

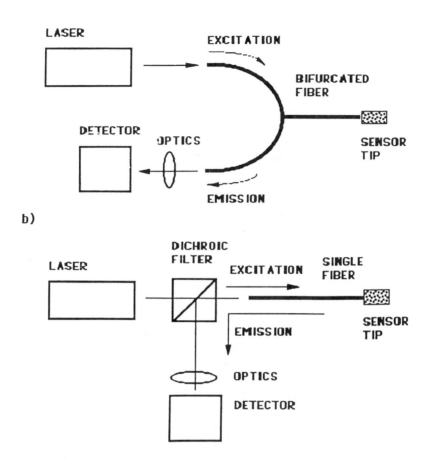

b)

FIGURE 1. Schematic diagram of bifurcated and single-strand fiber-optics sensors.

applications. Detection techniques for immunoassay can use a variety of processes including amperometric or potentiometric electrochemistry, piezoelectricity or microacoustics, and spectroscopic methods such as absorption, reflectance, or fluorescence. Although the basic detection principle is applicable to various techniques, the following discussions deal with fluoroimmunoassay (FIA) as the model system.

The simplest involves *in situ* FIS incubation followed by direct measurement of a naturally fluorescent analyte.[2] For nonfluorescent analyte systems, *in situ* incubation is followed by "development" in fluorophor-labeled second antibody. The resulting "antibody sandwich" produces a fluorescence signal that is directly proportional to bound antigen. The sensitivity obtained when using these techniques increases with increasing amounts of immobilized receptor. Enzyme immunoassays can further increase the sensitivity of detection of antigen-antibody interactions by the chemical amplification process whereby one measures the accumulated products after the enzyme has been allowed to react with excess substrate for a period of time.[3] A third detection scheme involves competition between fluorophor-labeled and unlabeled antigen. In this case, the unlabeled analyte competes with labeled analyte for a limited number of receptor binding sites. Assay sensitivity therefore increases with decreasing amounts of immobilized reagent.

In general, both labeled and unlabeled analyte molecules are premixed prior to measure-

NON FLUORESCENT SPECIES CAN ALSO BE MONITORED BY

USING COMPETITIVE BINDING AND FLUOROTAGGING METHODS.

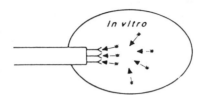

PLACE PROBE IN REFERENCE SOLUTION

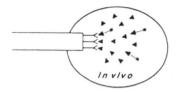 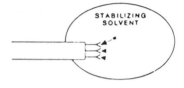

PLACE PROBE IN _IN VIVO_ SAMPLE   REMOVE PROBE AND MEASURE

FIGURE 2. Competitive-binding assay for fluorescence immunosensor. (A) Place the probe in a reference solution of labeled antigens; remove probe and measure fluorescence signal. (B) Place the probe in actual *in vivo* sample for predetermined period; unlabeled antigens compete with labeled antigens for binding sites on the probe. (C) Remove the probe; measure the diminished fluorescence signal.

ments in a competitive binding assay.[1] In an alternate procedure, the FIS device may function in the following manner: the FIS probe (i.e., fiber termini) with antibodies is placed in concentrated labeled antigen solutions.[1,4,5] After equilibrium has been established, the FIS probe is removed and washed to remove nonspecifically bound antigen. The probe prepared in this fashion is ready for use by placing it in a solution of antigens to be quantified. Fluorescence signals are measured before and after the fiber probe is placed in the sample. The antigens are allowed to react with the antibodies and displace the labeled antigens from the antibodies due to competitive binding. The resulting decrease in fluorescence signal, due to the displacement of fluorescent labeled antigens, will be related via a calibration plot to antigen concentration. The measurement procedures are schematically illustrated in Figure 2.

An essential requirement of a competitive binding assay is to be able to distinguish between bound and free antigens. In conventional FIA techniques, this is achieved by immobilizing one component, i.e., by linking the antigen to a substrate (plastic plate, beads), and subsequently, physically separating the substrate from the supernatant to measure the amount of bound fluorescent antigen. In many situations, one must resort to a separation procedure for labeled antigen (Ag*) and antibody-bound antigen (Ag*–Ab), e.g., precipitation, electrophoresis, chromatography, or physical adsorption onto a solid phase. In some cases, the formation of the (Ab–Ag*) antibody complex can be studied by fluorescence methods without the necessity for separation of free Ag* from bound antigen. Referred to as a homogeneous immunoassay, this procedure relies on techniques detecting differences in fluorescence polarization, energy transfer, quenching, and enhancement between the bound and unbound antigen.

The configuration of the sensing tip is one of the most important elements in the design of optimal FIS devices. This configuration is the determining factor for sensitivity, dynamics range, response time, specificity, and a useful lifetime of the sensor.[5-10] Typical configurations of the FIS probe are illustrated in Figure 3. For example, the antibodies can be bound either

POSSIBLE FIBEROPTIC TERMINI FOR FIS

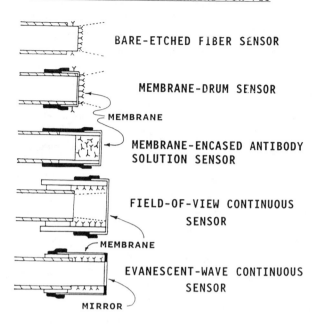

FIGURE 3. Different configurations for fiberoptics immunosensor probe.

to the termini of the optical fiber, or to the internal surface of a hollow cylinder, or to the wall of a microcuvette attached to the end of the fiber.

The appropriate immunoassay method for FIS devices is determined by the chemical characteristics of the antigen-antibody system of interest. For example, naturally fluorescent antigens and haptens may be analyzed by direct measurement. Nonfluorescent haptens may be detected via competitive binding or sandwich techniques, though the scarcity of hapten epitopes (antibody recognition sites) may diminish (or abolish) the sensitivity of sandwich analysis. An alternate means of gaining sensitivity in direct and sandwich assays entails increasing the absolute amount of immobilized receptor by increasing the fiber surface area. This can be accomplished with small-diameter fibers by utilizing evanescent-wave excitation of biomolecules bound to the circumference of the sensing tip. These sensors are characterized by large exposed surface areas and low evanescent-wave penetration depths. Their primary limitation is the fact that evanescently excited fluorescence is coupled less efficiently into the optical fiber than distal-face (vide infra) excitation.

III. ANTIBODY PREPARATION

A. ANTIBODY STRUCTURE

A brief and noninclusive overview of antibody structure is provided in this section. It is recognized that, while some readers may be immunologists, many are not and may consider discussion of antibodies and antigen-antibody complexes nonessential details compared to the truly interesting topic of analytical uses of immunosensors. Nonetheless, it is the belief of the authors that understanding of basic immunological/biochemical principles is essential for the analyst to progress in this area, and the better the understanding, the more satisfactory will be the design of the immunosensor. Much has been written describing antibodies; some basic information has been drawn from a variety of review articles and books.[11-20]

Antibody molecules all belong to the general class of proteins called immunoglobulins (Ig). The following description of Ig structure will be specific to immunoglobulin G (IgG), which comprises ~80% of the total Ig;[16] the differences between this and other Igs will be pointed out where appropriate. The molecular weight of IgG is 150,000 Da. The molecule consists of two light (L) and two heavy (H) chains, these chains simply being polypeptides composed of long segments of amino acids linked together.[16,18] The classic diagrammatic representation of the IgG molecule is shown in Figure 4. Although an oversimplification of an actually complicated folding pattern, this representation has some basis in physical reality, as revealed by electron microscopy of IgG molecules bound to antigen.[16] The L and H chains are held together by disulfide (S-S) bridges, and more importantly, by many hydrophobic interactions between the H and L chains.[16,19] The two heavy chains are also held together by disulfide linkages and hydrophobic interactions. The region from the joining point of the H and L chains to slightly beyond the S-S bonds between the heavy chains is called the hinge.[16] Due to its amino acid structure, the hinge region has a great deal of flexibility, but this same lack of complexity in structure renders it particularly susceptible to proteolytic (i.e., protein splitting) attack by enzymes.[19] Digestion by using the proteolytic enzyme papain produces cleavage at the indicated point (Figure 4) producing three IgG fragments: two identical ones denoted as Fab (antigen-binding fragment) and one called Fc (crystallizable fragment).[16] On the other hand digestion using the enzyme pepsin produces two fragments, one being essentially the Fc, while the other is a larger fragment containing the whole of the two L chains and part of each H chain, called F(ab')$_2$ (Figure 4).

The foregoing description of the fragments has brought out a highly important point in regard to antibody structure in relation to function. The regions on the end of the two arms of the "Y" in Figure 4 are the sites where antigen binding takes place. The Fab fragments are able to bind antigen because the antigen binding site is still intact on the fragments. An intact IgG molecule can, at most, bind two antigen molecules, since both arms have identical binding sites. Each Fab fragment can only bind one antigen, while the F(ab')$_2$ fragment, having both antigen binding sites, can bind two antigens.

The regions of both H and L chains near the end of the antibody arms are termed the variable regions or domains (Figure 4).[16] The polypeptide chains, whether H or L, actually consist of different numbers of linked globular domains (there are two types of domains, variable and constant), the length of each domain being ~108 amino acids.[18] The variable domain is so called because of the relatively large variability in amino acid sequence observed in this region when various primary amino acid sequences of antibodies are compared. In contrast, amino acid sequences in the constant domain are practically invariant, when different antibodies are compared. The L chain consists of one variable domain (starting at the H$_2$N-terminus) linked to one constant domain (Figure 4), while the H chains have a variable domain followed by three constant domains (for IgG, IgA, or IgD) or four constant domains (for IgM and IgE).[18] The structure of the variable domains of both H and L chains can be further subdivided into regions of hypervariability (i.e., more variation in amino acid sequence than seen in the variable region as a whole), separated by framework regions.[15] There are three hypervariable regions in each variable domain of the H and L chains (Figure 4). The polypeptide chains in the variable domain fold in such a manner as to bring the hypervariable regions into close juxtaposition, forming an overall surface which is in most intimate contact with the antigen.[15] Thus these hypervariable regions are called the complimentarity-determining residues (or regions). The framework regions serve to hold the complimentarity-determining residues in an overall similarity in three-dimensional structure, allowing enough flexibility to provide "fine-tuning" of the geometric structure of the antigen-binding site.[12] The key feature of the antibody molecule, therefore, can be seen as the ability to vary the primary amino acid structure in the complementarity-determining region, such that a vast number of different surface structures result which are complementary to the large array of antigenic surface features found in nature. (See also Section III.C.)

Although the antibody depiction in Figure 4 is useful to illustrate the overall structure, a

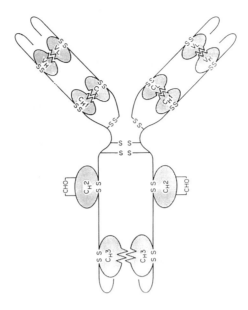

FIGURE 4. A representative IgG molecule, with two light (L) chains, containing variable (V_L) and constant (C_L) domains, and two heavy (H) chains, containing variable (V_H) and constant (C_H1, C_H2, and C_H3) domains. Points of cleavage by Papain and Pepsin are indicated along with the Fab and Fc fragments produced by Papain digestion. Details of the variable domains are also indicated: i.e., hypervariable regions (HV_1, HV_2, HV_3) and framework regions (FW_1, FW_2, FW_3, and FW_4). The free amino terminal end and free carboxylate end of the polypeptide chains are as indicated.

more accurate representation is shown in Figure 5. There are intrachain disulfide bridges on the H and L chains (as well as the interchain bridges in the hinge region) which form loops of folded polypeptides[16] which are the globular domains (variable and constant, as indicated previously). These globular domains interact (see Figure 5), mainly by hydrophobic interactions, and this interaction, along with the large number of disulfide bonds, contributes to the stability of the antibody molecule.[19] In contrast to the other constant domains, the C_H2 regions on adjoining H chains do not interact, because of carbohydrate structures attached to the amino acid backbone in these areas, which effectively block the domains from interaction.[16]

There are five distinct types of H chain, called G, A, M, D, and E, and depending on the type of H chain, an antibody may be classified into one of the five classes of immunoglobulins-IgG, IgA, IgM, IgD, and IgE (there are also subclasses, e.g., IgG_1, IgG_2, but these will not be further discussed).[16] The different Ig classes may be thought of as a biological system to provide antibody molecules with different response capabilities in relation to antigens. For example, IgA appears selectively in sero-mucous secretions (tears, saliva, lung fluids) and apparently has a role in defending external body surfaces against microbial assault; IgG appears to play a role in internal defense, as it diffuses more readily than other Igs into extravascular spaces.[16] IgM has a unique structure in relation to other Igs, as it consists of five "normal" Ig molecules, all linked together, and the total molecule has an antigen-binding capacity of ten. It is noteworthy that the Fc portion of the Ig molecule, which consists of the constant regions of H chains, has apparently little effect upon antigen-binding, but is highly important in directing the biological activity of the molecule.[16]

B. ANTIGENICITY CONSIDERATIONS

The unique property of antibodies which renders them potentially useful in immunosensors is their ability to combine with and bind a specific chemical structure which is called an antigen. In fact, the antibody originates as part of the immune system's response when this

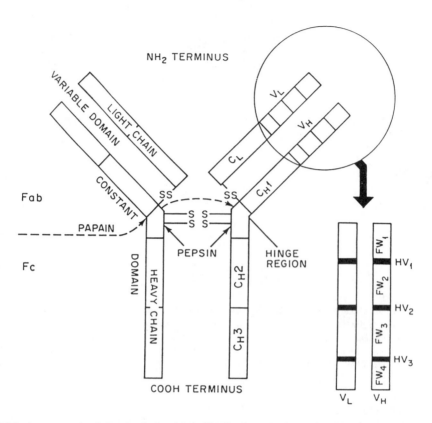

FIGURE 5. A representative IgG molecule showing the domains formed by intrachain disulfide bonds. The domains on adjoining heavy and light chains (or adjoining heavy chains) make contact through hydrophobic interactions. V_L, V_H = variable domain, light or heavy chain; C_L = constant domain, light chain C_H1, C_H2, C_H3 = constant domains, heavy chain; CHO — carbohydrate moieties attached to C_H2.

antigen is introduced into the organism. A brief overview of fundamental properties of antigens is provided in this section.[11,14,16,19,21] It will be useful to first define a variety of terms which will be used frequently. Immunogenicity is the ability of a substance to elicit or initiate an immune response (here the only aspect of the immune response of concern is antibody production). The substance which provokes the response is called an immunogen or antigen. A further distinction can be made between immunogenicity and antigenicity, however. Common antigens are large protein molecules. However, many small chemical substances can also elicit an antibody specific for this chemical, but *only* if the small chemical substance is part of a larger immunogenic protein (i.e., the chemical may be covalently coupled to amino acid residues of the protein). Such a small chemical molecule, which cannot by itself stimulate antibody formation but will combine with its specific antibody once formed, is called a hapten. If antigenicity is defined as the ability to be recognized by a molecule of the immune response (i.e., an antibody), then a hapten is antigenic but *not* immunogenic. Most common antigens are large proteins; antibodies elicited against them do not react with the whole protein, but with specific areas (sometimes quite restricted) called antigenic determinants. Epitope is a synonym for antigenic determinant. There can be many epitopes on a given macromolecular protein. The immunogenicity of haptens linked to proteins can now be explained, for in essence the hapten forms part of a new epitope on the protein's surface, which the immune system recognizes.

There appear to be multiple factors, some probably unknown, which govern immunogenicity. One highly important feature is that the antigen must be "foreign" to the host in which the antibody is made.[14,19] That is, the antigen must be recognized as "non-self" (or foreign to the circulation) by the immune system of the immunized organism. Quite often, the farther away the immunogen is from the immunized host in terms of phylogenetic relatedness (i.e., interspecies closeness), the stronger the immune response will be, but this is not invariably so.

Another key factor in immunogenicity is the molecular size of the antigen. Although there are a few exceptions, molecules of <1000 Da have proved to be nonimmunogenic;[14,22] most such structures can evoke antibodies if used as haptens. Molecules below 10,000 Da can be nonimmunogenic or only weakly so (example: glucagon — 3600 MW). Large protein molecules (serum albumins — 69,000 Da, globulin — 150,000 Da) in general evoke strong antibody responses, while even larger molecules (e.g., viruses and hemocyanins — 6 to 8 × 10^6 Da) give a very strong response.[22]

Another factor somewhat related to molecular size is complexity. Antigens must be of a certain complexity to evoke an immunogenic response.[19] Thus, microorganisms and large proteins are highly complex structures in three-dimensional terms, and evoke strong antibody responses. On the other hand, polymers of a single amino acid, no matter how large in molecular weight, are very poor immunogens.[14] One must be careful in predicting immunogenicity solely on the basis of complexity of structure, however. Dextrans, which are polymers of a single sugar, glucose, are immunogenic.

How many antigenic sites are on a given protein? Various estimates appear in the literature. Tijssen[19] indicates that proteins contain one epitope/40-80 amino acid residues, and suggests that the size of such an epitope can be quite small (5 to 7 amino acids). Shinnick and Lerner[23] say that there may be one antigenic site per 5000 Da (25 to 50 amino acids). Fudenberg et al.[14] list 5 epitopes for ovalbumin (42,000 MW) and as many as 40 for thyroglobulin (700,000 MW). From these numbers and estimates, it is apparent that there are a *limited* number of epitopes on a protein's surface, and that some portions of the surface are not antigenic.[11,23]

Although the preceding studies may suggest that only antibodies made against certain "immunodominant" epitopes will bind to a given protein, other studies suggest that the whole accessible surface may be a continuum of antigenic sites.[21] The reason only a limited number of antigenic sites manifest themselves on the intact protein may have more to do with steric hindrance associated with multiple antibody binding, rather than immunodominance of certain regions. Studies with the intact hemagglutinin protein of the influenza virus[23,24] have shown that the intact protein has only four antigenic sites. If the protein is digested into 20 peptide fragments and these fragments are used as immunogens, 15 of 20 antipeptide antibodies reacted with the intact protein with the same strength of binding as with the peptide they were raised against. Further, these peptides are not restricted to the same areas of the protein as the four antigenic sites on the native molecule, but extend over most of the primary sequence. These results suggest that fragments of proteins can serve to generate antibodies recognizing the whole protein.

In a remarkable investigation, Jackson et al.[25] found that two antigenic sites on a synthetic peptide (part of the influenza virus hemagglutinin) were separated by *only* three amino acid residues, a distance of 10 Å. Monoclonal antibodies to each antigenic site could bind simultaneously! This is an extremely close juxtapositioning of two large proteins, and probably points out that there is much still to be learned in regard to antibody recognition of antigenic epitopes. While the preceding study may provide a minimal spacing distance estimate for epitopes, a study by Hodges et al.[26] provides a minimal estimate of the size of an antigenic determinant. They found that an antibody raised against a peptide fragment of rhodopsin could bind a 4-amino-acid peptide (total MW ≅ 400 Da) with equal strength as it did the 18-amino-acid immunogen. It seems clear from this that antigenic epitopes can be quite small in length, and therefore very abundant over a large surface.

So far antigenicity has been discussed from a generalized viewpoint. The starting point for any antibody production procedure is achieving successful immunization. Immunization protocols abound in the literature. Above all else, one must avoid the induction of tolerance (i.e., no antibody response occurs following readministration of antigen). Tijssen[19] cautions against the use of high doses of antigen which may induce tolerance, and certain routes of exposure (i.e., intravenously for small or soluble antigens).

Successful immunization is still very much an empirical process, depending as it does upon so many factors (the immunogen, its physical form, route of administration, schedule of reimmunization, age, sex, and overall state of the immune system in the immunized animal). Antibodies to proteins can be obtained by using the protein (or peptide fragments?) as the antigen. Antibodies to haptens are obtained only when the small molecule is coupled to an immunogenic protein.[14] Ovalbumin, bovine serum albumin, and keyhole limpet hemocyanin are often used for this coupling — the chemistry to be used for covalent linkage to the protein depends upon the functional groups on the hapten. With a small molecule of doubtful immunogenicity, it will generally be found to be worthwhile to treat it as a hapten and couple it to an immunogenic protein.

Satisfactory immunization can often be achieved by injecting relatively small amounts (15 µg/kg body weight of animal) intramuscularly, subcutaneously, or intradermally.[19] Often, the antigen is administered as a mixture in Freund's complete adjuvant, which is a mineral oil plus emulsifier plus a dispersion of heat-killed *Mycobacterium tuberculosis*.[19] The presence of this particular organism often "boosts" the antibody response. The antibody response of the immunized animal should be monitored; several weeks may pass before the maximal response occurs. Usually, shorter immunization times will yield more specific antibodies, but the strength of avidity/affinity of the antibodies increases with time (and with larger intervals between "booster" injections),[19] so a tradeoff must often be made. It is common procedure to give "booster" injections with the antigen. Tijssen[19] suggests "boosters" should be given subcutaneously with no adjuvant, at intervals of about 2 weeks following the last immunization.

C. ANTIBODY-ANTIGEN INTERACTION

The basic antibody structural features which are involved in antigen binding (i.e., the variable domains, particularly the complimentarity determining regions) have been described. Due to the tremendous diversity which can be introduced into the antigen-combining region of the the antibody, the variety of binding sites is so great that at least some members of the total antibody repertoire can bind to any given cluster of atoms.[12] The details of how antibody-antigen interactions occur are being elucidated, at least for a small set of specific cases. Because such understanding is important in utilizing antibodies as analytical devices, the forces involved in the antigen-antibody interaction will be briefly discussed.

The antibody-antigen complex is not held together by covalent bonds; nevertheless, the strength of the interaction can be gauged from the often strikingly high affinity/avidity constants. It is generally agreed that there are four factors involved in the antigen-antibody interaction: (1) electrostatic forces, (2) hydrogen bonding, (3) hydrophobic attractions, and (4) van der Waals interactions.[16,22,27] Electrostatic or coulombic interactions occur between opposite electrical charges, and thus involve ionized sites (e.g., $-COOH$ and $-NH_2$ groups on amino acid side chains) or less strongly attracting dipoles. Hydrogen bonds involve interaction of a hydrogen atom, covalently bonded to a more electronegative atom and thus having a partial positive charge, with an unshared electron pair of a second electronegative atom. The interaction of water molecules is a classic example of hydrogen bonding; a variety of amino acid functional groups (particularly $-OH$ and $-NH_2$) could be involved in such interaction.[22] The hydrophobic interactions occur as a result of a strong tendency for apolar atomic groups (e.g., side chains of valine, leucine, phenylaline, proline, and tryptophan) to associate with one another in an aqueous environment; thus, their net interaction with water is decreased. Van der

Waals attractions are the result of external electron clouds of atoms forming dipole attractions, the dipoles being induced in a given atom by the very close approach of another atom which has a fluctuating dipole.[16,22] This last force becomes increasingly stronger as the interatomic distances decrease. In fact, the forces of hydrogen bonding, hydrophobic interaction, and van der Waals attraction all are relatively weak in binding strength and only become significant upon close approach of the pair of molecules. Studies of the few antigen-antibody interactions which have been dissected at the atomic level by X-ray crystallography have indicated that the predominant attractive forces in the antigen-antibody bond arise from a large number of hydrophobic and hydrogen bond interactions, as well as van der Waals forces arising from the close approach of the antibody to the antigen.[11,17,22,27] Thus the overall remarkable strength of the interaction is due to the extremely close fit of the molecular surfaces (1 to 2 Å),[22] the exquisite complementarity between antigen and antibody, and the formation of a large number of individual weak interactions, which become significant en masse.

The shape of the antigen binding site on the antibody has been the subject of a variety of studies. Obviously, from the previous discussion, one might expect it to vary with the shape of the antigen, yet certain overall features may remain the same. In the case of haptens, the antibody combining site can form a groove or pocket, much like an active site of an enzyme, and this groove accommodates the small antigen.[12,27] For a protein antigen, the groove concept probably does not generally apply. In the case of the chicken protein lysozyme, the antibody binding site was found to be a rather flat surface with knobs and valleys formed by the side chains of various amino acids in the combining site; the total size of the surface was 20×30 Å or 600 Å^2.[17,28,29] There were 17 antibody amino acid residues involved in interactions with the antigen, 11 of these interactions were hydrophobic, there were 12 hydrogen bonds, and no electrostatic interactions.[17] The antigen binding site also had a relatively high concentration of aromatic amino acids.[28] Evidence from the same lysozyme, anti-lysozyme system indicates that the antibody is sensitive to a single amino acid change in the antigenic determinant region of the antigen.[11,28] One further significant fact arising from study of the above antigen-antibody system was that there was no evidence of a conformational change in the antigen or antibody when binding occurred.[11,17,28] This fact supports the lock-and-key model of the antigen-antibody interaction, in which the antibody binding surface may be thought of as a "negative" image of the three-dimensional shape of the antigen. Study of another antigen-antibody system involving a viral neuraminidase and its antibody, however, has found evidence that antigen and possibly antibody undergo conformational changes upon binding.[17,30] The possibility of conformational changes upon antigen-antibody binding cannot be excluded.

A different aspect of antigen-binding site structure and specificity has been highlighted by Kabat.[27] He reviews amino acid sequence information from a number of antibodies from two different mouse strains, raised against polysaccharide antigens [α(1→6) dextrans]. Most of these antibodies show little difference in the ability to recognize and bind to the antigen. Yet the sequences of amino acids in the antibody complementarity regions show large differences between the two mouse strains (if the H chains are compared, 1 of 5 residues; 8 of 17, and 5 of 7 are the same in complimentarity regions 1, 2, and 3, respectively). The conclusion is that the antigen binding sites that recognize the same antigen can be made up of a very different array of amino acids, unless *only* those amino acids invariant between the two antibodies determine the antibody binding (which seems unlikely).

Further information regarding the unique antigenic specifications that structural features of the antibody can confer comes from a study in which site-directed mutagenesis of certain amino acids in the anti-lysozyme antibody binding site was carried out.[31] What this produced was an antibody which contained amino acid changes at specific positions, these positions being those felt to be important in overall interaction of the antibody with antigen. Changes of two amino acids in the antigen binding site increased the antibody's affinity for lysozyme eight- to ninefold, and improved its specificity for recognition of its specific antigen. Both the substituted amino acids were ionic in the original antibody structure, and the substitutions were uncharged although hydrophilic. Roberts et al.[31] speculate that the uncharged residues allowed a closer fit of the antigen-antibody surfaces, thereby increasing the number of hydrogen bonds. Substitution of glutamic acid by serine at another position effectively abolished lysozyme binding by the antibody, apparently by disruption of the interaction of H and L complementarity determining regions at this point. These results indicate the potential of genetic engineering to "improve upon nature" by reconstructing an antibody with desired properties.

The following examples show the amazing degree of structural detail the antibody is able to "see". Literally hundreds of examples could be cited to reinforce this point; only a selected few follow. Antibodies raised to m - aminobenzene sulphonate can distinguish between the m and p position of the two substituents on the benzene ring.[16] Substitution of the sulphonate group by a carboxyl group was also detected by the antibody.[16] An antibody against a complex glycolipid (sulphatide) was able to distinguish between the presence or absence of the sulfate group.[32] An antibody used in detecting 2-acetylaminofluorene was able to discriminate between the $-NH_2$ and the $-N-CO-CH_2$ structures at the 2-position of fluorene.[33] An antibody raised against digoxin was specific to the steroid structure of the molecule, did not recognize the sugar residues attached to the steroid-like nucleus of the molecule, and was found to be quite sensitive to relatively small changes in the steroid ring.[34] Such examples serve to underscore the potential for great molecular specificity inherent in the antibody's antigen binding site.

D. POLYCLONAL OR MONOCLONAL ANTIBODIES

In the development of immunosensors, the investigator must decide whether the antibodies desired for a particular application are to be derived from a polyclonal source or by monoclonal technology. Monoclonal antibodies are antibodies which are produced by the daughters of a *single* "B" cell. Since all the daughters are producing exactly the same antibody as the parent cell, a monoclonal antibody is completely homogeneous, all antibody molecules are the same, and all have been developed against a single antigenic determinant. Polyclonal antibodies, by contrast, are antibodies circulating in serum of animals immunized with a specific antigen. Because these antibodies have arisen from clones of a number of separate "B" cells, the antibodies are heterogeneous and different antibodies in this mixture react with different antigenic determinants. Therefore, polyclonal antisera will always be a mixture while monoclonals will never be (unless deliberately mixed in the laboratory).

The question of whether monoclonal or polyclonal antibodies should be utilized should not be necessarily viewed as a mutually exclusive proposition. In many instances, it may be instructive and useful to prepare polyclonal antisera for use in a proof-of-principle experiment, followed by production of monoclonal antibodies with specific desired characteristics. In this section the authors provide some comparative comments about the advantages and disadvantages of the two technologies, simply as a guide to the fledgling investigator.

Polyclonal antibodies, if prepared properly, can have high specificity and high avidity.[19] The quality of the antibodies obtained depends very much upon the purity of the immunogen used to elicit the antibody response. An impure immunogen preparation will result in a mixture of antibodies some of which will be against the various impurities. Immunization protocols can be adjusted and modulated to optimize the avidity (usually will increase with total time of immunization and interval between booster shots) of the antisera and keep cross-reactivity to an acceptable level.[19] Purification of the desired antibodies from the antisera will almost always be required since the majority of the Ig present in the antisera will be irrelevant.

A major advantage of polyclonal antibodies is the relative ease with which they may be obtained, i.e., by immunizing an appropriate experimental animal and then taking an appropriate quantity of blood. Rabbits are frequently used; larger animals would provide a larger supply of immune serum, but cost of animal care may be prohibitive. If the necessity for a high degree of immunogen purification is also taken into account, the whole procedure loses some of its apparent simplicity. There are two major disadvantages in regard to polyclonal antibodies. First, these antibodies will always have multiple specificity (even if it is possible to purify only the antibodies to a *single* antigen from the antisera, if the antigen has more than one epitope, the antibodies will be a mixture of antibodies recognizing different epitopes), and hence can never provide the monospecificity of a monoclonal antibody. Second, because the antibodies arise from bleeding an immunized animal at some point in the immunization protocol, different batches of antisera taken at different times will inevitably have a somewhat different antibody composition.[19]

Monoclonal antibodies can, at least theoretically, provide the solution to the problems of polyclonal antisera listed above. Kohler[35] lists the following advantages associated with monoclonal antibodies: (1) each hybrid cell line produces only one unique antibody, (2) there is potentially an unlimited antibody supply, (3) immunization with an impure antigen can still lead to an antibody against *only* the antigen of interest, (4) potentially all specificities (i.e., antibodies against all antigenic epitopes) can be obtained, (5) it is possible to manipulate monoclonal antibodies by genetic engineering techniques, and (6) the technique is very general, in terms of what antigen can be used and desired properties of the antibody. Since each successful hybridoma is the result of a fusion of a myeloma cell with a "B" cell reacting with one epitope of the antigen, it can be seen that an antigen preparation that contains a number of impurities can still provide good results.[36] This fact alone recommends the monoclonal antibody technology for antigens which can only be obtained in very small amounts and/or in an impure state. King and Morrow[37] provide an interesting discussion of procedures to select the monoclonal antibody of interest from such a "shotgun" immunization.

Balanced against the many advantages of monoclonal antibodies are some significant disadvantages which must be carefully considered. First, hybridoma cell lines are inherently unstable;[19] a hybrid producing an antibody of interest can lose this ability as the cell line is expanded. This potential problem requires the preservation of frozen cultures to which the investigator can return if the further propagated line loses antibody production. Even this is no guarantee of ultimate success. Secondly, the monoclonal antibody may not be of the correct antibody class (or subclass) for the desired application.[13] For example, the monoclonal antibody with the desired specificity may be an IgM, a less suitable choice for some applications than an IgG. Thirdly, and perhaps of paramount concern, monoclonal antibody technology is very labor intensive and time consuming.[19] Strong expertise in cell culture techniques and familiarity with modern high-volume immunoassay procedures (e.g., ELISA) are more or less minimal requirements to even begin a monoclonal antibody effort.

One final point needs to be made in regard to potential problems with monoclonal antibodies. In spite of their homogeneity, monoclonals do not always have as high an affinity and specificity as one might expect and desire.[19] This is particularly true if a weak immunogen is used in immunization.[13] In fact, polyclonal antisera may have better avidity and specificity for a given antigenic determinant than a monoclonal antibody. This can be most readily understood by considering the affinity/avidity "bonus"[16] which polyclonal antibodies have as an inherent property. It must be remembered that a collection of polyclonal antibodies recognizes *different* epitopes on the same antigen. If a given antigen has two epitopes, for example, and the polyclonal antisera contain antibodies against both epitopes, then the *apparent* strength of the avidity of antigen-antibody binding will be significantly stronger than expected based upon the affinity of either of the two antibodies for their epitope. This is because both antibodies can bind to the same antigen, each at its appropriate epitope. The antigen-antibody complex breaks up when the antibody and antigen dissociate (usually the half-life of the formed complex is quite short — on the order of a few seconds). But in the case where two antibodies are attached to the antigen, the complex dissociates only when *both* antibodies detach simultaneously. The probability of this happening is considerably less than either one detaching separately. Thus the overall antigen-antibody complex appears to stay intact, even though there may be fewer antibody molecules attached to the antigen. This same fundamental mechanism can be invoked to explain the apparent specificity "bonus" sometimes obtained with polyclonal antibodies. The antibodies against a given antigen in a polyclonal system will be a spectrum of antibodies, each having a different affinity for the immunogen *and* a different spectrum of cross-reactivity with other antigens.[16,19] This broad spectrum of cross-reactivity tends to "dilute out" the cross-reactivity toward any *particular* irrelevant antigen. The monoclonal antibody does not share in this benefit, and if it demonstrates cross-reactivity with another antigenic epitope, there is no way to eliminate the interference.[19] It can thus be seen that in spite of the many advantages of monoclonal antibodies, a good polyclonal antisera can still be very useful. (An interesting example where a polyclonal antisera preparation proved to have more desirable properties than monoclonal antibodies prepared for the same assay was described by Brown et al.[38])

E. ANTIBODY ACTIVITY ASSAYS AND ANTIBODY PURIFICATION

Purity of the antibody is the critical parameter for immunosensor specificity. Once an antibody preparation, whether monoclonal or polyclonal, is available, some means must be employed to determine whether this preparation has antibodies which recognize the desired antigen, the affinity/avidity of the interaction, the extent of cross-reactivity, etc. Some objective assay of antibody activity must be employed. The authors will briefly discuss a suite of assay procedures with which they have had personal experience and have found to be relatively amenable to even those with no prior immunological background. This is not intended to be an exhaustive compendium of immunoassay procedures; the authors have tried to focus particularly on assays which can be adapted to detect antibody rather than antigen, as they assume most readers are interested in preparation of antigen-specific *antibodies* for use in immunosensors. The four general categories to be discussed are (1) precipitin reactions, (2) agglutination assays, (3) solid-phase immunosorbent assays (generally enzyme-linked), and (4) RIAs. The available literature on each of these topics is vast; the references are only to general methodology texts, from which the reader may obtain appropriate references to specific applications.

1. Precipitin Reactions

Good discussions of this topic can be found in Garvey et al.,[39] Roitt,[16] Fudenberg et al.,[14] and Rose and Bigazzi.[40] Precipitin reactions occur when antibody and antigen combine in such proportions (called equivalence) that a complex network or lattice of antibody-antigen aggregates forms and precipitates. With either antibody or antigen excess, such precipitable aggregates do not form. This precipitation reaction can be observed in solution; a very commonly used modification is the Ouchterlony technique, or double immunodiffusion in agar. In this procedure, antigen and antibody diffuse toward each other from antigen and antibody wells punched in semisolid agar, and at the point in the agar where the two concentrations reach equivalence, a precipitin line forms. This technique is easy (the agar plates with punched holes can be bought commercially) and can provide useful information as a qualitative screening test for presence of antigen-specific antibody. A reasonable qualitative idea of antibody purity and cross-reactivity can be obtained by examining the precipitin lines (see above references for details). The technique can also be used to determine the *relative* strength of various antibody preparations, by making serial dilutions of the preparations, and testing them against single (or a few) antigen concentrations. The dilution at which the precipitin line disappears provides a crude idea of the strength of the antibody. The Ouchterlony technique is useful, but it has distinct limitations. Its sensitivity is not great, particularly in relation to other procedures to be mentioned. A second major limitation is that the test is useless for antigen-antibody interaction where no precipitation occurs. This is the case with hapten-antibody reactions, and also with some low molecular weight antigens. Other procedures must be utilized in these cases.

2. Agglutination Techniques

General discussion of this assay's principles and practice can be found in Rose and Bigazzi,[40] Fudenberg et al.,[14] Garvey et al.,[39] and Roitt.[16]

In agglutination reactions, the general principle is the same as in precipitin reactions. In this case, a lattice-like aggregate of antibody and antigen again forms, but this time the antigen is adsorbed or covalently linked to particles or cells. The result of the aggregate formation is a clumping of particles, and the assay can be examined visually (or microscopically) for evidence of such clumping. There are two basic types of agglutination reactions: direct and indirect or passive. In the direct techniques, a cellular or particulate antigen reacts directly with an antibody. The classic example of this is the typing of red blood cell surface antigens (A, B, or O) using appropriate antisera to each of the blood cell antigen types. The passive techniques will probably be more useful to most investigators interested in immunosensor applications. In this procedure, the test antigen is coated or adsorbed onto red blood cells, latex particles, glass spheres, etc., and the coated particles are used to assay antisera preparations for agglutination activity.

The key to the success of this technique is being able to coat the cells or particles with the antigen. The most commonly used indicator particles are red blood cells; besides their ready availability, they have a variety of functional groups on their surface, lending themselves to a wide range of chemical modifications. In some instances, antigens can be simply adsorbed to the cell surface. A common technique to increase this adsorptivity is pretreatment of the cells with tannic acid. Other chemical treatments, including covalent coupling with water-soluble carbodiimides, etc., have also been used to affix the antigen to the cell surface. Once the antigen-coated cells (or particles) are prepared, they are dispensed into individual wells of microtiter plates (the authors prefer the "V" well bottom configuration, although a "U" bottom is almost as good) which also contain serial dilutions of antisera being tested for the presence of antibody. The extent of dilution of the antiserum which still produces agglutination can be taken as a measure of the amount of antibody in the antiserum preparation. This assay can also be used to monitor for cross-reactivity. Agglutination assays are more difficult to perform than precipitation assays, and are again only semiquantitative, but they have the definite advantage of increased sensitivity — i.e., levels of detection of submicrogram quantities of antibody per milliliter.[14]

3. Enzyme Immunoassays

Extensive information concerning these techniques can be found in the book by Tijssen,[19] these assays are also discussed in monographs concerning monoclonal antibodies[41] and in the series Immunochemical Techniques, Part B, Vol. 73[42] and Part E, Vol. 92 in *Methods in Enzymology*.[43]

The essence of the procedure is the immobilization of either antigen or antibody on the plastic surface of wells in a microtiter plate. Subsequently different reagents are added, depending on the assay system used, but because the reagents of interest are directly (or indirectly) immobilized to the solid surface, washings can take place between each step to remove extraneous molecules. This eliminates much of the nonspecific binding which might otherwise occur. If specific antigen is first immobilized to the wells, addition of antisera containing the corresponding antibody will result in antibody attaching to the immobilized antigen. The addition of an anti-immunoglobulin antibody for the species of animal the first antibody was prepared in (i.e., the second antibody, made in a goat against rabbit IgG, recognizes and binds to the first antibody, derived from a rabbit), and subsequent binding completes the immune "sandwich". There is one additional factor which accounts for the tremendous sensitivity of the assay. The second (or detector) antibody has an enzyme co-valently coupled to it; if any second antibody binds in the well, enzyme is therefore present and addition of an appropriate substrate will result in a color being produced in the well. Theoretically, one antibody molecule in a single well could, under appropriate conditions, be detected. Although this limit is not practically achieved, this fact of "enzyme amplification" (one enzyme molecule reacts with thousands of substrate molecules) gives the technique extreme sensitivity. The basic series of steps indicated above are susceptible of many permu-tations. A monospecific antibody (sometimes called a "capture" reagent) can be the first molecule immobilized to the well. Subsequently impure antigen can be added and simple washing may result in a single antigen being immobilized. Another antibody to this antigen (provided the antigen has more than one epitope) can then be added, followed by the enzyme-linked anti-immunoglobulin antibody as previously described. Modifications of the detector system (i.e., biotin/avidin systems) can be instituted to improve sensitivity. A variety of enzymes (e.g., alkaline phosphatase, horseradish peroxidase, urease) can be chosen depending upon the analyst's desire for a specific assay. Fluorescent substrates can, in some cases, be chosen, further enhancing the sensitivity.

These enzyme immunoassays (EIA or ELISA-enzyme-linked immunosorbent assays) have tremendous inherent advantages recommending them to the investigator. First, they are highly sensitive. Second, they are relatively easy to perform, given some minimal equipment (micro-plate washers, multitip micropipettors). Third, as a result, many assays can be run at one time. (In fact, without ELISA, it is difficult to imagine how most hybridoma screenings for monoclonal antibodies could be done, although RIA can also be used.) Fourth, the assays are (or can be) highly reproducible, and fifth, they are generally applicable to a wide variety of antigen-antibody systems. Where suitable, EIA systems have much to recommend them. Nevertheless, a few caveats must be pointed out. Hapten-antibody systems may not be suitable for EIA analysis. The hapten may not bind to the plastic surface, or if it binds, the hapten-specific antibody may not be able to interact with it. (In the authors' experience, they were unable to adapt an EIA procedure to detect an anti-benzo(a)pyrene antibody). In some measure, this problem may be overcome by using a hapten-protein conjugate. A second caveat is that, because of the large number of separate assay reagents and assay steps, truly optimizing an EIA system can be a laborious undertaking. Fortunately, for routine use, such painstaking optimization may be unnecessary.

4. Radioimmunoassay

General discussion of this procedure is in Fudenberg et al.,[14] Garvey et al.,[39] and Kennett et al.,[41] The *Methods in Enzymology* series on Immunochemical Techniques has many ex-amples.[42-44] This procedure, because of its excellent sensitivity and quantification capabilities, has proven to be widely used and popular. The usual RIA assay is applied to determining the concentration of antigen (e.g., a hormone) in a sample of interest. The same general principles can be applied to detection of an antigen-specific antibody using a radiolabeled antigen, with slight modification. The basic principle of the assay depends upon reaction of antibody with radiolabeled antigen, followed by a step to separate free from antibody-bound antigen. The effectiveness of this separation is essential to the success of the assay. The antibody may be adsorbed on plastic, thus effecting separation by washing. An anti-immunoglobulin second antibody which reacts with the first antibody may be added to precipitate the antigen-first antibody complex. Alternatively, various substances (activated charcoal, talc, zirconyl phos-phate) may be added to adsorb the free antigen, leaving the antigen-antibody complex in solution. The authors have had considerable experience with the dextran-coated activated charcoal procedure[45] and find it to be convenient and reproducible, in the cases where they have used it.

One obvious necessity for a successful RIA is to have a radiolabeled antigen. With protein antigens, this is usually not a problem as iodination with ^{125}I or ^{131}I is normally successful. With small hapten antigens, radiolabeling can be a very serious problem. As already mentioned, the RIA has the advantage of excellent sensitivity and good reproducibility. It is also amenable to processing large numbers of samples. There are, nevertheless distinct disadvantages. Perhaps the greatest of these entails working with radioactive materials in general, including the problem of expensive counting equipment, and the problem of waste disposal. Also, the availability of the radiolabeled antigen (possible expense for custom synthesis) must be investigated.

F. ANTIBODY PURIFICATION

Once a suitable antibody has been prepared, some purification procedure will almost always be found to be necessary. In general, for most immunosensor configurations, the highest density of antigen-specific antibody attainable at the sensor tip will be desirable. The presence of irrelevant proteins in the sensor tip will not only dilute the available specific antibody, but may also produce positive interferences. The purification process can conveniently be divided into two phases: (1) separation of immunoglobulin fraction from all other proteins present in the initial antiserum (or ascites or culture supernatant) preparation, and (2) separation of the antigen-specific antibody from all other irrelevant immunoglobulins.

The old, tried and true method of ammonium sulfate purification, followed by DEAE-cellulose chromatography (e.g., see Tijssen;[19] Garvey et al.[39]) will be found to be still satisfactory, although somewhat slow and laborious. Since the advent of monoclonal antibodies, techniques for purification of IgGs, in particular, have proliferated greatly, most such procedures being based on column chromatography, using ion-exchange or gel filtration (size separation) as the basic principles.[19,46] Only a partial list of the various separation procedures available is provided: J. T. Baker has Bakerbond ABx and Bakerbond MAb for antibody purification; AMF Laboratory Products has Zetachrom 60 disks; Phenomenex also supplies MEMSEP chromatography cartridges; DuPont has HPLC gel filtration columns (GF-250, GF-450, and corresponding XL columns). BioRad supplies a number of column materials such as HPHT MAPS, DEAE Affi-Gel Blue, and CM-Affi-Gel Blue. Yet another general procedure for separation of IgGs from most other proteins relies on affinity chromatography principles. An anti-IgG antibody against IgG from a given animal species can be immobilized on a chromatography packing material, and will bind to the appropriate IgG. A more generic method utilizes immobilized Protein A or Protein G. Both of these bacterial proteins bind to the Fc region of IgG, and by passage through such an affinity column, IgG is separated from other nonbinding proteins. Protein G appears to bind a wider spectrum of IgG subclasses from various species than does Protein A.

The separation of an antigen-specific antibody from other types of antibody is a far more selective process than what has been described in the foregoing. In fact, there is only one method of choice, that being affinity chromatography. The antigen is immobilized on a suitable matrix, and the antibody mixture is passed through the column. Antigen-specific binding of antibody occurs, and the antibody of interest is then recovered after dissociation from the antigen using some appropriate elution scheme.[19,39] Although the procedure is straightforward theoretically, the following technical problems can occur. There can be nonspecific binding of other proteins to the column; these can sometimes be eliminated by alternating high salt (e.g., NaCl-0.3M) and no salt washes. The antibody of interest can sometimes be difficult to elute from the column. The usual procedure involves changing the eluant pH, ionic strength (or combination thereof), or adding chaotropic reagents (e.g., SCN^-). Anti-hapten antibodies can often be eluted by using hapten-containing solutions. Finally, it should be recognized that an antigen-affinity column will inevitably produce a certain "sorting" of antibodies based inversely on affinity. Thus the lowest affinity antibodies will first be eluted, followed by those of higher affinity. The highest affinity antibodies (perhaps the most desirable) will be the most difficult to elute.[19]

IV. DIFFERENT TYPES OF IMMUNOSENSORS

The design of an FIS device depends on the type of immunoassay (homogeneous or heterogeneous assays) and on the choice of the detection technique. The instrument development is also based on the selection of the fiberoptics sensor design (e.g., light transmission onto the distal end with covalently bound antibodies or microcavity, or excitation and collection of light via the evanescent-field method). This section illustrates the different combinations of FIS designs and immunoassay procedures.

A. FLUORESCENCE IMMUNOSENSORS
1. Bifurcated Fiberoptics Instrument
An example of a portable device based on a bifurcated fiberoptics waveguide has been developed for immunofluorescence assays.[3] Figure 6 shows the schematic diagram of the device. The sensing probe, mounted at the common leg of the bifurcated optical waveguide, is designed to monitor small amounts of sample introduced through the aperture into the probe head. As little as 5 µl of liquid sample can be monitored by the instrument.

The schematic design of the sensing probe is illustrated in Figure 7. The probe is designed to be adaptable to various configurations of sensing heads. With the configuration of Figure 7(a), the internal wall surface of the probing head is coated with the immobilized bioreagent of interest; with this configuration, the sample is introduced inside the sensing probe and the front-surface fluorescence detected directly by the fiberoptic waveguide. With configuration (b), the external surface of the probing head is coated with immobilized bioreagent and the back-scattered fluorescence is detected; configuration (b) requires material for the head to be optically transparent to the UV excitation and the luminescence emission. With configuration (a), the instrument can be a multipurpose tool using a sensing probe adapted to commercially available bioreagent-coated microplates with multiple bioassay microwells.

The microwells can also be used as interchangeable sensor heads. The utility of the fiber-optic biosensor is illustrated with a model enzyme-linked immunosorbent assay (ELISA) used to measure varying amounts of rabbit immunoglobulin G (IgG) from 10 pg to 1 µg. The sensitivity of this assay is three orders of magnitude better than that obtained with a conventional ELISA absorption spectrometric assay using a commercially available reader.[3] The ELISA technique used in this procedure with the fiberoptics instrument is based on the quantitation of a fluorescent product, 4-methylumbelliferone (MUB), resulting from the enzymatic cleavage of the appropriate substrate by the enzyme alkaline phosphatase (Figure 8). This substrate, consisting of MUB covalently attached to a phosphate group, has only a very weak fluorescence. The antigens to be measured (i.e., rabbit IgG) are first bound to the microwells, which serve as the sensor heads. Enzyme-conjugated antibodies (E-Ab) specific for the bound antigen (i.e., antirabbit IgG antibody) are then allowed to react with the antigen. Since excess amounts of E-Ab are washed off, the quantities of E-Ab retained are proportional to the antigen (i.e., IgG) bound. The following step involves the addition of weakly fluorescent enzyme substrate, i.e., 4-methylumbelliferyl phosphate (MUB-P), that will be cleaved by enzymatic reaction producing the fluorescent MUB.

2. Single-Fiber Devices
Several single-fiber FIS instruments were developed previously.[2,4,6] The details of the device are provided elsewhere.[3,4] The sensor tip can have different designs having antibodies which are either covalently attached to the fiberoptics sensor tip[2,4,5] or enclosed in a membrane-drum cavity.[6-9]

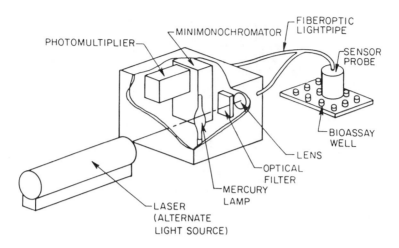

FIGURE 6. Schematic diagram of the bifurcated fiberoptics monitor for bioassays. (From Vo-Dinh, T., Griffin, G. D., and Ambrose, K. R., *Appl. Spectrosc.*, 40, 696, 1986. With permission.)

a. Covalently-Bound Antibody Sensor

The experimental procedures used to prepare optical fibers with covalently bound antibodies for fluorescence measurements were as follows: quartz fibers were stripped of their cladding for a length of 7 to 8 mm; bare fibers were then derivatized with 3-glycidoxypropyltrimethoxysilane (GOPS) using described procedures.[3,4] Following oxidation with periodic acid, the fibers were incubated for 36 to 48 h in solutions containing 2 mg/ml of rabbit IgG (for the control fibers) or 2 mg/ml crude IgG fraction from sera of rabbits immunized with BP-BSA (for the BP-antibody fibers). Phosphate buffered saline (PBS) was used as the diluent for the IgG preparations. After covalently linking the IgG protein to the fibers, the final step was reduction with sodium borohydride ($NaBH_4$). The fibers were then rinsed with PBS and ready for use.

The development of a fiberoptics chemical sensor based on the principle of competitive-binding fluorescence immunoassay has been reported.[4] Rabbit immunoglobulin G (IgG) is covalently immobilized on the distal sensing tip of a quartz optical fiber. The sensor is exposed to FITC labeled and unlabeled anti-rabbit IgG. The 488-nm line of an argon-ion laser provides excitation of sensor-bound analyte. This results in fluorescence emission at the sensing tip of the optical fiber. Sensor response is inversely proportional to the amount of unlabeled anti-IgG in the sample. Limits of detection (LOD) vary with incubation time, sample size, and measurement conditions. For 10 μl samples, typical LOD are 25 femtomoles of unlabeled antibody in a 20-min incubation period.

b. Membrane-Enclosed Sensor

Figure 9 shows an FIS sensing tip having a membrane-enclosed cavity constructed with 200- to 300-μm diameter core/cladding plastic-clad fused silica fiber.[6,8,9] Very fast cellulose dialysis membrane (7-μm thick, molecular weight cutoff = 10,000) was stretched across the face of a piece of plastic heat-shrink tubing and positioned with a band of heat shrink. The tip was assembled so that it could slide on and off the fiber and, when in place, a tight seal would form between the fiber jacket and the plastic tubing. Approximately 2 to 3 mm of bare fiber core was exposed. The plastic heat-shrink tip was tapered, resulting in an inner diameter of roughly 300 μm. The distance between the membrane and the fiber face was adjusted to about 0.5 mm, yielding an approximate sensor volume of 40 nl.[6]

The excitation sources were either an Omnichrome model 3112 10-mW He:Cd Laser or an

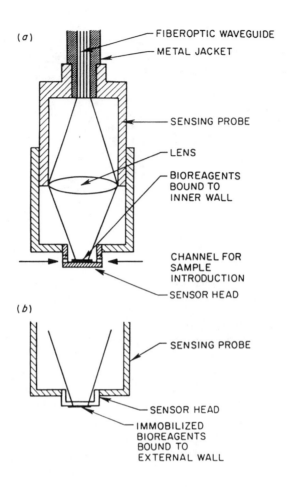

FIGURE 7. Configuration of the sensing probe of a portable fluorosensor. (From Vo-Dinh, T., Griffin, G. D., and Ambrose, K. R., *Appl. Spectrosc.*, 40, 696, 1986. With permission.)

8-mW Liconix model 4230 He:Cd Laser. Both lasers were operated at 325 nm and directed through a laser line filter having 20% transmission. Selection of the fluorescence from antigen was accomplished using a 40-nm bandpass filter having 40% transmission centered at 400 nm. A Hamamatsu model R760 photomultiplier tube biased at −1000V DC was used to detect the fluorescence emission. The photocurrent was processed with a picoammeter (Keithley model 485) and displayed on a strip chart recorder. A shutter, programmed to open for 1 s every 90 s, was used to provide controlled-duration laser excitation. This procedure helped minimize photothermal/photochemical effects.

B. TOTAL INTERNAL REFLECTION/EVANESCENT-FIELD IMMUNOSENSORS

Optical fibers utilize total internal reflection (TIR) to achieve long propagation length with very low loss, but another complementary feature of TIR can be used in optical sensing for surface sensitivity

Evanescent-field spectroscopy technique is an extension of the well-known internal reflection method. The evanescent field technique uses an optical waveguide as the transmitting medium in place of the crystal medium often used in internal reflection spectroscopy. When light is reflected at a dielectric interface, i.e., at the interface between two materials of different refractive indices (Figure 10), the energy associated with the light is not totally confined to

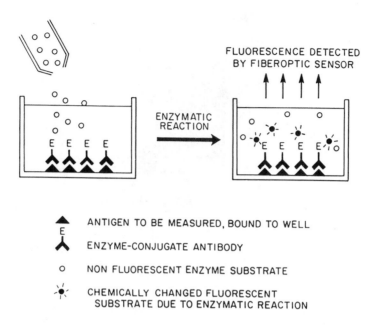

FIGURE 8. Principle of the substrate-labeled fluoroimmunoassay. (From Vo-Dinh, T., Griffin, G. D., and Ambrose, K. R., *Appl. Spectrosc.*, 40, 696, 1986. With permission.)

the material in which the incident and reflected waves are propagated. There is a drastic decrease of energy away from the reflected point into the second material. This field is known as the evanescent field, because energy cannot be propagated in this direction. It decays within a distance comparable with the wavelength of the light.[47-50]

The operating principle of evanescent-field sensors can be described by considering a light beam striking the interface between two transparent media (Figure 10). The light beam is assumed to strike from the medium with the greater refractive index ($n_1 > n_2$), total internal reflection occurs when the angle of reflection θ is larger than the critical angle θ_c.

$$\theta_c = \sin^{-1}\left(\frac{n_2}{n_1}\right) \tag{1}$$

In this case the evanescent wave penetrates a distance (x_p), on the order of a fraction of a wavelength, beyond the reflecting surface into the medium of refractive index n_2. According to Maxwell's equations, a standing sinusoidal wave, perpendicular to the reflecting surface, is established in the denser n_1 medium (Figure 10). Although there is no net flow of energy into the nonabsorbing, rarer n_2 medium, there is an evanescent, nonpropagating field in that medium, the electric field amplitude (E) of which is largest at the surface interface (E_0) and decays exponentially with distance, d, from the surface:[48,50]

$$E = E_0 \cdot \exp\left(\frac{-d}{x_p}\right) \tag{2}$$

The depth of penetration (x_p) defined as the distance required for the electric field amplitude to fall to exp (−1) of its value at the surface, is given by:

$$\frac{\lambda/n_p}{x_p} = 2\pi\left[\sin^2\theta - \left(\frac{n_2}{n_1}\right)^2\right]^{1/2} \tag{3}$$

MEMBRANE TIP SENSOR

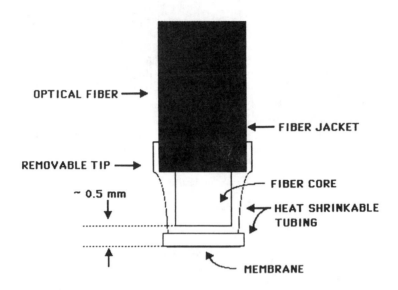

SENSOR VOLUME ~40 nL W/200-μm DIAMETER FIBER

FIGURE 9. Membrane-enclosed sensor head. (From Tromberg, B. J., Sepaniak, M. J., Alarie, J. P., Vo-Dinh, T., and Santella, R. M., *Anal. Chem.*, 60, 1901, 1988. With permission.)

As θ varies from 90°, and approaches θ_c, x_p becomes infinitely large and, at a fixed angle, increases with closer index matching (i.e., as $n_2/n_1 \to 1$). Also, because x_p is proportional to wavelength, it is greater at longer wavelengths.

Thus, one can select x_p to induce optical interaction mainly with compounds close or adjacent to the interface and minimize interferences from bulk solution, by an appropriate choice of the refractive index n_1 of the transparent support material, of the incident angle, and of the wavelength.

The assay sensitivity may be improved by combining the evanescent-wave principle with multiple internal reflections. The number of reflections (N) is a function of the length (L) and thickness (T) of waveguide and angle of incidence θ:

$$N = LT \cdot \cot \theta \qquad (4)$$

Kronig and Little[48] first suggested that this property of light reflection could be exploited to design a fluorescence-based immunosensor with few or no washing steps. An antibody is immobilized on the waveguide. A displacement, a competition, or a sandwich assay is then done with the analyte and fluorescently-labeled analogue of the analyte or fluorescently-labeled second antibody. The fluorescence is excited by the evanescent field which, as the field decays rapidly away from the interface, excites predominantly the fluorescent label in the immobilized antibody-analyte complex and excites very little of the label attached to analyte analogue or second antibody free in solution. Thus the fluorescence emitted, some which will be coupled back into the waveguide, is a measure of the amount of analyte bound. Such an optical fiber in a capillary tube using evanescent-field immunoassay has been developed by Hirschfeld.[49] The sensor has an optical rod positioned in the center of a slightly larger capillary tube. Antibodies are covalently immobilized onto the sensing surface and preloaded with fluorescently labeled antigens. The liquid sample is introduced by capillary action, and a competitive assay occurs between the preloaded labeled antigens and the sample antigens.

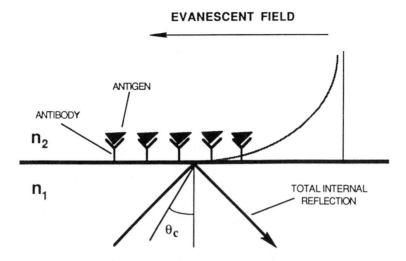

EVANESCENT FIELD

FIGURE 10. Evanescent wave at the interface between two media of refraction indices n_1 and n_2. Antibodies are shown covalently bound to the surface of the core fiber of index n_1.

Sutherland and coworkers[50] described an optical technique for detecting and monitoring antibody-antigen reactions at a solid-liquid interface. The antibody is covalently immobilized onto the the surface of either a planar or cylindrical waveguide made of fused quartz.[50] The reaction of immobilized antibody with antigen in solution is detected through use of the evanescent wave component of a light beam, which has a characteristic depth of penetration of a fraction of a wavelength into the aqueous phase, thus optically interacting primarily with substances bound or located very close to the interface and only minimally with the bulk solution. This resulting *in situ* spatial separation of the antibody-bound from free antigen precludes a formal separation step and allows the reaction to be monitored kinetically. An immunoassay for methotrexate by absorption spectrometry achieved a detection limit of about 270 nmol/l; binding of methotrexate by immobilized antibody was monitored by the decrease in transmittance at 310 nm. An immunofluorometric assay for human IgG could detect as little as 30 nmol/l; binding of fluorescein-labeled antibody was monitored by the increase in signal above 520 nm using excitation at 495 nm.

C. SURFACE PLASMON SENSORS (SPR)

Immunosensors can be based on the application of SPR principle. It is possible to arrange a dielectric/metal/dielectric sandwich such that when light impinges on a metal surface, a wave is excited within the plasma formed by the conduction electrons of the metal.[51,52] A surface plasmon is a surface charge density wave at a metal surface. When a plasmon resonance is induced in the surface of a metal conductor by the impact of light of a critical wavelength and angle, the effect is observed as a minimum in the intensity of the light reflected off the metal surface. The critical angle is naturally very sensitive to the dielectric constant of the medium immediately adjacent to the metal, and it therefore lends itself to exploitation for immunoassay. For example, the metal can be deposited as, or on, a grating; on illumination with a wide band of frequencies, the absence of reflected light at the frequencies at which the resonance matching conditions are met can be observed.

The dispersion relation for a surface plasmon is given by:[51]

$$k_{sp} = \frac{\omega}{c}\left(\frac{1}{\varepsilon} + \frac{1}{\varepsilon_m}\right)^{-1/2} \tag{5}$$

where ε_m is the real part of the dielectric constant of the metal at the given frequency, and ε is the dielectric constant for the dielectric medium outside the metal. The frequencies of interest are those where ε_m is negative. When incident light impinges on the surface at an angle θ, the wave vector of the light parallel to the surface is

$$k_x = \frac{\omega}{c} \varepsilon^{1/2} \sin \theta \qquad (6)$$

It can be derived from Equations 5 and 6 that for a negative ε_m the parallel wave vectors for the incident light (k_x) and the surface plasmon (k_{sp}) are not equal for any values of ω and θ. However, if the metal is in the form of a thin film on, e.g., a glass substrate, then the dielectric constants on the two sides of the metal are different, i.e., ε is larger in Equation 6 than in Equation 5. If the film is thin enough, light incident on the glass/metal interface will also affect the metal/air interface. In this case there exists an angle of incidence θ where Equations 5 and 6 match each other. At this θ_R surface plasmon resonance will occur.

In FIS devices the surface plasmon resonance is detected as a very sharp decrease of the light reflectance when the angle of incidence is varied. The resonance angle is very sensitive to variations in the refractive index of the medium just outside the metal film. Since the electric field probes the medium within only a few hundreds of nanometers from the metal surface, the condition for resonance is very sensitive to variations in thin films on this surface. Changes in the refractive index of about 10^{-5} are easily detected.

The surface plasmon wave penetrates in both directions normal to the interface; consequently, the incident angle or frequency at which resonance is observed is dependent on the refractive index of dielectric at the interface. Liedbert et al.[51] have shown that surface plasmon resonance can be used as the basis of a genuine reagentless immunosensor if large analytes are to be monitored. The antibody is immobilized on the metal. When a large antibody binds, it displaces solution (with a refractive index or around 1.34) by, for example, protein (with a refractive index of 1.5). The effective refractive index of the dielectric adjacent to the metal is thus changed in proportion to the amount of analyte bound and the surface plasmon resonance (incident angle or resonance frequency) is shifted accordingly. Flanagan and Pantell[52] have shown that the amount of analyte bound can be directly related to the resonance shift even when the resonance curve is distorted by scattering caused by surface roughness, thus relieving one of the constraints of precise control of metalization which would be unattractive in the mass production of cheap sensors.

D. OTHER TYPES OF IMMUNOSENSORS

Aizawa and coworkers reported the development of a luminescence-based electrochemical immunosensor.[53] The electrochemical luminescence immunoassay of the sensor was based on the following principle. The antigen, e.g., human serum albumin (HSA), was labeled with an aromatic hydrocarbon molecule such as aminopyrene. The aminopyrene-labeled HSA emits luminescence by an electrochemical technique, while the labeled HSA complexed with antibody generates less luminescence depending on the antibody concentration. This effect may be caused by the steric hindrance to the access of the aromatic hydrocarbon making it less accessible to the electrochemical luminescence detection. The sensor consisted of a potential-controllable platinum electrode which is attached to an optical fiber. The luminescence signal was detected by a photomultiplier through the optical fiber electrode. The device used the luminol label system for detection. Woodhead and Weeks described the use of chemiluminescent molecules as nonisotopic alternatives to radioactive labels in immunoassays.[54]

V. ANTIBODY UTILIZATION IN FIS DEVICES

A. IMMOBILIZED ANTIBODIES

To make a useful immunosensor, the antibody of interest must be, in some way, retained at the sensor tip. A variety of strategies can be used to this end. Whatever procedure is involved, one requirement is that the antibody, as much as is possible, retains its antigen-binding activity. Perhaps the easiest and most satisfactory procedures enclose the antibody in solution, within a semipermeable membrane cap which fits over the end of the sensor.[6,9] The analyte solution is kept separate from the antibody by the semipermeable membrane, through which the analyte of interest diffuses and then interacts with the antibody. Obviously, such an arrangement only works for relatively small analytes (antigens) which can diffuse through the semipermeable membrane (whose pores must not allow the antibody to pass through). Other potential problems could arise from diffusion limitations or absorption on the membrane. Nevertheless, the authors have found the "membrane-drum" type sensor to perform well for detection of the metabolite of benzo(a)pyrene, BPT (r-7, t-8, 9, c-10-tetrahydroxy-7,8,9,10-tetrahydrobenzo(a)pyrene) at ultra-trace levels in aqueous solution.[6,9]

There are a wide variety of procedures by which antibody may be adsorbed/linked to the fiber itself. Although simple adsorption on quartz/glass (or better, plastic) is possible, most investigators prefer a more firm anchorage, particularly when multiple washes may be envisioned. A variety of covalent linkages may be utilized — the important caveats being to (1) try to avoid denaturing the antibody during linkage, so it does not lose antigen-binding activity, and (2) try to avoid linking at the antigen-binding site, because such linkage may provide steric hindrance to antigen binding.

In the authors' laboratory a comparative study using several different procedures to attach antibody to silica beads has recently been completed.[55] The beads are first derivatized with GOPS; GOPS can also be used to derivatize quartz optical fibers. The use of this reagent introduces diol groups on the surface of the spheres. After this initial treatment, different techniques were utilized to attach antibody. In one method, HIO_4 was used to oxidize the diols to aldehyde groups, and upon addition of antibody, covalent coupling occurred through formation of the Schiff base with free primary amino groups present in the antibody protein (e.g., ε*-amino of lysine). Obviously the site of attachment on the antibody cannot be controlled. The Schiff base linkage is subsequently reduced with sodium borohydride to stabilize the linkage. In another procedure, the GOPS-derivatized beads were treated with 1,1'-carbonyldiimidazole (CDI), followed by antibody. The linkage was again through a free primary amino group on the antibody. (Note that cyanogen bromide and N-hydroxysuccinimide are also frequently used as coupling agents for binding through primary amino groups).

Another method of linkage investigated by Alarie et al.[55] involved utilizing free-SH groups on the antibody molecule. To generate these, $F(ab')_2$ fragments were prepared and the S–S bonds in the hinge region were reduced with dithiothreitol. Silica beads were derivatized with GOPS, activated with 2-fluoro-1-methylpyridinium toluene-4-sulfonate, and subsequently reacted with the reduced Fab fragments. The linkage of antibody, in this case, occurs at the SH groups in the hinge; the antigen binding site should therefore be unhindered. Alarie et al.[55] also investigated a procedure where antibody is linked to the beads through protein A binding. Silica beads having protein A on the surface were incubated with antibody, and the resulting complex was stabilized by cross-linking the antibody covalently to the protein A with dimethylsuberimidate. In this case, protein A is known to bind antibody in the Fc region, so again the antigen binding site should be free. For all different immobilization procedure, the total amount of antibody immobilized and the amount of active immobilized antibody was determined, using two antigen-antibody systems.

Not surprisingly, linkage via protein A was found to preserve antibody activity, although the amount of antibody bound was rather low in comparison to other procedures. Somewhat surprising was the fact that CDI produced large amounts of antibody bound and reasonable retention of antibody activity. The linkage via the SH group of the Fab fragment was approximately equivalent to CDI coupling. The direct linkage via GOPS was least satisfactory as there were large losses of antibody activity. A conclusion that might be drawn from this study is that random linkage on the antibody, while unattractive on theoretical grounds, may in actual practice be acceptable, probably because there are many available amino groups on the antibody surface, a large proportion *not* being in the antigen-binding site. It may still be worthwhile, however, to attempt coupling in the Fc region, simply because one should be assured of retaining the bulk of the antibody activity. In this regard, Little et al.[56] provide the interesting idea of covalent coupling at the carbohydrate moieties attached to the C_H2 portion of the heavy chain. They found enhanced antigen binding with this procedure compared to random coupling through amino groups.

Liu and Schultz[57] described a fluorescence-based optical fiber system for monitoring reactions between ligands (e.g., haptens) and their receptors (e.g., antibodies). Competitive binding reactions take place within a hollow dialysis fiber which is attached to the tip of the optical fiber. The results indicate that assay sensitivity is a function of valence (antibody and hapten), the spacing between immobilized antibody sites, and the nature of the carrier portion of a multivalent hapten.

B. AFFINITY-AVIDITY CONSIDERATIONS

The performance of immunosensors is dependent on the affinity/avidity of an antibody for its antigen. The affinity of an antibody will determine the overall sensitivity (i.e., the limit of detection will increase with increases in affinity) and specificity (specificity increases with larger differences in antibody affinity for specific and nonspecific antigen) of an analytical system based on this antibody.[19] A distinction should be made between the terms affinity and avidity. Affinity is a thermodynamic parameter based on the strength of the interaction of one antibody site with one site on the antigen.[19,41] Avidity on the other hand, is a term which expresses the strength with which an *antiserum* (a collection of antibodies with different affinities) binds an antigen (usually with multiple epitopes). It is proper to speak of affinity when referring to a monoclonal antibody which recognizes a single epitope; avidity is appropriate to polyclonal antibodies in an antiserum. Throughout the following discussion, the authors will use the term affinity as the general term for describing the strength of the antibody-antigen interaction; this should be understood as avidity in the case of polyclonal antisera.

Because the forces holding antibody and antigen together are noncovalent, there is a continuous process of association and dissociation between antigen and antibody, during which antibody and antigen may become separated. This fundamental reversibility of the antibody-antigen interaction must be grasped, i.e., that there is a constant separation and reattachment of antibody and antigen molecules as the two species interact in solution. This reaction can be written (where Ab = antibody and Ag = antigen): Ab· + Ag ⇌ Ab · Ag. The law of mass action as applied to this reversible reaction produces the following equation for the equilibrium constant: K = [Ab · Ag]/[Ab] [Ag]. This equilibrium constant K is the affinity constant. If there is strong interaction between Ab and Ag, the equilibrium will favor the [Ab · Ag] complex, the affinity constant will be relatively large, and the antibody can be said to show strong (or high) affinity. Conversely, a smaller affinity constant will mean a shift toward greater concentrations of free antibody and antigen, and a correspondingly lower affinity for the antibody.

A variety of methods may be employed to experimentally determine the affinity constant, including such procedures as equilibrium dialysis,[39] radioimmunoassay,[41] or solid-phase immunoassay[58] to list a few of many adopted procedures. Analysis of such experimentally derived data produces a calibration graph from which constants can be derived. The determination of affinity constants can be very useful when it is desirable to compare the *relative* affinities of a suite of antibodies, in order to select one with the appropriate properties. An interesting technique to compare affinities is that of van Heyningen et al.,[59] which uses a Michaelis-Menten (used in studying enzyme kinetics) type of procedure in which antibody-antigen complex concentration is plotted against antibody concentration. The concentration of antibody required to achieve 50% of the maximum antibody-antigen complex concentration is indicative of the affinity of the antibody, i.e., the lower the antibody concentration required to achieve 50% binding, the higher the affinity.

In the special case where the antigen is a hapten [H] with a single antigenic determinant, the equilibrium constant is: $k = [Ab \cdot H]/[Ab][H]$, and at a certain free hapten concentration where half the antibody sites are bound by hapten, $[Ab \cdot H] = [Ab]$ (free) and $k = 1/[H]$ (free). The affinity constant in this case is therefore the reciprocal of the free hapten concentration where half the antibody sites (whatever the antibody concentration) are saturated. Antibodies with high affinity bind hapten strongly, and require only a low hapten concentration to achieve half saturation of the antibody. As an example, very strong antibodies may have affinity constants as high as 10^{11} 1/mol;[16] the concentration of hapten required to half-saturate this antibody would be 10^{-11} mol/l. It is obvious then that antibodies with a higher affinity/avidity constant will be more sensitive in an antibody-based sensing device, in terms of limit of detection for a given antigen. Tijssen[19] provides data that illustrate this fact, as well as demonstrating that an increasingly larger amount of antigen is required to saturate an antibody to a given level as the affinity constant decreases (amount of antigen required to achieve 50% saturation increases 18-fold when K decreases from 10^8 to 10^6 1/mol). Absolom and van Oss[22] consider K values of $<10^4$ 1/mol to be weak, values of 10^4 to 10^6 to be of low to medium strength, values of 10^6 to 10^8 1/mol to be relatively strong, and values $>10^8$ to be strong to very strong.

The equilibrium constant of the antigen-antibody interaction equation can also be expressed in terms of rate constants of each half reaction, i.e.,

$$Ab + Ag \underset{ass}{k} Ab \cdot Ag$$

$$Ab \cdot Ag \underset{diss}{k} Ab + Ag$$

$$\text{So } K = \frac{k_{ass}}{k_{diss}}$$

Using appropriate methods, these rate constants have been measured in a few cases. It has been found these reactions occur extremely rapidly.[22] The k_{ass} for most haptens studied has been found to be approximately in the range 10^7 to 10^8 1/mol/s. This rate is very close to the limiting value based on diffusion processes, and says in effect that as rapidly as diffusion processes bring antigen and antibody together, reaction occurs. Protein antigens have an initial k_{ass} much lower (~100-fold) because their bulkiness inhibits diffusion.[19] The k_{diss} shows much greater variation than the k_{ass} (i.e., from 10^{-5} to 10^3/s for a variety of haptens).[22] Thus the overall equilibrium constant, K, largely depends on the k_{diss} at least for small antigens.[19] Antibodies with strong binding can have k_{diss} of 10^{-4}/s, while antibodies with low affinity can have k_{diss} of 10^3/s. The half-life of the interaction (i.e., the time required for dissociation of 1/2 the initial amount of Ab·Ag) can be calculated from the k_{diss} by the equation $t1/2 = 0.693/k_{diss}$. For high affinity antibodies (i.e., K ? 10^8 1/mol, $k_{diss} = 0.005$/s), the $t1/2 = 140$ s, while for low affinity antibodies (i.e., K = 10^5 1/mol, $k_{diss} = 80$/s), the $t1/2 = 0.009$ s.[22] It is therefore clear that a major reason for the relatively weak binding of low affinity antibodies is the extremely short lifetime of the Ab·Ag complex. A change in antibody valency (i.e., from univalent to divalent for Fab compared to IgG, for example) significantly slows the rate of dissociation of the Ab·Ag complex, thus contributing to a larger affinity constant.

An understanding of the significance of the interrelationship of the k_{diss} and the equilibrium constant is important when developing an immunosensor. As the sensing device is washed during various stages of the procedure, free antigen is removed, and the Ab·Ag complex will dissociate to some extent to reestablish equilibrium conditions. The extent of this dissociation will have important effects on the sensitivity of the device. Tijssen[19] presents theoretical data on the effect of multiple washes on the extent of antibody saturation, as a function of affinity constant, K. For antibodies with high K (10^9), two washes reduce 90% saturation to 80%. Starting with the same initial saturation, antibodies with $K = 10^7$ show a reduction to 20% saturation, while antibodies with $K = 10^6$ are reduced to <1% saturation. For these low affinity antibodies, therefore, there is little relation between the initial antigen concentration and what remains associated with the antibody after several washes.

The affinity/avidity of an antibody not only determines the strength of the antigen-antibody interaction but has important consequences with regard to antibody specificity. Antibodies are often spoken of as having cross reactivity. What this essentially means is that the antibodies (or a subpopulation of antibodies in a polyclonal antiserum) react with an antigenic epitope other than the one which induced their formation.[14] This situation is particularly apt to occur when the antibody preparation is heterologous, with many antibodies of differing combining site geometry (i.e., an antiserum). The antibodies involved in cross-reactivity identify and bind to antigenic sites structurally similar to the antigenic determinant used for immunization. Because these cross-reacting antigenic sites are only *similar* to the antigenic epitope the antibody was raised against, it is to be expected that there will be less complementarity between the antibody site and the similar antigen, and therefore the affinity constant will be lower than for the antibody-immunizing antigen interaction. (There is a theoretical possibility that antibodies with low affinity for the immunizing antigenic epitope may have higher affinity for other antigenic epitopes.) Thus the specificity of an antibody preparation can be defined in terms of its affinity constant for the immunizing antigen, compared to its affinity constant for cross-reacting antigen epitopes. Tijssen[19] defines a highly specific antibody as one for which the ratio of these two affinity constants is 1000 or greater. Although cross-reactivity is often thought of as a problem for polyclonal antisera, monoclonal antibodies can also demonstrate cross-reactivity. Also, polyclonal antisera can be highly specific, despite the fact there are subpopulations of antibodies having differing affinities.

Besides cross-reactivity, Tijssen[19] also points out that antibodies may exhibit shared reactivity. This occurs when two antigens have one or more identical epitopes. A subpopulation of antibodies from an antiserum would be expected to react with equal affinity to both antigens. If a monoclonal antibody recognized this epitope of identity, it would be unable to distinguish between the two antigens. In most cases where antibodies are being selected for use in immunosensing devices, careful selection of the appropriate immunogen structure should essentially eliminate shared reactivity (particularly with monoclonal antibodies). Cross-reactivity, however, must always be taken into account when an immunoassay is being developed.

C. REGENERABLE SENSORS/MULTIPLE MEASUREMENTS

The performance of immunosensors is determined by the Ag-Ab binding constants. A relatively low Ag-Ab constant results in reversibility but low sensitivity of the sensor. Many antibodies bind antigens very strongly and specifically, thus providing a selective but also irreversible tool for detection. There is a great interest in developing reversible or pseudo-reversible sensors rather than a "one-shot" dosimeter. One method suggested by Andrade and

CAPABILITIES OF FIS FOR
MULTIPLE MEASUREMENTS

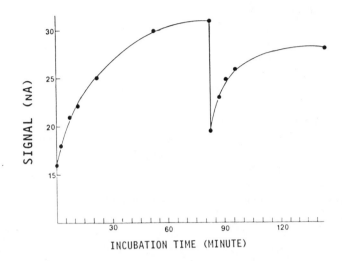

FIGURE 11. Multiple measurements with a fiberoptics immunosensor. (Taken from Vo-Dinh, T., Griffin, G. D., Ambrose, K. R., Sepaniak, M. J., and Tromberg, B. J., in *Polyaromatic Hydrocarbons: A Decade of Progress*, Cooke, M. and Dennis, A. J., Eds., Battelle Press, Columbus, OH, 1988, 885. With permission.)

coworkers[60] involves regulation of the ligand-receptor (Ag-Ab) binding constant. Also one may require a fast response time to allow continuous and semicontinuous measurements. Ag-Ab systems with high binding constants exhibit dissociation rates which are very low.[61] Sensors using these Ag-Ab systems have a very slow response time or are essentially irreversible. In order to regenerate and to reuse these sensors, it is desirable to regulate and modify the Ag-Ab binding strength to allow the complex to dissociate in a short time period.

There are several methods that allow modification of Ag-Ab binding constants. One method consists of decreasing the contribution of the hydrophobic effect to ligand binding. Reduction of this hydrophobic effect can be achieved by reducing the polarity of the bulk solvent.

Another approach is based on regulating the temperature at the Ag-Ab binding site. Thermodynamic studies of antifluorescyl antibodies have shown that temperature perturbation can be used to modify antigen-antibody binding strength.[62] Affinities increased by about 300-fold when the temperature decreased from 70 to 2°C. The effect appeared to be reversible over this temperature range.

Andrade and coworkers suggested an approach based on the regulation of the microenvironment near the Ag-Ab binding site.[60] This approach is an extension of the method of modifying the properties of the bulk solvent surrounding the Ag-Ab system. A synthetic photosensitive polymer containing azobenzene groups is coimmobilized with an antibody at a solid surface. The trans-cis isomerization of azobenzene-containing polymers can result in significant changes in the shape and size of the polymer molecule. Radiative energy from light induces a conformational change in the synthetic polymer which produces a coil expansion. The expanded polymer changes the solution microenvironment in the vicinity of the Ab binding sites, thus regulating the Ag-Ab binding process.

FIS devices with covalently bound receptors have the potential for multiple measurements. Following a first measurement, the sensor tip can be "cleaned" for future reuse by disrupting the antibody-antigen binding using an appropriate chaotropic reagent. An initial demonstration of the multiple-measurement capabilities of an FIS is shown in Figure 11. Rabbit IgG was immobilized on the fiber surface.[5] The sensor was then incubated with 1% Bovine Serum Albumin (BSA), in order to minimize the effects of nonspecific binding, followed by 1.8 mg/ml anti-Rabbit IgG labeled with FITC. Measurements for each incubation time were made in PBS after a 30-s rinse. After approximately 85 min, the sensor was placed in 3 M KSCN/PBS chaotropic reagent to disrupt the antigen-antibody association. At this point, the fluorescence signal fell nearly to baseline. Continuation of the incubation in 1.8 mg/ml anti-IgG-FITC produced a steady rise in fluorescence signal to nearly the same level as that produced by the first incubation. Limitations of the use of chaotropic reagents include the potential denaturation of antibody activity following many cleaning procedures.

D. CONTINUOUS IMMUNOSENSORS

Certain medical applications of immunosensors require *in vivo* continuous measurement of metabolic and therapeutic substances to monitor patients in critical care situations or to assess a patient's physiologic or pharmacokinetic response. Anderson and Miller recently described a fiberoptics-based sensor using a homogeneous fluorescence energy transfer immunoassay, which operates in a continuous and reversible manner to monitor the drug phenytoin.[63] B-phycoerythrin-phenytoin and Texas Red labeled antiphenytoin antibody were sealed inside a short length of cellulose dialysis tubing which was cemented to the distal end of an optical fiber. When the sensor was placed into a solution of phenytoin, the drug crossed the dialysis membrane, displaced a fraction of the B-phycoerythrin-phenytoin from the antibody, and produced a change in fluorescence signal which was measured with a fiber optic fluorometer. The sensor had a concentration response of 5 to 500 µmol/l phenytoin with a response time of 5 to 15 min. The chemical kinetics of the antibody-hapten indicator reaction were modeled mathematically and simulation showed that response time in the minutes range can be achieved when the dissociation rate constant is greater than approximately 10^{-3} s^{-1}. The ratio of the labeled and unlabeled hapten dissociation rate constants influences the analyte concentration range to which the sensor will respond.

E. TIME-RESOLVED FLUORESCENCE: STRATEGY FOR IMPROVED SENSITIVITY

In many FIA procedures, the sensitivity of detection is often limited by the background fluorescence of the biological samples. A common method adopted to minimize background fluorescence consists of isolating the fluorescence of interest by means of optical filters or monochromators. An additional approach relies on distinguishing the fluorescence emission of interest on the basis of its temporal characteristics. The fluorescence associated with many body fluids such as serum proteins has lifetimes on the order of approximately 10 nsec. The lifetimes of the fluorescent labels commonly employed are also of the order from 50 to 100 nsec. In contrast, the fluorescent lanthanide chelates, such as those of europium and terbium, provide a unique class of fluorescent labels because of their long fluorescence decay times, which are in the 10^3 to 10^4 nsec range, thus allowing easy time resolution between the background emission and the analyte fluorescence. The relatively long fluorescence lifetimes of the lanthanides provides the basis for the development of a simple time-resolved luminescence monitor.[64]

FITC is the most commonly used labeling fluorophore in fluoroimmunoassays. Though FITC has a reasonably large fluorescence quantum efficiency, it has the disadvantage of possessing a relatively small Stokes shift (approximately 30 nm). Hence, simple spectral rejection of backscatter radiation is not very efficient and large optical background levels have been observed in our previous FIS work, particularly when intense laser radiation was employed for excitation.

An alternative type of fluorescent label, rare-earth metal chelates, offers some unique and important characteristics for fluoroimmunoassays. The metal chelates which are used most often in fluoroimmunoassays are those of Eu(III) and Tb(III) complexed with β-diketones and EDTA derivatives.[64] Excitation with these labels involves absorption in the near-ultraviolet spectral region by the ligand. This is followed by energy transfer from the ligand's excited singlet state, through its triplet state, to the resonance levels of the rare-earth ion. The metal-ion emission exhibits a narrow band in the green-red visible spectral region. Concomitant with this energy transfer process is an unusually large Stokes shift (typically 200 to 300 nm) and an extremely long emission lifetime (typically 0.1 to 1.0 msec). The former characteristic facilitates efficient spectral rejection of the backscattered radiation, while the latter characteristic permits virtual elimination of optical background, including sample matrix fluorescence, by the use of time-resolved detection.

Kuo and coworkers[65] reported the use of a time-resolved fluoroimmunoassay procedure in which the long-lived fluorescence from a rare-earth chelate is measured directly in serum-containing samples. The terbium label is attached to the protein antigen via the bifunctional chelating agent (1(*p*-benzenediazonium)-EDTA. The labeled antigen is allowed to react with an immobilized antibody in a classical nonequilibrium competitive-binding immunoassay. By measuring the long-lived fluorescence of the labeled antigen remaining free in solution, could quantitatively determine the concentration of antigen present in the original serum sample. This method is illustrated by application to immunoglobulin G (IgG), a commonly occurring protein that is present in normal human serum at a typical concentration of 12 g/l (80 μmol/l).

Time-resolved detection is very important in fiber-optic fluorimetry, a technique that ordinarily exhibits large signal backgrounds from back-scattered radiation. With time-resolved detection to reject the backscattered radiation, the limit of detection for Eu 2-naphthoyltrifluoroacetonate is 10^{-12} *M*, nearly three orders of magnitude lower than for the fiber-optic measurement of the most common fluorescent label, FITC. Commercially available reagents labeled with a europium chelate were used to demonstrate the potential utility of time-resolved fluorimetry in fiber-optic immunoassays. Rabbit immunoglobulin G (IgG) was covalently bonded to the distal end of quartz optical fibers prior to exposure to anti-rabbit IgG labeled with europium chelate. The limit of detection for the assay was approximately 0.1 μg/ml.[64]

VI. EXAMPLES OF MEASUREMENTS USING IMMUNOSENSORS

A. IMMUNOSENSOR FOR THE CARCINOGEN BENZO(A)PYRENE

The application of the basic principles of immunology to a practical case can be illustrated by the authors' experience in developing a fluorimmunosensor for the polynuclear aromatic hydrocarbon, benzo(a)pyrene (BP).[2,66] Benzo(a)pyrene was selected as the model PNA compound because this important compound is found in many industrial and residential environments (chemical, petroleum, coke oven, and synfuel industries; woodburning fireplace and cigarette smoke) and because it is known to be carcinogenic. Since BP is known to be nonimmunogenic when injected, it was treated as a hapten and linked to a larger protein structure. To do this it was necessary first to prepare a chemically reactive derivative of BP, since BP is relatively inert, chemically speaking. This goal was accomplished by nitration of BP, subsequent reduction, and reaction with phosgene to form the isocyanate at the 6 position of BP.[67] The reactive BP isocyanate was then covalently bound to free amino groups of BSA and the resulting conjugate (abbreviated BP-BSA) was used to immunize rabbits. (Spectral analysis indicated seven or eight BP molecules were attached to each BSA molecule.) The immunization schedule involved initial intramuscular injections of BP-BSA in Freund's complete adjuvant (0.3 to 0.5 mg/kg body weight), followed at 2- to 3-week intervals by subcutaneous injections of BP-BSA in Freund's incomplete adjuvant (0.1 to 0.15 mg/kg body weight).

To follow the course of immunization and the development of polyclonal antisera, two assays of relative simplicity and rapidity were used. These were the Ouchterlony double immunodiffusion technique and a passive hemagglutination assay in which BP-BSA was adsorbed to the surface of sheep red blood cells.[68] The Ouchterlony assay was not able to distinguish between BSA and BP-BSA, although the hemagglutination assay did provide evidence for the differential development of antibodies to BP, as opposed to antibodies for BSA, since the antisera samples showed a higher titer against BP-BSA than BSA alone. By monitoring the anti-BP-BSA antibody levels during the course of immunization, it was found that a plateau of antibody titer was reached. Large blood samples (~30 mls) were taken from each rabbit at this point.

An ELISA assay for anti-BP-BSA activity was carried out on each separate pool of antisera, since the ELISA technique was found to be much more sensitive than the hemagglutination procedure. The antisera with highest antibody levels were subsequently assayed for anti-BP activity using a radioimmunoassay.[45] This assay employed the charcoal procedure for separation of antibody-bound and free radiolabeled BP and was found to provide a rapid, reproducible, and highly efficient (i.e., ~98 to 99% of free BP was adsorbed to the charcoal) assay methodology. Since radiolabeled BP *alone* (not conjugated to BSA) was used in the assay, positive results assured that the antibody recognized the free hapten. Furthermore, rabbit serum from nonimmunized animals showed very low BP binding in the assay, indicating that BP was not simply adsorbing to the lipid fraction of serum.

It was also possible, using the radioimmunoassay, to investigate the specificity of the antisera for the BP chemical structure. Using conditions of antigen excess, a series of assays were set up in which radiolabeled BP was increasingly diluted with a mixture of 16 different polynuclear aromatic hydrocarbons. The hydrocarbon mixture contained naphthalene, acenaphthylene, acenaphthene, anthracene, benz(a)anthracene, benzo(a)pyrene, fluoranthene, pyrene, fluorene, phenanthrene, chrysene, benzo(b)fluoranthene, benzo(k)fluoranthene, benzo(ghi)perylene, dibenz(a,h)anthracene, and indeno(1,2,3-cd)pyrene. Figure 12 shows the results of this study. It is clear from this figure that the BP antibodies present in the antisera display considerable specificity for the molecular structure of BP. If the antibodies had no specificity for BP in relation to other PAHs, then one would predict that at a molar ratio of 1:1 (total PAH:BP), the picomoles of antibody-bound BP would drop to one half the amount observed when no competing PAHs were present (i.e., to ~22 pmol). The fact that this does not occur suggests that most (or all) of the other PAHs in the competing mixture do not fit in the antigen binding site on the antibody as effectively as does BP. Even at molar ratios where there are ten times as many molecules of competing PAHs as there are molecules of BP, the amount of BP bound is still 63% of the value for BP binding in the absence of competing PAHs. This may be contrasted with a predicted decrease to 9 to 10% of the initial value (dashed curve in Figure 12), if the antibody had no specificity for BP. It is therefore concluded that the antibody produced upon immunization with a BP-protein conjugate has strong specificity for the BP molecular structure.

The IgG fraction from the antisera with the greatest anti-BP activity was isolated by ammonium sulfate precipitation and DEAE-cellulose chromatography. Attempts to further purify the anti-BP antibody fraction using an affinity (e.g., BP-BSA bound to cyanogen bromide-activated sepharose) column were unsuccessful because, although an antibody fraction bound to the column, very poor recovery resulted from attempts at elution. Antibodies to BP were produced by polyclonal techniques and were covalently attached to a fiberoptics sensing probe of an FIS. A helium-cadmium laser was used as the excitation source.

Figure 13 shows the schematic diagram of the fiberoptics immunosensor.[2] The instrument consists of an optical fiber having antibodies immobilized at the sensor tip; excitation radiation from a helium cadmium laser (325 nm) is sent through a beamsplitter onto the incidence end of the optical fiber. The laser radiation is transmitted inside the fiber onto the

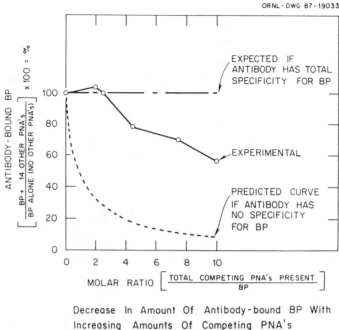

ORNL-DWG 87-19033

Decrease In Amount Of Antibody-bound BP With Increasing Amounts Of Competing PNA's

FIGURE 12. Specificity studies of antibodies against benzo(a)pyrene.

sensor tip, where it excites the analyte molecules [BP] bound to the antibodies. The excited antigen fluorescence is collected and retransmitted back to the incidence end of the fiber, directed by the beamsplitter onto the entrance slit of a monochromator, and recorded by a photomultiplier. The intensity of the fluorescence signal measured is proportional to the amount of antigen bound to the sensor tip.

In this study the measurement of BP molecules trapped on the fiberoptics tip involved a three-step procedure. (1) The fiber-optics sensor tip was immersed into a 5-μl drop of sample solution containing BP; during the incubation time, set at 10 min, the BP molecules, which diffused toward the sensor tip, were bound to the antibodies immobilized on the sensor tip. (2) Following incubation the sensor tip was removed from the sample and rinsed with a PBS solution; this operation took about 10 s. (3) To conduct the measurement, laser excitation radiation was directed to the sensor tip by opening a shutter; the fluorescence from the BP molecules excited by the laser radiation was measured for a few seconds.

Figure 14 shows the temporal response of the FIS with covalently bound Ab following incubation in a BP solution. The plateauing of the sensitivity curve after 1 h may be due to saturation of the antibodies by the BP molecules and indicated that steady-state conditions were reached. The results in this figure were obtained with a $(2 \times 10^{-7} M)$ solution of BaP in phosphate buffer/1% ethanol. After the first 10 min, the signal reached 50% of its maximum value. During the next 50 min, the fluorescence signals increased only by another 50% of the maximum value.

B. IMMUNOSENSOR FOR DNA ADDUCT PRODUCTS

With membrane-drum sensors, measurements can be obtained using a sequential and a stepwise procedure.[6,9] Sequential measurements were performed by filling the membrane sensor tip with antibody solution (typically 0.3 mg/ml) for each sample. After each measurement the sensor head was refilled with fresh antibody. Each FIS was incubated in 1-ml stirred antigen solution for a given time interval and rinsed in PBS solution for about 5 min. During this period, fluorescent antigen diffused across the membrane and was conjugated to its

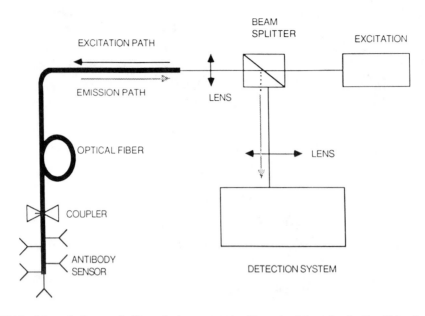

FIGURE 13.   Schematic diagram of a fiberoptics immunosensor with covalently bound antibodies. (Taken from Vo-Dinh, T., Tromberg, B. J., Griffin, G. D., Ambrose, R. K., Sepaniak, M. J., and Gardenhire, E. M., *Appl. Spectrosc.*, 41, 735, 1987. With permission.)

specific antibody. When the sensor was rinsed in PBS, unbound antigens and/or interfering substances were dialyzed out of the sensing tip. Antibody-bound material remained, confined by the membrane to the fiber's viewing region. Signal was obtained either from the slope of the signal rise or from the difference between pre- and post-incubation signals in blank PBS solutions. Signal-to-noise (S/N) values were calculated using the peak-to-peak noise of blank PBS solutions.

Due to the large amount of antibody that remains unbound during these nonequilibrium dialysis measurements, stepwise calibration data were also obtained. These assays involved loading the sensor tip with antibody and performing fixed-time incubations. Rinsing the sensor in PBS served to mark the endpoint of the measurement and the new baseline for the next analysis. Both the slope and the signal difference were used to obtain data. Sensor response was independent of the measurement sequence of antigen solutions (i.e., signals obtained going from high to low concentrations were the same as those obtained going from low to high concentrations).

Figure 15 is an illustration of the calibration information obtained with a membrane-drum sensor. Antigen solutions containing r-7,t-8,9,c-10-tetrahydroxy-7,8,9,10-tetra-hydrobenzo(a)pyrene or "BPT" were prepared in PBS. Concentrations ranged between 1.6×10^{-9} and 1.1×10^{-7} M. The sensor was filled with 0.3 mg/ml monoclonal anti-BPT antibody. Results obtained with both stepwise and sequential measurements are represented in Figure 15.

The fluorescence quantum efficiency for Ab:BPT complex was determined to be approximately the same as that for free BPT. Accordingly, the concentration of Ab:BPT was determined by comparing the postincubation FIS signal to bare fiber signals in free BPT. Each data point was obtained using a 15-min sample incubation period. The slope of the calibration plot curve, d[Ab:BPT]/d[BPT] = ~2, represents the sensor's concentrating effect after 15 min. Since the lowest detectable Ab:BPT concentration visible to the fiber is 1×10^{-9} M, the limit of detection (LOD) using 15-min incubations is simply determined by 1×10^{-9} $M/2$ ~5×10^{-10}

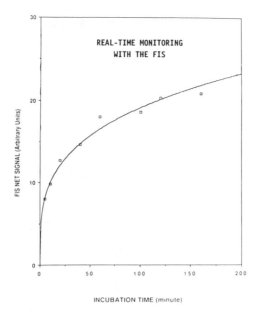

FIGURE 14. Temporal response of the fiberoptics immunosensor with covalently bound antibody. (Taken from Vo-Dinh, T., Tromberg, B. J., Griffin, G. D., Ambrose, R. K., Sepaniak, M. J., and Gardenhire, E. M., *Appl. Spectrosc.*, 41, 735, 1987. With permission.)

M BPT. Lower LODs are attainable with longer incubations due to the fact that longer incubations allow the sensor to retain more Ab:BPT complex within its viewing region. The absolute LOD is 1×10^{-9} M $\times$ 40 nl $\sim 4 \times 10^{-17}$ mol Ab:BPT, or 40 attomol.

In order to determine the specificity of the membrane-tip FIS, two structurally similar haptens were added to a standard 1.6×10^{-8} M BPT sample.[6] The standard BPT sample was "spiked" with three different amounts of 3,4-dihydroxy-1,2-epoxy-1,2,3,4-tetrahydrobenz(a)anthracene (ADE) and a single concentration of 7,8-dihydroxy-7,8-dihydrobenzo(a)pyrene or "BPD". Fluorescence signals from FIS and bare fiber measurements performed in three pure samples and four spiked samples are illustrated in Figure 16. All FIS measurements were conducted with high antibody concentrations in order to minimize the possibility of analyte competition for a limited number of binding sites.

The bare fiber intensities displayed in Figure 16 reflect the extent of spectral interference in each of the mixtures. In an attempt to differentiate between ·BPT and interferents, FIS assays involved 15 min incubations followed by 4.5 min PBS rinses. The results in Figure 16 indicate that, in the two-component BPT/ADE samples, the FIS was able to chemically discriminate between BPT and higher ADE concentrations up to two orders of magnitude (Spike 3). The absence of ADE binding was further confirmed in measurements of pure ADE. Note that ADE exists mainly in tetrol forms in aqueous solution. Pure BPD, however, did exhibit binding to the anti-BPT antibody. This cross-reactivity probably contributed to the elevated FIS signal observed in the three-component BPT/ADE/BPD sample. These results indicate that, even when spectral selectivity is not used, membrane-tip sensors can resolve multi-component samples. Furthermore, FIS resolving power is ultimately a function of the specificity of antibody:hapten complex formation.

VII. CONCLUSION

There is tremendous potential for immunosensors in a wide variety of areas. As fiberoptics

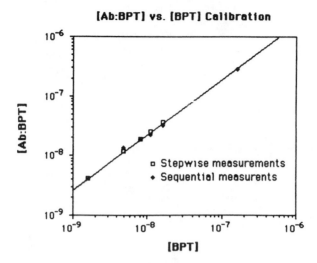

FIS incubation time = 15 minutes.

Slope = 2 = concentrating factor at 15 minutes.

LOD at 15 minutes = 5×10^{-10} M BPT.

FIGURE 15.   Calibration curve of the fiberoptics immunosensor with membrane-drum tip. (Taken from Vo-Dinh, T., Tromberg, B. J., Sepaniak, M. J., Griffin, G. D., Ambrose, K. R., and Santella, R. M., in *Fluorescence Detection II*, Menzel, E. R., Ed., SPIE, Bellingham, WA, Vol. 910, 1988, 87.)

immunosensors pass form the conceptual into the development phase, future work will be market-driven. Several areas that will benefit from the development of immunosensors include:

* Biomedical applications
* Environmental analyses
* Biotechnology
* Food and agriculture
* Industrial/military applications

   Although many technical problems yet need to be overcome, immunosensors offer powerful tools for detecting chemicals and studying biological systems. Due to their high sensitivity, immunosensors are well suited to the analysis of trace contaminants in environmental samples. Sensitivity reported for fiberoptics immunosensors are in the 10^{-8} to 10^{-12} M range. In clinical chemistry, the volume of sample available is usually small and the presence of high concentrations of proteins require exquisite specificity only available in immunological assays. Fiberoptics immunosensors can use very small amounts of sample ranging from 40 nL to a few microliters to detect attomole amounts of carcinogen-DNA adducts for early cancer diagnosis.[6,8,9] For the health care industry, fiberoptic immunosensors may bring about significant changes in the future. They may be used for continuous monitoring during critical care of patients. Immunosensors might be used to monitor therapeutic levels of drugs and give early warning of abnormal physiological changes. These devices can also be used to monitor fermentation processes and antibody production in biotechnology-based industries. They can be used in agriculture and the food industry for measuring important chemical, physiological, and biological parameters. Finally, immunosensors may be used to detect toxic chemicals in order to protect the health of workers and personnel at industrial and military installations.

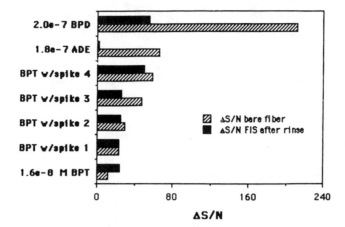

FIS INTERFERENCE STUDY

Incubation time = 15 min. Rinse time = 4.5 min.

SPIKE 1: 1.8×10^{-8} M ADE.

SPIKE 2: 1.9×10^{-7} M ADE.

SPIKE 3: 1.9×10^{-6} M ADE.

SPIKE 4: 1.9×10^{-6} M ADE and 4×10^{-8} M BPD.

FIGURE 16.　Interference studies using a fiberoptics immunosensor. (Taken from Tromberg, B. J., Sepaniak, M. J., Vo-Dinh, T., and Griffin, G. D., in *Optical Fibers in Medicine III*, Katzir, A., Ed., SPIE Publishing, Bellingham, WA, 1988.)

REFERENCES

1. **Smith, D. S., Hassan, M., and Nargessi, R. D.,** Principles and practice of fluoroimmunoassay procedures, in *Modern Fluorescence Spectroscopy*, Vol. 3, Wehry, E. L., Ed., Plenum Press, New York, 1982, chap. 4.
2. **Vo-Dinh, T., Tromberg, B. J., Griffin, G. D., Ambrose, K. R., Sepaniak, M. J., and Gardenshire, E. M.,** Antibody-based fiberoptics biosensor for carcinogen benzo(a)pyrene, *Appl. Spectrosc.*, 41, 735, 1987.
3. **Vo-Dinh, T., Griffin, G. D., and Ambrose, K. R.,** A portable fiberoptic monitor for fluorimetric bioassays, *Appl. Spectrosc.*, 40, 696, 1986.
4. **Tromberg, B. J., Sepaniak, M. J., Vo-Dinh, T., and Griffin, G. D.,** Fiberoptic chemical sensors for competitive binding fluoroimmunoassay, *Anal. Chem.*, 59, 1226, 1987.
5. **Vo-Dinh, T., Griffin, G. D., Ambrose, K. R., Sepaniak, M. J., and Tromberg, B. J.,** Fiberoptics immunofluorescence spectroscopy for chemical and biological monitoring, in *Polyaromatic Hydrocarbons: A Decade of Progress*, Cooke, M. and Dennis A. J., Eds., Battelle Press, Columbus, OH, 1988, 885.
6. **Tromberg, B. J., Sepaniak, M. J., Alarie, J. P., Vo-Dinh, T., and Santella, R. M.,** Development of antibody-based fiber-optic sensors for detection of a benzo[a]pyrene metabolite, *Anal. Chem.*, 60, 1901, 1988.
7. **Sepaniak, M. J., Tromberg, B. J., and Vo-Dinh, T.,** Fiber optic affinity sensors in chemical analysis, *Prog. Analyst Spectrosc.*, 11, 481, 1988.
8. **Tromberg, B. J., Sepaniak, M. J., Vo-Dinh, T., and Griffin, G. D.,** Development of fiberoptics chemical sensors, in *Optical Fibers in Medicine III*, Katzir, A., Ed., SPIE Publishing, Bellingham, WA, 1988.
9. **Vo-Dinh, T., Tromberg, B. J., Sepaniak, M. J., Griffin, G. D., Ambrose, K. R., and Santella, R. M.,** Immunofluorescence detection for fiberoptics chemical and biological sensors, in *Fluorescence Detection II*, Menzel, E. R., Ed., Proceedings of SPIE, Bellingham, WA, Vol. 910, 1988, 87.
10. **Ives, J. T., Lin, J. N., and Andrade, J. D.,** Fiber-optic fluorescence immunosensors, *Am. Biotechnol. Lab.*, 3, 10, 1989.

11. **Alzari, P. M., Lascombe, M. B., and Poljak, R. J.,** Three-dimensional structure of antibodies, *Annu. Rev. Immunol.*, 6, 555, 1988.

12. **Davies, D. R. and Metzger, H.,** Structural basis of antibody function, *Annu. Rev. Immunol.*, 1, 87, 1983.

13. **DePinho, R. A., Feldman, L. B., and Scharff, M. D.,** Tailor-made monoclonal antibodies, *Ann. Intern. Med.*, 104, 225, 1986.

14. **Fudenberg, H. H., Stites, D. P., Caldwell, J. L., and Wells, J. V.,** *Basic and Clinical Immunology*, Lange, Los Altos, CA, 1976.

15. **Kabat, E. A.,** Antibody complementarity and antibody structure, *J. Immunol.*, 141 (Suppl. 7), S25, 1988.

16. **Roitt, I.,** *Essential Immunology*, 4th ed., Blackwell Scientific, Oxford, 1980, chaps. 1 to 5.

17. **Sutton, B. J.,** Antigen, recognition by B cells: antibody-antigen interactions at the atomic level, *Immunol. Suppl.*, 1, 31, 1988.

18. **Taussig, M. J.,** Molecular genetics of immunoglobulins, *Immunol. Suppl.*, 1, 7, 1988.

19. **Tijssen, P.,** Practice and theory of enzyme immunoassays, in *Laboratory Techniques in Biochemistry and Molecular Biology*, Vol. 15, Burdon, R. H. and van Knippenberg, P. H., Eds., Elsevier, Amsterdam, 1985, 39.

20. **Tonegawa, S.,** The molecules of the immune system, *Sci. Am.*, 253(4), 122, 1985.

21. **Berzofsky, J. A.,** Intrinsic and extrinsic factors in protein antigenic structure, *Science*, 229, 932, 1985.

22. **Absolom, D. R. and van Oss, C. J.,** The nature of the antigen-antibody bond and the factors affecting its association and dissociation, *CRC Crit. Rev. Immunol.*, 6(1), 1, 1986.

23. **Shinnick, T. M. and Lerner, R. A.,** Predetermined antibody specificity, in *Proceedings of BioTech 84, USA*, Online Publications, Pinner, UK, 1984, 639.

24. **Green, N., Alexander, H., Olson, A., Alexander, S., Shinnick, T. M., Sutcliffe, J. G., and Lerner, R. A.,** Immunogenic structure of the influenza virus hemagglutinin, *Cell*, 28, 477, 1982.

25. **Jackson, D. C., Poumbourios, P., and White, D. O.,** Simultaneous binding of two monoclonal antibodies to epitopes separated in sequence by only three amino acid residues, *Mol. Immunol.*, 25(5), 465, 1988.

26. **Hodges, R. S., Heaton, R. J., Parker, J. M. R., Molday, L., and Molday, R. S.,** Antigen-antibody interaction. Synthetic peptides define linear antigenic determinants recognized by monoclonal antibodies directed to the cytoplasmic carboxyl terminus of rhodopsin, *J. Biol. Chem.*, 263(24), 11768, 1988.

27. **Kabat, E. A.,** Antibody combining sites: how much of the antibody repertoire are we seeing? How does it influence our understanding of the structural and genetic basis of antibody complimentarity, *Adv. Exp. Med. Biol.*, 228, 1, 1988.

28. **Mariuzza, R. A., Phillips, S. E. V., and Poljak, R. J.,** The structural basis of antigen-antibody recognition, *Ann. Rev. Biophys. Biophys. Chem.*, 16, 139, 1987.

29. **de la Paz, P., Sutton, B. J., Darsley, M. J., and Rees, A. R.,** Modeling of the combining sites of three anti-lysozyme monoclonal antibodies and of the complex between one of the antibodies and its epitope, *EMBO J.*, 5(2), 415, 1986.

30. **Colman, P. M., Air, G. M., Webster, R. G., Varghese, J. N., Baker, A. T., Lentz, M. R., Tulloch, P. A., and Laver, W. G.,** How antibodies recognize virus proteins, *Immunol. Today*, 8(11), 323, 1987.

31. **Roberts, S., Cheetham, J. C., and Rees, A. R.,** Generation of an antibody with enhanced affinity and specificity for its antigen by protein engineering, *Nature* (London), 328, 731, 1987.

32. **Fredman, P., Mattsson, L., Andersson, K., Davidsson, P., Ishizuka, I., Jeansson, S., Mansson, J.-E., and Svennerholm, L.,** Characterization of the binding epitope of a monoclonal antibody to sulphatide, *Biochem. J.*, 251, 17, 1988.

33. **Silbart, L. K., Nordblom, G., Keren, D. F., Wise, D. S., Jr., Lincoln, P. M., and Townsend, L. B.,** A rapid and sensitive screening method for the detection of anti-2-acetylaminofluorene immunoglobulins, *J. Immunol. Methods*, 109, 103, 1988.

34. **Ehrlich, P. H., Moustafa, Z. A., Justice, J. C., Harfeldt, K. E., Gadi, I. K., Sciorra, L. J., Uhl, F. P., Isaacson, C., and Ostberg, L.,** Human and primate monoclonal antibodies for in vivo therapy, *Clin. Chem.*, 34(9), 1681, 1988.

35. **Köhler, G.,** Derivation and diversification of monoclonal antibodies, *Science*, 233, 1281, 1986.

36. **Morrison, S. L., Canfield, S., Porter, S., Tan, L. K., Tao, M.-H., and Wims, L. A.,** Production and characterization of genetically engineered antibody molecules, *Clin. Chem.*, 34(9), 1668, 1988.

37. **King, S. W. and Morrow, K. J., Jr.,** Monoclonal antibodies produced against antigenic determinants present in complex mixtures of proteins, *Biotechniques*, 6(9), 856, 1988.

38. **Brown, R. C., Aston, J. P., St. John, A., and Woodhead, J. S.,** Comparison of poly- and monoclonal antibodies as labels in a two-site immunochemiluminometric assay for intact parathyroid hormone, *J. Immunol. Methods*, 109, 139, 1988.

39. **Garvey, J. S., Cremer, N. E., and Sussdorf, D. H.,** *Methods in Enzymology*, 3rd ed., W. A. Benjamin, Reading, MA, 1977.

40. **Rose, N. R. and Bigazzi, P. E.,** *Methods in Immunodiagnosis*, John Wiley & Sons, New York, 1973, 1-10; 45-70.

41. **Kennett, R. H., McKearn, T. J., and Bechtol, K. B., Eds.,** *Monoclonal Antibodies. Hybridomas: A New Dimension in Biological Analyses,* Plenum Press, New York, 1980.

42. **Langone, J. J. and Van Vunakis, H., Eds.,** *Methods in Enzymology, Vol. 73. Immunochemical Techniques Part B,* Academic Press, New York, 1981.

43. **Langone, J. J. and Van Vunakis, H., Eds.,** Monoclonal antibodies and general immunoassay methods, in *Methods in Enzymology, Vol. 92. Immunochemical Techniques Part E,* Academic Press, New York, 1983, 1.

44. **Langone, J. J. and Van Vunakis, H., Eds.,** *Methods in Enzymology, Vol. 84. Immunochemical Techniques Part D,* Academic Press, New York, 1982.

45. **Herbert, V., Lau, K.-S., Gottlieb, C. W., and Bleicher, S. J.,** Coated charcoal immunoassay of insulin, *J. Clin. Endocrinol.,* 25, 1375, 1965.

46. **Boonekamp, P. M. and Pomp, K.,** Cation-exchange chromatography-A one-step method to purify monoclonal antibodies from ascites fluid, *Science Tools,* 33(1), 5, 1986.

47. **Wolfbeis, O. S.,** Fiber optical fluorosensors in analytical and clinical chemistry, in *Molecular Luminescence Spectroscopy: Methods and Applications, Part II,* Schulman, S. J., Ed., John Wiley & Sons, New York, 1988.

48. **Kronig, M. N. and Little, W. A.,** A new immunoassay based on fluorescence excitation by internal reflection spectroscopy, *J. Immunol. Methods,* 8, 235, 1975.

49. **Hirschfeld, T. E.,** Fluorescent immunoassay employing optical fiber in capillary tube, U.S. Patent 4,447,546, 1984.

50. **Sutherland, R. M., Dahne, C., Place, J. F., and Ringrose, A. S.,** Optical detection of antibody-antigen reactions at a glass-liquid interface, *Clin. Chem.,* 30, 1533, 1984.

51. **Liedberg, B., Nylander, C., and Lundstrom, I.,** Surface plasmon resonance for gas detection and biosensing, *Sensors Actuators,* 4, 299, 1983.

52. **Flanagan, M. T. and Pantell, R. H.,** Surface plasmon resonance and immunosensors, *Electron. Lett.,* 20, 968, 1984.

53. **Aizawa, M., Ikariyama, Y., Tanaka, M., and Shiushara, H.,** Electrochemical luminescence-based optical immunosensor, *Proceedings Symposium on Chemical Sensors,* England, 362-9, 1987.

54. **Woodhead, J. S. and Weeks, I.,** Chemiluminescence immunoassay, *Pure Appl. Chem.,* 57, 523, 1986.

55. **Alarie, J. P., Sepaniak, M. J., and Vo-Dinh, T.,** Evaluation of antibody immobilization techniques for fiber optic-based fluorimmunosensing, *Anal. Chem.,* (submitted) Jan. 1989.

56. **Little, M. C., Siebert, C. J., and Matson, R. S.,** Enhanced antigen binding to IgG molecules immobilized to a chromatographic support via their F_c domains, *BioChromatography,* 3(4), 156, 1988.

57. **Liu, B. L. and Schultz, J. S.,** Equilibrium binding in immunosensors, *IEEE Trans. Biomed. Eng.,* BME-33, 133, 1986.

58. **Schots, A., Van der Leede, B. J., DeJongh, E., Egberts, E.,** A method for the determination of antibody affinity using a direct ELISA, *J. Immunol. Methods,* 109, 225, 1988.

59. **Van Heyningen, V., Brock, D. J. H., and Heyningen, S.,** A simple method for ranking the affinities of monoclonal antibodies, *J. Immunol. Methods,* 62, 147, 1983.

60. **Andrade, J. D., Liu, J. N., Herron, J., Reichert, M., and Kopeck, K.,** Fiber optic immunosensors: sensors or dosimeters, fiber optic and laser sensors IV, SPIE, Vol. 718, 280, 1986.

61. **Sela, M., Ed.,** *The Antigens,* Vol. 6, Academic Press, New York, 1982.

62. **Kranz, D. M., Herron, J. N., and Voss, E. W.,** *J. Biol. Chem.,* 257, 6987, 1982.

63. **Anderson, F. P. and Miller, W. G.,** Fiber optic immunochemical sensor for continuous, reversible measurement of phenytoin, *Clin. Chem.,* 34, 1417, 1988.

64. **Petrea, R. D., Sepaniak, M. J., and Vo-Dinh, T.,** Fiberoptic time-resolved fluorimetry for immunoassays, *Talanta,* 35, 139, 1988.

65. **Kuo, J. E., Milby, K. H., Hinsberg, W. D., Poole, P. R., McGuffin, V. L., and Zane, R. N.,** Direct measurement of antigens in serum by time-resolved fluoroimmunoassay, *Clin. Chem.,* 31, 50, 1985.

66. **Griffin, G. D., Ambrose, K. R., Thomason, R. N., Murchison, C. M., McMannis, M., St. Wecker, P. G. R., and Vo-Dinh, T.,** Production and characterization of antibodies to benzo(a)pyrene, in *Polynuclear Aromatic Hydrocarbons, Tenth International Symposium,* Cooke, M. and Dennis A. J., Eds., Battelle Press, Columbus, OH, 1988, 329.

67. **Hirata, A. A. and Brandiss, M. W.,** Passive hemagglutination procedures for protein and polysaccharide antigens using erythrocytes stabilized by aldehydes, *J. Immunol.,* 100, 641, 1968.

68. **Creech, H. J.,** Isocyanates of 3,4-benzpyrene and 10-methyl-1,2-benzanthracene, *J. Am. Chem. Soc.,* 63, 576, 1941.

Chapter 18

ORIGIN, CONSTRUCTION, AND PERFORMANCE OF AN *IN VIVO* OXYGEN SENSOR

John I. Peterson and Einar Stefansson

TABLE OF CONTENTS

ABSTRACT

The origin of the work on fiber optic chemical sensor development will be outlined, including a brief discussion of the original pH sensor. A description of the original PO_2 sensor based on fluorescence quenching will follow, including the theory, instrumentation, and a research application.

I. HISTORICAL AND THEORETICAL BACKGROUND

A. ORIGIN OF FIBER OPTIC SENSOR DEVELOPMENT
1. The pH Sensor

At the beginning of 1976, this author and Goldstein of the Biomedical Engineering and Instrumentation Branch, National Institutes of Health (NIH), were considering ways to meet the request of an NIH researcher interested in respiratory physiology, to measure the blood "gases" (pH, PCO_2, and PO_2) in the muscle of an exercising person. The electrode technology of the time did not seem to offer promise, and Goldstein asked if I thought optical fibers could be used for this purpose. I said it might be worth trying, so we embarked upon a project to develop fiber optic chemical sensors, with me working on the sensor and Goldstein developing instrumentation for it. I started work on a pH sensor, since I felt that it would be the easiest, and if I could not make one, then I would not continue further.

The pH sensor worked well[1] and was based upon the idea of a dye indicator package at the end of a pair of optical fibers, forming a miniature spectrophotometric cell. In order to fix and support the dye, it was immobilized onto polyacrylamide gel microspheres in a copolymerization. The microspheres were easily made in a batch large enough for many sensors, were hydrophilic and water and ion permeable, and the microsphere form allowed easy packing of

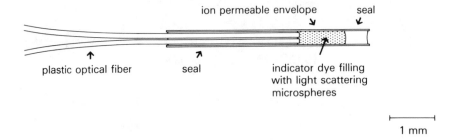

FIGURE 1. Construction of the pH sensor.

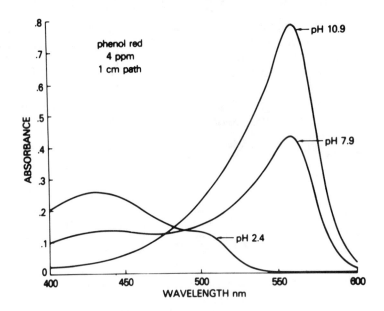

FIGURE 2. Principle of dye measurement of pH.

the sensor. Light scattering microspheres were included in the larger gel microspheres to return the light back along a fiber to the measuring instrument. The microspheres were enclosed by an ion-permeable cellulosic tube fastened to the end of the plastic optical fiber, and sealed at the end. The interior of the sensor equilibrated with the pH of the exterior by migration of hydrogen ions through the cellulosic membrane. Plastic fibers were used for reasons of safety, flexibility, and ease of use. Figure 1 shows the design of the pH sensor.

The principle of the measurement was an adaptation of the same method that had been used classically for spectrophotometric absorption measurement of pH with dyes, and applies equally to the measurement of metal ions with dye indicators. The dye in solution exists in two (or more) tautomeric forms, with the equilibrium concentration of each form depending on the pH. Each form has a different spectral absorption, for example the acid form absorbs blue light and the base form absorbs green light. In this sensor, phenol red was the dye, and the absorption of green light was measured as a function of the pH. Figure 2 illustrates the principle of using the dye to measure the pH.

The original instrument[2] was based upon illuminating the sensor through one of the fibers with green light from a lamp through a filter wheel. The green light was absorbed by the dye as a function of pH, and returned through the other fiber for intensity measurement. The relation of intensity to pH was processed by the instrument. Red light was also passed to the sensor and its returning intensity measured and used to form a ratio of green and red, to correct for optical variations, since the red was not absorbed by the dye.

The pH sensor and instrumentation were further refined and converted to needle form by Markle[3] and used at NIH. The general concept and form of the sensor provided a basis for further sensor development. Vurek at NIH demonstrated the use of the pH sensor for PCO_2 measurements.[4] The ion permeable membrane was replaced with one of silicone rubber, which made the sensor gas permeable and insensitive to ions. It was filled with phenol red and bicarbonate solution, to give a dye equilibrium determined by the PCO_2.

2. The PO$_2$ Sensor

a. Theory

This author started development work on an equilibrium, or reversible, oxygen sensor based on a dye indicator, along the same general form of construction as the pH sensor, starting in 1978. Initially, an effort was made to find a suitable optical absorbance indicator, but this approach was abandoned after it became evident that there was little hope of finding one with sufficient stability.

The phenomenon of quenching of the fluorescence of dyes by oxygen was known since the turn of this century, and had been studied as a means of oxygen measurement in gases by Kautsky and Hirsch in the 1930s.[5] It is a useful approach to oxygen analysis, having been applied to oxygen measurement in several cases,[6-15] and offers one of the few ways available for oxygen analysis. Although fluorescent dyes can be quenched by various things, especially in solution, oxygen is unique among the common ambient gases in its quenching effect. It has an unpaired electron structure which is sensitive to conversion to a paired, or singlet, structure by the absorption of quite low level of energy, and dyes activated by visible light have sufficient energy to transfer to oxygen to result in its activation.

The quantitative relation for this process had been set forth by Stern and Volmer in 1919.[16] This relation results from the assumption that the dye molecules are excited at a given rate by the illumination, and that the excited molecules lose their energy by several routes, each with its characteristic rate constant, which sum to give an overall rate of energy loss equal to the rate of activation. One of the routes of energy loss is by collision with oxygen. Energy is transferred to oxygen, converting it to singlet, or excited, oxygen, with the collision rate determined by the concentration of oxygen. The rate of energy loss by transfer to oxygen, in competition with energy loss by other routes, predominantly fluorescent decay, results in the oxygen pressure-sensitive decrease in fluorescence described by the equation. Fluorescent decay of dye molecules follows an exponential relation, like radioactive decay, with a mean time constant. The longer the mean time of the excited state, the higher the probability of collision with oxygen, and thus the more sensitive is the quench effect to oxygen concentration. The presence of oxygen decreases the apparent mean lifetime, so the Stern-Volmer equation can be stated in terms of either the ratio of emitted intensity or the ratio of apparent time constant of the unquenched dye to that of the quenched dye. For instrumentation purposes, the relation can be stated as :

$$PO_2 = G \left((U/Q)F - 1 \right)$$

where U and Q are the intensities or lifetimes of the unquenched and quenched dye, and G and F are calibration factors representing the gain and zero offset. In practice, the relation is curved and a correction must be applied.

In general, all aromatic compounds and related dyes, and some other organic compounds, show the effect of fluorescence quenching by oxygen. The simplest aromatic, benzene, has one of the highest quench constants, but it is not visually observed since the emission is in the ultraviolet. With nonfluorescent dyes, the absorbed energy is thermally dissipated before fluorescent decay can occur, the excited state lifetime being very short. With most fluorescent dyes, the oxygen quenching effect is not noticed either because of a short mean fluorescent lifetime, fluorescent decay occurring preferentially to the activation energy being lost by molecular transfer, or else the dye is not in a situation where it is exposed to efficient oxygen collision.

The latter is a common situation, which allows high brightness of fluorescent dyes in most applications. In a liquid or solid, both the solubility of oxygen and its diffusion rate are low, combining to give a low oxygen permeability, and a very low collision rate with dye molecules, thus quite weak quenching. In the classic studies of Kautsky and Hirsch, the dyes were usually supported on an inorganic adsorbent such as silica gel, in the gas phase. This provided a very sensitive effect, with the dye spread monomolecularly on a surface exposed to free gaseous diffusion. It is important that the dye be exposed to oxygen in a high permeability medium in order to observe sensitive quenching.

II. CONSTRUCTION OF FIBER OPTIC PO₂ SENSOR AND INSTRUMENTATION

A. SENSOR CONSTRUCTION

The general method of construction of the sensor was similar to the pH sensor: an oxygen permeable container at the end of an optical fiber pair, filled with a reversible dye indicator fixed on a support.

An alternate form of construction was tried, with the dye contained in a solid or liquid solution at the end of an optical fiber. This method was not suitable because the limited permeability of oxygen in solids or liquids resulted in insufficient sensitivity. The only solid material with sufficient oxygen permeability to be useful is silicone rubber, but it does not have sufficient dye solubility. The problem of making a dye with suitable solubility in silicone rubber and quench sensitivity was later solved by Marsoner et al.[17] and additional dyes were suggested by Wolfbeis and Carlini.[18]

About 70 dyes were tested to find one which met the requirements of visible light excitation, stability to aging and light exposure, minimal toxicity, good quench sensitivity, and hydrophobicity, suitable for dispersion on a hydrophobic support. Perylene dibutyrate was found to be the best available at the time.

An important requirement was the use of an organic, high surface area adsorbent material for fixing the dye in a dispersed form. Classically, inorganic adsorbents had been used, but their attraction for water made them unsuitable for use in an aqueous environment, and the quenching effect was humidity sensitive. A support which would contain a high concentration of dye was needed to have enough dye for sufficient brightness, and the dye must be dispersed monomolecularly to make it susceptible to oxygen collision. The best material was found to be a polystyrene adsorbent, XAD4 (Rohm and Haas). An oxygen sensitive packing material for the sensor was made by evaporating a solution of the dye in methylene chloride onto the adsorbent.

A suitable envelope material for containing the packing was CelGard (Celanese). This is porous polypropylene, and combines the features of high gas permeability, hydrophobicity, and mechanical strength.

The sensor was made by gluing a length of CelGard tubing onto the end of a pair of optical fibers, filling it with the packing described, and sealing the end. More extensive details are given in the original paper.[19] A current version of the sensor is based on similar construction using attachment to a single 1/2-mm diameter fiber. The single fiber sensor is enclosed in a 3/4-mm outside diameter needle, with a side window about 1 mm long near the end, for physiological applications. Figure 3 illustrates the construction of the PO₂ sensor.

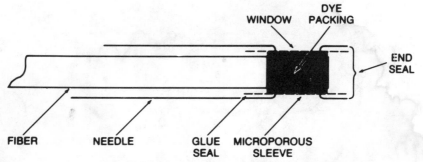

FIGURE 3. Construction of the PO2 sensor.

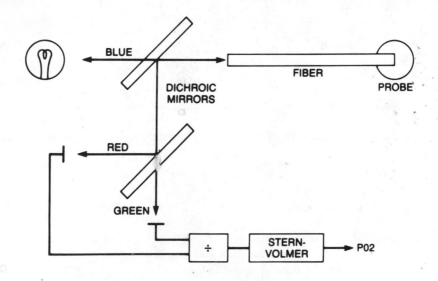

FIGURE 4. Diagram of the opto-electronic system.

B. INSTRUMENTATION

The current version of the instrumentation is based upon a computer operated system, consisting of a Macintosh of computer connected to an optoelectronic box to which the fiber optic cable connects. The computer provides power for the photodetection and light source system, and automatic control and data collection, including calibration. Figure 4 shows a diagram of the optoelectronic system.

The light from a lamp is filtered to select the blue region and directed into the fiber through a dichroic mirror. A fluorescent dye system at the sensor end returns green light, which is varied in intensity by the oxygen quenching effect, and red light, which provides optical compensation for instrumentation variations and f____ __nding. The green and red light signals are reflected by the first dichroic mirror and ___ _____ by a second dichroic mirror, and each is then filtered and converted t_ an an___ ___ __ photodetector/amplifier. The electronic signals are digitized by the compu___ __ _____ ed.

The signals are sampled a large number of ti___ ____ _ging to eliminate instrumentation noise and ambient light fluctuation. The green ___ ___ signals are sampled with the light source on and with it off, to provide a correctio___ ___ _posure of the sensor to ambient light. The combined signals are then converted by the Stern-Volmer relation using calibration factors, corrected for curvature in the response, and the experimental data are stored in a file, printed, and displayed graphically. Signal collection and processing for a data point requires 2 s.

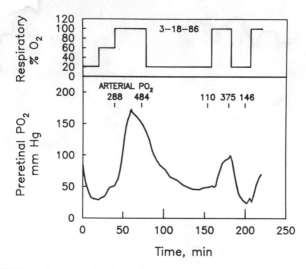

FIGURE 5a.   Response of a fiber optic PO_2 sensor to arterial PO_2 in the dog eye.

III. SENSOR PERFORMANCE AND RESEARCH APPLICATION

The system is designed to provide PO_2 readings over the 0 to 200 torr range, to the nearest torr. The sensor can be sterilized by any of the methods which do not require a temperature above about 60°C, including ethylene oxide, since higher temperatures will damage the components. A plastic fiber cable length of up to 20 m can be used. The response to a change in oxygen partial pressure is exponential, with a time constant (63% response to a step change) of about a three fourths of a minute, and about 1 min when the sensor is enclosed in a needle. A temperature sensitivity of 0.6% decrease in PO_2 indication per C increase is observed. The sensor can be used at elevated hydrostatic pressure equivalent to a water depth of 20 m. A sensor has a useful life of only a year.

The sensor has been used for measurement of the PO_2 level in dog eyes, with the sensor placed in the preretinal vitreous near the retina, or in the anterior chamber of the eye. The oxygen tension was simultaneously measured with conventional polarographic electrodes, either in the same eye or in the corresponding location in the fellow eye. The fiber optic oxygen sensor showed oxygen tension levels similar to those measured polarographically. Both sensors respond to changes in inspired oxygen concentration, showing similar levels and response times. The fiber optic sensor shows a higher oxygen level, as would be expected, since the polarographic electrode consumes oxygen and will deplete it from the surrounding volume. Figure 5a shows a typical oxygen tension in the preretinal vitreous in a dog. Figure 5b shows a similar measurement in comparison with a polarographic electrode. The inspired oxygen tension variation is shown at the top of the figure. Further details of these experiments are described in References 20 and

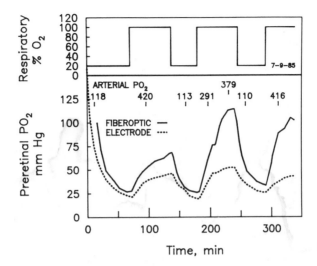

FIGURE 5b. Response of a fiber optic PO$_2$ sensor to arterial PO$_2$ in the dog eye, in comparison with a polarographic electrode.

REFERENCES

1. **Peterson, J. I., Goldstein, S. R., Fitzgerald, R. V., and Buckhold, D. K.,** Fiber Optic pH Probe for Physiological Use, *Analytical Chemistry,* 52, 864, 1980.
2. **Goldstein, S. R., Peterson, J. I., and Fitzgerald, R. V.,** A Miniature Fiber Optic pH Sensor for Physiological Use, *Journal of Biomechanical Engineering,* 102, 141, 1980.
3. **Markle, D. R., McGuire, D. A., Goldstein, S. R., Patterson, R. E., and Watson, R. M.,** "A pH Measurement System for Use in Tissue and Blood, Employing Miniature Fiber Optic Probes", in "1981 Advances in Bioengineering", presented at the Winter Annual Meeting of the American Society of Mechanical Engineers, Washington, D. C., November 15-20, 1981, ed. D. C. Viano, The American Society of Mechanical Engineers, United Engineering Center, 345 West 47th Street, New York, NY 10017, 1981, pp. 123-4.
4. **Vurek, G. G., Fuestel, P. J., and Severinghaus, J. W.,** A Fiber Optic PCO$_2$ Sensor, *Ann. Biomed. Eng.,* 11, 499, (1983).
5. **Kautsky, H.,** Quenching of Luminescence by Oxygen, *Transactions of the Faraday Society,* 35, 216, 1939.
6. **Pollack, M., Pringsheim, P., and Terwoord, D.,** A Method for Determining Small Quantities of Oxygen, *J. of Chem. Phys.,* 12, 295, 1944.
7. **Bowen, E. J.,** Fluorescence Quenching in Solution and in the Vapour State, *Transactions of the Faraday Society,* 50, 97, 1954.
8. **Orban, G., Szentirmay, Z.   .d Patko, J.,** A Highly Sensitive Luminescent Oxygen Detector, Proc. of the International Conference on Luminescence, 1966, pp. 611-13.
9. **Shaw, G.,** Quenching by Oxygen Diffusion of Phosphorescence Emission of Aromatic Molecules in Polymethyl Methacrylate, *Transactions of the Farada Society,* 63, 2181, 1967.
10. **Bergman, I.,** Rapid-response Atmospheric Oxygen M⬤⬤sed on Fluorescence Quenching, *Nature,* 218, 396, 1968.
11. **Jones, P. F.,** On the Use of Phosphorescence Quen⬤⬤rmining Permeabilities of Polymeric Films to Gases, *Polymer Letters,* 6, 487, 1968.
12. **Stevens, B.,** Instrument for determining oxygen quan   easuring oxygen quenching of fluorescent radiation, U.S. Patent 3,612,866, 1971.
13. **Knopp, J. A. and Longmuir, I. S.,** Intracellular measurement of oxygen by quenching of fluorescence of pyrenebutyric acid, *Biochim. Biophys. Acta,* 279, 393, 1972.
14. **Lübbers, D. W. and Opitz, N.,** The PCO2-/PO2-optode; a new probe for measurement of PCO2 or PO2 in fluids and gases, *Z. Naturforsch.,* 30c, 532, 1975.

15. **Wolfbeis, O. S., Posch, H. E., and Kroneis, H. W.,** Fiber optical fluorosensor for determination of halothane and/or oxygen, *Anal. Chem.*, 57, 2556, 1985.

16. **Stern, O. and Volmer, M.,** Über Die Abklingungszeit Der Fluoreszenz, *Phys. Z.*, 20, 183, 1919.

17. **Marsoner, H. J., Kroneis, H. W., and Wolfbeis, O. S.,** Sensorelement zur Bestimmung des O_2-Gehaltes einer Probe sowie Verfahren zur Herstellung desselben, European Patent Appl. 109,959, 1984.

18. **Wolfbeis, O. S. and Carlini, F. M.,** Long-wavelength fluorescent indicators for the determination of oxygen partial pressures, *Anal. Chim. Acta*, 160, 301, 1984.

19. **Peterson, J. I., Fitzgerald, R. V., and Buckhold, D. K.,** Fiber optic probe for in vivo measurement of oxygen partial pressure, *Anal. Chem.*, 56, 62, 1984.

20. **Peterson, J. I. and Stefansson, E.,** Fiber Optic Oxygen Sensor, in "Biosensors International Workshop 1987", GBF Monographs Vol. 10, Schmid, R. D., Ed., Gesellschaft für Biotechnologische Forschung mbH, Braunschweig-Stöckheim, West Germany, 1987, 235.

21. **Stefansson, E., Peterson, J. I., and Wang, Y. H.,** Intraocular oxygen tension measured with a fiber optic sensor in normal and diabetic dogs, *Am. J. Physiol.*, 256, (4 Part 2), H1127-33, 1989.

Chapter 19

BIOMEDICAL APPLICATIONS OF FIBER OPTIC CHEMICAL SENSORS

Otto S. Wolfbeis

TABLE OF CONTENTS

I. INTRODUCTION

The most obvious advantage of optical fibers in medicine and bioanalytical sciences is the possibility of having access to otherwise inaccessible regions, be it for imaging purposes (as in endoscopy) or as a light-guiding system (as in laser surgery). A quite different field of application is the use of fibers for sensing clinically important parameters such as temperature, cardiac output, pressure, pulse, and chemical para␣␣ ␣ ␣ch as pH and oxygen. This chapter focuses on the optical determination of chemica␣ ␣ ␣ ␣mical species of interest in clinical chemistry. Various reviews on the use of fibers ␣ ␣ ␣ical sciences are available,[1-4] but none of them covers the present day clinical che␣ ␣ ␣ ␣ y aspect in sufficient detail.

There are two main groups of chemical sensors: in plain fiber sensors, fibers are employed as mere lightguides to convey light from the source to the sample, and collect diffusely reflected, transmitted, or emitted light to guide it to a photodetector (see Chapter 3, Section II). Obviously, the species to be determined must have a measurable optical parameter which, for instance, may be the intrinsic color of blood. However, most parameters of interest, including oxygen and the proton, do not display intrinsic optical properties that can be exploited in plain fiber sensing.

In the second group, a sensing chemistry is placed at the end of the fiber or on its core. Interaction of the analyte with the reagent phase leads to a change in the optical properties of the reagent phase. Numerous indicator-mediated sensing schemes are known (Chapter 3, Section III) and offer a wide variety of analytical procedures for *in vivo* applications. However, only a few of those have been realized in practice so far. Both the plain fiber ("bare-ended") optrodes and the indicator-mediated optrodes will be discussed in this chapter. Unlike in many physical sensors, variations in the signal detected by chemical sensors and biosensors usually occur independently of the fibers. Changes in amplitude, phase, or frequency may be monitored.

II. PLAIN FIBER SENSING

Plain fiber sensors are more simple in design and manufacturing than are indicator-phase sensors. However, they are mostly less selective and suffer from interferences by (1) any substance that adsorbs at the same analytical wavelength, (2) even small changes in the optical parameters such as refractive index at the sensing tip, and (3) ambient light. They therefore usually are operated at at least two wavelengths and/or background subtraction.

Indicator-mediated optrodes frequently have an optical isolation at the fiber tip to prevent ambient light and sample to interfere in the optical system, and a fairly constant tip chemistry with its almost invariable refractive index. However, like in plain fiber sensors, additional discriminations such as pulsed excitation plus electronic background subtraction or sequential excitation are useful techniques in order to improve selectivity.

Plastic, glass, and quartz fibers can be used, depending on the analytical wavelength applied. Plastic fibers have a larger aperture, are flexible and easy to work with, but have poor transmission below 420 nm and do not tolerate multiple heat sterilization. Glass fibers are available in small size and have low attenuation, but have a small aperture and are suitable only down to about 380 nm. Quartz fibers transmit in the UV, but the aperture is not better than glass. Both glass and quartz fibers are fragile and present a risk when breaking *in vivo*. However, plain fiber sensors usually have sufficient biocompatibility because a broad variety of materials with proven biocompatibility is available. This contrasts the indicator-mediated sensors where material constraints often limit compatibility. One way to improve the *in vivo* performance is to cover the sensor surface with covalently bound heparin.

A. FIBER OXIMETERS

The fact that hemoglobin (Hb) and oxyhemoglobin (OxyHb) have quite different absorption spectra has resulted in the development of a widely known class of sensors called oximeters. These devices measure the oxygen *saturation* (OS) of blood by exploiting the difference in the spectra of Hb and OxyHb. This contrasts the oxygen optrodes described in Chapter 10 which measure oxygen *partial pressure*. OS is defined as the amount of oxygen carried by the Hb in erythrocytes relative to the maximum carrying capacity. Typical arterial blood is 95% saturated and venous blood may be 75%. Figure 1 shows the absorption spectra of Hb and oxyHb. Another substance whose absorption of light in the near infrared changes corresponding to its reduced and oxidized state is cytochrome aa$_3$, the terminal member of the respiratory chain.

Measurement of saturation and oxygen consumption have been used to determine cardiac output. The presence of unusually high oxygen saturation in blood samples taken from the right heart may indicate congenital abnormalities of the cardiovascular anatomy. Measurement of mixed venous saturation indicates the effectiveness of the cardiopulmonary system; low saturation may indicate reduced ability of the system to deliver oxygen from the lungs, compromised cardiac output, or reduced oxygen-carrying capability or capacity of the blood.

The development of a fiber optic technique incorporating instantaneous as well as continu-

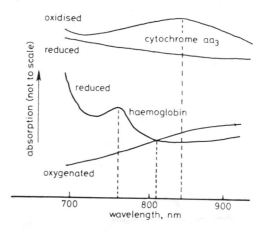

FIGURE 1. Absorption spectra of hemoglobin, oxygenated hemoglobin, and the two forms of cytochrome aa$_3$. (From Martin, M. J., Wickramasinghe, Y. A. B. D., Newson, T. P., and Crowe, J. A., *Med. Biol. Eng. Comput.*, 25, 597, 1987. With permission.)

ous measurement of OS without withdrawal of blood samples constituted a major step forward since former methods for measurement of Hb saturation suffered from numerous drawbacks. In the first version,[5] OS was obtained from the light intensities at two wavelengths (805 and 660 nm) diffusely reflected by unhemolyzed blood. Unhemolyzed blood is a highly scattering substance due to the difference in the refraction indices of red cells and plasma. The reflectivity is the same for Hb and OxyHb at 805 nm, so that it can serve as a reference wavelength where approximately 10% of the incident light are reflected. A linear relationship exists between OS and ratio of backscattered light, but the numerical values of the constants in an equation that relates OS to reflected light intensity were found to differ substantially from the data obtained in conventional reflectometry using a spectrophotometer[6] (see also Equation 1).

In a related study,[7] an instrument consisting of a rotating filter wheel in front of the fiber was used, with a photomultiplier alternately measuring the reflectivities at 640 and 805 nm. It was shown that both the respiration rate and heartbeat are clearly recorded on the output trace of the *in vivo* fiber oximeter. Also, temperature and flow velocity have a distance effect on the scattering properties of blood. In an examination of intracardiac OS in 31 patients, a standard error of only 1.1% saturation was found[8] with an instrument having a response time of 1.5 s. The time lag between the two measurements may, however, be an important source of error, so that later versions were operated at much higher alternating frequencies.

To reduce some of the noise due to the flow effects and artifacts produced by the catheter tip hitting the inner wall of the blood vessel, some "mechanical" methods have been proposed. Without seriously compromising the dynamic response, flow effects could be reduced[10] by inserting the fiber into a catheter, so that its tip is just a few millimeters from the tip of the catheter through which blood is sampled at a constant rate. To prevent artifacts produced by close contact between catheter tip and endocardium tripode-like protector on the tip of the catheter has been developed.[11]

A more detailed study on the reflectivity of Hb and OxyHb revealed an isosbestic region between 840 and 900 nm and led to the design of a new two-photocell instrument whose optical components were described in detail.[12] The apparatus has been tested for static and dynamic imbalance, and various sources of errors due to blood flow and hematocrit were discussed.

Following the advent of solid state optoelectronic components, an all solid state fiber oximeter equipped with light-emitting diodes as light sources was designed, calibrated *in vitro*, and put into intial clinical operation.[13] It was found to independent of blood temperature

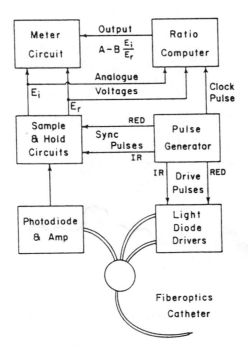

FIGURE 2. Block diagram of an LED-based fiber optic oximeter. (From Johnson, C. C., Palm, R. D., Stewart, D. C., and Martin, W. E., *J. Assoc. Adv. Med. Instrum.*, 5, 77, 1971. With permission.)

variations and hematocrit above 40%. The calibration is shifted approximately 1% per 0.1 pH unit. Figure 2 shows a block diagram of the oximeter. The pulse generator develops a 200-Hz square wave clockpulse to control the repetition rate of the oximeter, a pair of 8-μs pulses to drive the LED on alternate half-cycles of the clockpulse, and a pair of shorter pulses to control the sample and hold circuits. Light from the LEDs is tightly coupled into a branch of fibers. Reflected light is coupled to a photodiode, amplified, and delivered to the sample and hold circuits which read the pulse heights whose ratio is computed.

Figure 3 shows a typical calibration graph with a linear relationship between reflectance ratio and OS. Pulsatile flow variations of the reflectance were observed (Figure 4) and do not cancel out completely in the ratio mode. The pulsatile curves are not easy to reproduce and irregular waveforms are obtained, as can be seen in Figure 4, curves A and B, with the catheter pointed downstream. For upstream flow, the waveforms are quite different (curve C).

An extremely small fiber optic probe has been developed[14] which can be inserted through a teflon sheath percutaneously placed in a peripheral artery. A similar solid state instrument was later used for *in vivo* studies.[16] Arterial and venous OS was determined in 47 patients, and cardiac output in 130 cases, using indocyanine green as an injected dye. By using a device equipped with two photocells and operated at 640 and 920 nm, the standard deviation over the entire saturation range was 3.1% (n = 152, pig blood). Intra-arterial OS measurements have been applied mainly in the case of the critically ill.[17]

Rather than invasively, OS may as well be measured transcutaneously by registering Hb spectra via fibers. Keller and Lübbers have used a commercially available light guide oximeter that measures the reflection spectra of the skin over the 520 to 600 nm range. The microprocessor-controlled instrument consists of a rotating interference filter and the usual optical components. A fiber bundle is used to guide light to the skin and collect reflected light. The spectra of Hb and OxyHb mixtures have two peaks, the distance of which depends on OS in the 40 to 80% OS range. At an OS below 40%, the spectra flatten and a single peak of Hb develops.

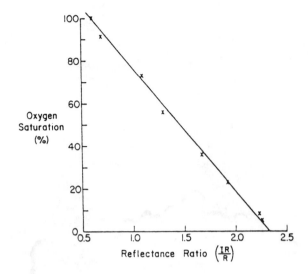

FIGURE 3. Typical calibration curve obtained with a fiberoptic catheter in human blood by measuring the ratio of reflectances at 685 nm (red) and 920 nm (infrared). Ratios of 0.62 and 2.33 were obtained at 100 and 0% oxygen saturation, respectively. (From Johnson, C. C., Palm, R. D., Stewart, D. C., and Martin, W. E., *J. Assoc. Adv. Med. Instrum.*, 5, 77, 1971. With permission.)

A major field of application of transcutaneous fiber oximetry is in neonatal intensive care. A pulsatile arterial bed (such as the patient earlobe or toe) is sandwiched between a pair of LEDs (emitting at 660 and 925 nm). A fiber bundle which sends transmitted light of both wavelengths to a photodetector. The intensity of light reaching the photodiode is determined by the pigmentation and thickness of the skin and the absorption of venous and arterial blood in the tissue. The absorbance of skin and venous blood is constant; that of arterial blood changes because of pulsatile blood flow. The constant background can be subtracted from the arterial signal in the usual way.

Ear oximeters are commercially available, e.g., from Hewlett Packard (model no. 47201 A). Since these tests are performed without fibers, they shall not be discussed here, although they are of great significance for fiber sensing as well. A list of useful references is found in a paper by Mendelson et al.[18]

Rather than measuring at two fixed wavelength, it has been shown[19] that acquiring the whole spectrum of blood using a diode array spectrometer yields even more information. Instead of reflection, the absorption is measured. It is said that derivative spectra reduce systematic error, signal averaging minimized noise, and least squares calculations allow the maximum information to be extracted from the spectra. Relative amounts of OxyHb, Hb, and carboxy-Hb can be determined within 30 s with a precision of about 15%. Alternatively, OS may be measured with a guided-wave spectroscopic sensor fabricated from low-cost optical components.[20]

In a more recent version,[21] a spectrometer has been coupled to a CCD array camera, thus permitting the recording of highly time-resolved transmittance and reflectance spectra via a quartz fiber optics. Fibers of 100 to 400 μm thickness were used in single fiber, double fiber, and fiber bundle configuration. The respective reflectance spectra vary considerably. Because blood tends to clot at the distal end of the fiber, experimental animals were heparinized (250 units per kilogram weight). In contrast to former methods for calculation of OS from reflectance data which are based on Equation 1:

$$OS = a + b \cdot R_{ref}/R \tag{1}$$

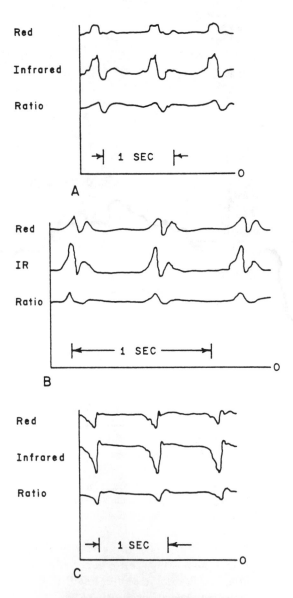

FIGURE 4. Oximeter output wave forms showing response to blood flow pulsation for fresh dog blood of Hct 66 and 100% oxygen saturation. A, catheter pointed downstream in tubing; pH 7.05. B, catheter pointed downstream; pH 7.18. C, catheter pointed upstream; pH 7.18. (From Johnson, C. C., Palm, R. D., Stewart, D. C., and Martin, W. E., *J. Assoc. Adv. Med. Instrum.*, 5, 77, 1971. With permission.)

where a and b are empirical parameters, R is the measured reflectance at the shorter wavelength, and R_{ref} the reference intensity measured at the isosbestic wavelength, the optical multichannel analyzer method described above was found to be highly dependent upon Hb concentration, the relation between a and b and Hb concentration being

$$a = a_0 + a_1 Hb + a_2 Hb^2 \tag{2}$$

and

$$b = b_0 + b_1 Hb + b_2 Hb^2 \qquad\qquad (3)$$

where a_{0-2} and b_{0-2} are true coefficients that can be obtained in a calibration procedure applying least square fits.

Several problems, such as hematocrit dependence, inaccuracy at low OS, and the need for frequent recalibration, have limited the clinical and research applications of fiber oximeters based on the simple relationship given in Equation 1. An empirical three-wavelength technique to partially compensate for these errors has been introduced,[22] and several transcutaneous oximeters which determine OS from the ratio of the magnitudes of the pulsatile components of the diffusely reflected light intensities at red and infrared wavelengths have been described.[23,24] The isosbestic reference wavelength, in this case, also compensates for tissue absorption due to skin pigmentation, skin thickness, and vasculature.

The theory, design, and initial evaluation of an intravascular sensor for measurement of OS and hematocrit (Hct) has been presented by Schmitt et al.[24] A model based on the diffusion of light in an anisotropically scattering medium is developed and used to predict the effects of physiological parameters and the source/detector configuration on the diffuse reflectance of blood. Unlike in former instrumentation, three light sources are used operating at 660, 820, and 820 nm. An implantable integrated circuit sensor with hybrid optoelectronics on a glass-capped silicon substrate is used. Figure 5 shows a sketch of the intravascular optical sensor, while Figure 6 demonstrates the effect of varying Hct.

Fiber oximeters are commercially available from Oximetric Inc. (1212 Terra Bella Ave., Mountain View, CA 94034) and BTI (4765 Walnut Street, Boulder, CO 80301). The Oximetrix 3 instrument uses three LEDs, emitting at different wavelengths, to send light down a single fiber in a disposable catheter. The reflected light returns through a receiving fiber to the instrument, is detected, and then analyzed by a microprocessor in real time. The state of the art up to 1983 has been reviewed.[25] More recently, Domanski et al.[26] have developed devices for simultaneous determination of Hct and OS using three LEDs and a specially designed optoelectronic head in contact with blood drawn from the body using a syringe catheter. A simple method for Hb determination was presented[27] which is based on the knowledge of Hct. It was examined in hundreds of patients.

Finally, it should be mentioned that another important application of fiber oximeters is the measurement of dye concentration in blood. In this technique, a dye is injected into the circulatory system at a given instant and at a given location. The concentration of the dye as a function of dye (the "dye dilution curve") is recorded. The dye (e.g., indocyanine green) is usually chosen such that it absorbs only at one wavelength, so that the signal at the second wavelength serves as an internal reference. Typical wavelength are 805 and 930 nm, and the ratio of the reflectivities at these two wavelengths (R_{900}/R_{800}) was shown to be linear up to 20 mg/l indocyanine green. Obviously, however, the injected dye interferes (at levels above 1 mg/l) in the determination of OS at these wavelengths. More recent instrumentation is designed to make measurements at 660, 805, and above 900 nm, so that both OxyHb and dye concentration can be measured simultaneously by proper interpretation of the light signals. Rather than by reflectometry, the dye may be determined via its intrinsic fluorescence. A discussion of this subject is, however, beyond the scope of this book.

B. MISCELLANEOUS PLAIN FIBER SENSORS FOR CHEMICAL AND BIOCHEMICAL SPECIES

The intensity of reflected or transmitted light at specific wavelengths in the visible and near-infrared can provide useful information concerning the metabolic state of tissue.[28] Fiber optics can play an important role in this research area. Metabolism can also be studied at an early stage in the respiratory chain by monitoring the redox state of NADH. The redox

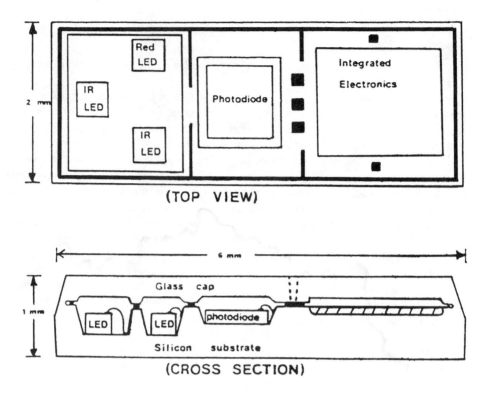

FIGURE 5. Top and side view of an LED-based implantable oximeter. (From Schmitt, J. M., Meindl, J. D., and Mihm, F. G., *IEEE Trans. Biomed. Eng.*, 33, 98, 1986. With permission.)

potential E_h of the NAD$^+$/NADH system at pH 7.0 is known to be $0.320 \pm 0.03 \times \log([\text{NAD}^+]/[\text{NADH}])$ V and reflects the redox status of the tissue. NADH displays strong fluorescence at 455 nm when excited at around 340 to 350 nm through a quartz fiber. NAD$^+$, in contrast, is nonfluorescent. Very small areas of tissue can be examined so that metabolism at a localized level can be followed.[29-31]

NADH is favorably excited by a nitrogen laser system in combination with single-strand fibers.[31] In a typical experiment, the 337-nm nitrogen laser output is split into two beams: one beam is focused onto the incident end of the optical fiber and serves to excite NADH. The other is used to pump a dye laser whose output at 805 nm is also focused onto the incident end of the fiber. The intensity of the reflected 805-nm line (and, in later work, the 586-nm line) can be used to correct for blood-induced disturbances of the rat heart. The metabolic state of the moving heart was determined in a similar way. Applications of this procedure include (1) pharmacological studies by monitoring the protective effect of drugs (such as pentobarbital) in myocardial ischemia, (2) evaluation of revascularization procedures by monitoring NADH for estimation of the reversibility of ischemic injury, and (3) myocardial protection during cardiopulmonary bypass.

A fiber sensor for continuous monitoring of NADH in culture broths of bioreactors is available from Ingold AG (Switzerland). Another class of substances whose absorption of light in the near-infrared changes corresponding to its reduced or oxidized state are in cytochromes, the terminal members of the respiratory chain (Figure 1), and the oxidized and reduced flavoproteins.

An elegant way to suppress straylight in plain fiber fluorimetry of biological samples is to make time-resolved measurements in the nanosecond time regime. In a typical arrangement,[32] light pulses from a spark source are selected in an optical filter and passed through a beam splitter. The exciting beam then enters a light guide, the distal end of which contacts the

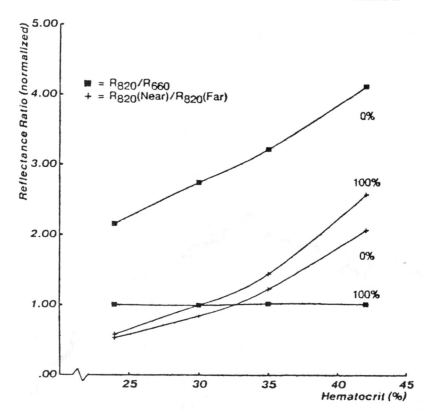

FIGURE 6. Plots of experimental data which show the effect of Hct on the infrared and near far-infrared reflectance ratios as a function of oxygen saturation (OS). Ratios are normalized at 100% OS and Hct 30%. (From Schmitt, J. M., Meindl, J. D., and Mihm, F. G., *IEEE Trans. Biomed. Eng.*, 33, 98, 1986. With permission.)

sample. Fluorescence, scattered light, and reflected light is conveyed back to the beam splitter where part of it is reflected into the detector and registered after selection in a secondary filter. The time difference between light source and detector pulses is measured with a delayed coincidence system. The method is not applicable to concentration determination of fluorescent biomolecules of fluorescent biological species such as NADH or FAD, but to probe minor changes in the microenvironment of such species.

Plain fibers also have been used to perform *in vivo* measurements of drugs in minute volumes of body fluid by using conventional methods of absorptiometry[33] and fluorometry,[34] but also of more sophisticated methods such as two-photon-excited fluorometry, and sequentially excited fluorometry.[34] Detection limits of approximately 0.5 μM were determined for the drug adriamycin (doxorubicin) in interstitial fluid and whole blood and are comparable for the three techniques. In constructing the sensor, the protective coating and optical cladding of a 200-μm quartz fiber was stripped off to about 2 cm from its terminus. Then, a piece of a close fitting glass capillary tubing was slid over the stripped fiber and attached to a modified 20-gauge needle with epoxy. The available capillary volume was about 200 nl. Using this sensor, virtually all fluorescence emanated from the first millimeter of the solution in front of the fiber tip. In an alternative version, the fiber may be threaded into a blunt-ended 26-gauge needle of a syringe. When the fiber is withdrawn a certain distance, a defined sample chamber is created within the needle. Figure 7 shows a schematic of the experimental arrangement. The sensor was also used for pharmacokinetic studies.

Similarly, bilirubin and methotrexate can be determined in serum with a fiber optic system terminated in a 19-gauge hypodermic needle and a reflective cap at its end so to produce a small absorbance cell.[35] Although these methods usually do not yield absolute analyte concen-

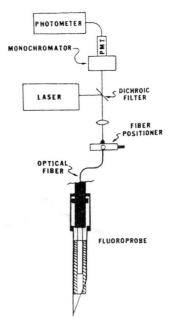

FIGURE 7. Experimental arrangement for performing *in vivo* fluorometry with a plain single-strand fiber inserted into the body through a standard needle catheter. The volume "seen" by the fiber in around 200 nL. (From Sepaniak, M. J., *Clin. Chem.*, 31, 671, 1985. With permission.)

trations due to background absorption or fluorescence of serum (see Chapter 3, Figure 30), they do sufficiently correctly reflect relative concentration changes.

A concept of the fiber optic based absorbance sensor is shown in Figure 8. Again, the fiber is threaded in a standard catheter, thus allowing its insertion into tissue or body fluids. A piece of aluminum foil is attached to the end of the inner needle (which contains the optical fiber). Fluids can be drawn into the sample irradiation cavity by aspiration, the volume between foils and fiber being filled through the hole shown in the figure. Typical pathlengths (twice the distance from fiber tip to foil) are 0.5 to 4.3 mm.

Further applications of plain fibers for biochemical *in vivo* analysis involve direct fluorimetry of the animal brain[36] and a device for measuring calcium(II) transients in the cerebral cortex of cats.[37] The calcium-binding enzyme aequorin is introduced through a thin channel into the extracellular space in a reservoir-type optrode (see Chapter 3, Section V.G). The chemiluminescence emitted by aequorin after interaction with calcium can be monitored with an optical fiber and is related to the actual Ca(II) activity. The probe was used to study epileptic seizures.

A plain fiber system has been described[38] for analysis of carbon monoxide, carbon dioxide, and oxygen bound to Hb or dissolved in blood. The fiber is placed at a critical point such as the aorta, and infrared light of wavelength near 5.13, 4.3, and 9.0 μm, respectively, is passed through a small dimension of blood. Measuring the respective transmission at these wavelengths (which are at the NIR overtones of the respective IR absorptions) gives a parameter for CO, CO_2, and O_2, respectively. Similarly, small concentrations of heavy water (D_2O) can be detected in the bloodstream by inserting a fiber and measuring the absorption by D_2O of the 4-μm infrared light which is fairly specific for D_2O and is not absorbed by H_2O and blood components.[39]

III. INDICATOR-MEDIATED SENSORS

Plain fiber sensors can only be applied to species that have an intrinsic optical property that can be sensed via fibers. Unfortunately, most species of interest in diagnosis are not susceptible to direct photometry. Particularly pH, oxygen, carbon dioxide, potassium, and glucose

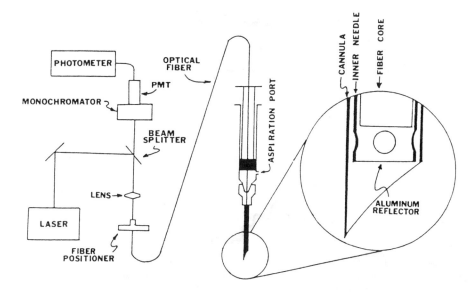

FIGURE 8. Fiber optic catheter for invasive absorbance measurement, and associated instrumentation. (From Coleman, J. T., Eastham, J. F., and Sepaniak, M. J., *Anal. Chem.*, 56, 2246, 1984. With permission.)

are not directly measurable so that indicator-mediated optosensors have to be used. The following is a review on the use of such sensors in biomedical sciences.

A. RESPIRATORY GAS ANALYSIS

Analysis of inspired and expired gases plays an important role in respiratory physiology, in lung function diagnosis, and in supervision of the critically ill. It is known that severe trauma, shock, or sepsis result in abnormal metabolic activity, leading to modifications in oxygen consumption and carbon dioxide production. Monitors to make on-line determinations of gas exchange therefore are desirable clinical tools. Usually, oxygen consumption and CO_2 production are measured in an open respiratory circuit.

Monitoring inspired and expired gas requires fast-responding devices. Oxygen has been determined by mass spectrometry, paramagnetic resonance, or by absorptiometry. Fluorescence-based oxygen optrodes (Chapter 10) offer a promising alternative. Kroneis[40] has developed a method for on-line monitoring gases using a highly oxygen-sensitive dye [benzo(ghi)perylene] dissolved in a very thin silicone rubber membrane that is placed on a planar glass plate (Figure 9). Separation of excitation light and fluorescence emission is accomplished by means of total internal reflection. The system has a very fast response (t_{90} in the order of 30 ms; see Figure 10) for both directions, and its accuracy compares favorably with mass spectroscopic methods, with a correlation coefficient of $r = 0.9998$. There is no interference by water vapor, carbon dioxide, nitrous oxide, enflurane, and ethrane. Halothane, in contrast, heavily interferes in acting as a quencher as well. As a matter of fact, this effect has been used to determine both oxygen and halothane, using a two-sensor technique.[41]

Figure 11 shows the results obtained in breath-by-breath analysis of oxygen and CO_2. Because CO_2 optrodes at present are too slow for application in respiratory gas analysis, an IR-based CO_2 monitor (Datex Normocap) was used. A small portion of the inspired and expired gas coming from a mouthpiece was bypassed through both sensor systems at a flow rate of 150 ml/min. Obviously, the profiles are identical for both detection systems, indicating the favorable properties of the optrode system. The signal depends on the pulse. The standard deviation of the sensor is 0.02% oxygen (n = 12) in the 0 to 21% oxygen range, and 0.04% in the 21 to 50% range. For an analysis of errors in determination of respiratory gas exchange, see the excellent paper of Ultman and Burzstein.[42]

Other experiments using oxygen optrodes in respiratory analyzer have been reported by

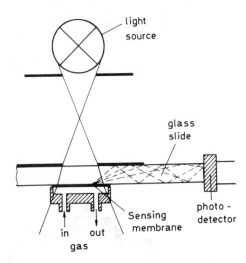

FIGURE 9. Schematic of a fast responding oxygen sensor for respiratory gas analysis. The light source excites the fluorescence of a sensing membrane placed in the gas flow-through cell. Emitted fluorescence, which is quenched by oxygen, is guided to the photodetector by total internal reflection and thereby efficiently separated from the exciting light whose angle of incidence does not match the condition of total reflection.

Bacon and Demas[43] and Barnikol et al.[44,45] Response times as fast as 20 ms have been reported,[45] with a dead space of 100 ml and almost no cross-sensitivity from the usual gases. It is even said to be insensitive towards water vapor (which usually interferes in surface-adsorbed dye sensors) and even halothane.[45] The sensor construction is simple in that a dye such as tetraphenyl-porphyrin (with its longwave excitation and emission wavelengths) is adsorbed on hydrophobic particles which then are placed on a gluing tape in a single layer. The influence of humidity was eliminated by using hydrophobic supports for the dye and heating the sensing layer to 45°C. The device has been applied to measure the bronchial volume of guinea pigs and to study the behavior of the bronchial system when applying drugs.[46]

B. TRANSCUTANEOUS AND TISSUE SURFACE ANALYSIS OF OXYGEN AND CARBON DIOXIDE

Rather than taking blood samples and performing gas analyses, or inserting a sensor into blood vessels, it has been shown to be possible to measure oxygen and carbon dioxide pressure transcutaneously.[47] Since, however, conventional methods for oxygen determination consume oxygen (and therefore may introduce undesired pressure gradients), optical methods based on fluorescence quenching are potentially superior because there is no analyte consumption being involved. Generally, measurements are performed at elevated skin temperatures, typically 41 to 43°C, so to improve gas permeability.

It is surprising to see that experiments on the use of oxygen optrodes for transcutaneous oxygen measurement have not been published yet, although some experiments have been performed.[48] Both pCO_2 and pO_2, however, have been determined on the surface of the isolated guinea pig heart.[49] A thin indicator layer containing an oxygen-sensitive fluorescent indicator and covered with a 6-μm teflon membrane was brought into close contact with the heart. Hermetic sealing was achieved using a tissue glue (Figure 12). The area that could be sensed was around 6 mm in diameter. The oxygen-sensitive material was a viscous solution of pyrenebutyric acid in dimethylformamide stabilized with agarose particles. The CO_2 optrode was a solution of 4-methylumbelliferone in bicarbonate buffer, with agarose added as mechanical stabilizer. Both sensing materials were placed on a quartz window and covered with black teflon. Response times as short as 2 to 3 s (t_{90}) for oxygen and 3 to 4 s for CO_2 were reported.

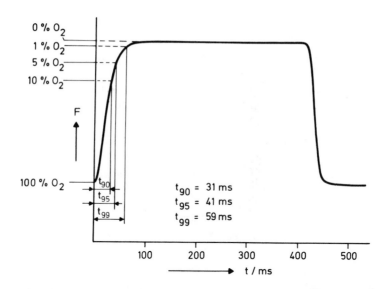

FIGURE 10. Response time of the oxygen sensor as used in analysis of respiratory gases. (From Kroneis, H. W., Unpublished results, 1983.)

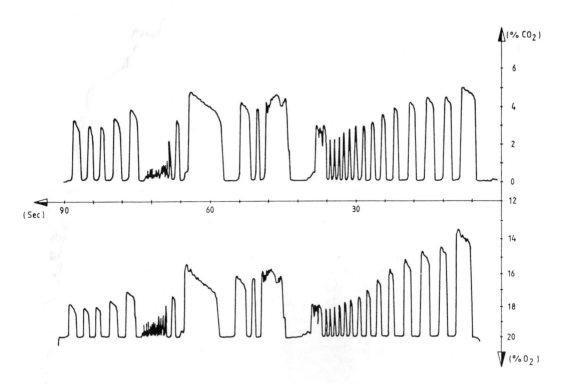

FIGURE 11. Typical trace for on-line analysis of oxygen (bottom, measured with an oxygen optrode) and carbon dioxide (top, via ir spectroscopy). (From Kroneis, H. W., Unpublished results, 1983.)

Both pCO_2 and pO_2 could be monitored at the tissue surface using the two-wavelength referencing method to account for interferences by small positional changes and some stray-light. The performance of the oxygen sensor compares favorably with that of an oxygen electrode placed in the right ventricle of the heart, and both calibration curves showed linearity

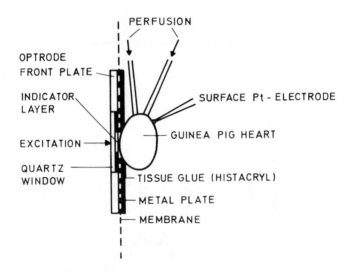

FIGURE 12.   Schematic of the experimental arrangement for optically measuring surface pO_2 of the isolated guinea pig heart. (Redrawn from Opitz, N., Weigelt, H., Barankay, T., and Lübbers, D. W., in *Oxygen Transport to Tissue*, Vol. 3, Silver, I. A., Erecinska, M., and Bicher, H. I., Eds., Plenum Press, New York, 1978, 99. With permission.)

over a wide range. Obviously, tissue (background) fluorescence does not seriously interfere. This may probably be further improved by choosing indicators with analytical wavelengths in the visible where tissue fluorescence is much weaker. Model experiments have also been performed to compensate for movements of the heart which causes a periodical modulation of the optical signal.[50]

C. *IN VIVO* SENSING OF pH, pO_2, AND pCO_2: EARLY EXPERIMENTS

The considerable interest in optrode-type invasive sensors results from their minute size, spatial flexibility, and the lack of reference cell and electrical links to the body. Needle electrodes mostly have large shafts and can cause traumatic damage to the tissue, especially during exercise. In addition, they require implantation of a reference electrode, are fairly expensive when used as disposables, and are more likely to cause clotting than are optrodes which can be fabricated from biocompatible materials. One alternative to in-dwelling sensors, which is withdrawing many blood samples and performing measurements *in vitro*, requires skillful technique to avoid artifacts and is far more labor-intensive. Moreover, blood withdrawal as a means of studying tissue is not practical in poorly perfused tissues such as muscle.

Plastic fibers offer the possibility of going into the body. They have a breaking strength in the order of 1000 kg cm^{-2}, or 0.5 kg for a 250-μm fiber. Glass fibers are stronger. The allowable bend radius is limited by loss in short bends. An important feature of most fiber sensors is that the indicator system is reversible, i.e., it is in an equilibrium, as opposed to diffusion-rate dependent measurements upon which some membrane-selective electrodes such as the oxygen electrode rely. This provides long-term stability of calibration.

1. *In Vivo* pH Sensing

Blood pH measurements present unique challenges due to the lack of a method to determine accuracy[51-53] although an international recommendation for blood pH measurements exists.[54] Blood pH, by convention, is defined as the pH indication of a glass electrode, calibrated in standard buffer solution, when in contact with blood. While it remains uncertain whether blood indeed has this pH, *the glass electrode is industry standard for the measurement of in vitro blood pH* and is the source for the existing body of data on clinical blood pH. It is used in the commercial clinical pH analyzers.

Unlike the measurement of the partial pressures of oxygen and carbon dioxide, where accuracy can be verified by tonometry of the blood with primary gas standards, there is no method available to determine the accuracy of a blood pH measurement. This is the result of the inherent difficulty of computing or measuring the hydrogen ion activity except in simple dilute solutions. Therefore, the clinical goal in pH measurements has been to achieve precision and consistency in the measurements in terms of a practical acidity scale rather than to lay claim to accuracy. This is strived for by the pragmatic use of an "operational definition of pH" obtained by the use of accepted pH standard buffer solutions.

Because pH optrodes do not require a reference element and a variable liquid junction potential is avoided, it appears that it is possible to measure absolute pH values. This, however, is not correct. The inherent inaccessibility of absolute chemical activities of single ions is buried in the effects of variable ionic strength on the pK_a of the indicators. A variable Donnan potential across the membrane separating the indicator layer from the sample also represents a source of error.

At about the time when the first experiments with living exposed tissue were performed without the use of fibers (see the previous section), Peterson et al.[55,56] demonstrated the feasibility of measuring pH *in vivo* by placing the fiber in the body (see also Chapters 8 and 18). A 0.4-mm diameter probe, constructed from two plastic fibers, remotely senses the color change of a dye indicator contained within an acutely implanted sealed cellulose hollow fiber permeable to hydrogen ions (Chapter 8, Figures 4 and 5). A supporting electronic module provides tungsten filament illumination, light measurement with a photodiode plus operational amplifier, analog and digital circuitry to provide appropriate signal averaging and processing, and a mechanical assembly to enable the optical density measurements to be made both at 560 nm and, for normalization purposes, in the red.

Over the physiological range from pH 7.0 to 7.4, the optrode agrees with a standard glass pH electrode to within 0.01 pH units. The temperature coefficient is 0.017 pH per C. Figure 13 shows the results of an early study on the performance of a pH optrode in sheep, and a comparison with independently obtained data. Although not identical with the values obtained on withdrawn blood, the agreement between numerical data and trends is very good. According to the authors, it was not possible to say which of the measurements was most correct. Obviously, the fiber optic method gives as good an indication of blood pH levels as electrode methods.

Tait et al.[57] have enclosed such a sensor in a 20-gauge needle with a cut-away side. The optrode was calibrated by three pH measurements, two at the same temperature to establish the slope and intercept, and a third at a different temperature to establish the temperature coefficient. The probe was found to have an accuracy of ±0.05 pH units and was used to compare the performance of the optrode with that of a miniature glass electrode in sensing extracellular acidosis during regional ischemia in the working dog heart. There was no statistically significant difference between the two types of pH sensor. The time constant of the optrode (t_{95} 12 to 15 min) is considerably longer than that previously reported.[56] This is probably due to the fact that the mechanical support necessary to make the probe sufficiently robust to implant in the beating heart resulted in a large reduction in the exposed surface area of the proton-permeable chamber.

A more precise version of Peterson's sensor was specially designed for application in small blood vessels[58] and consisted of a 25-gauge hypodermic needle with an ion-permeable side window and 75-μm fibers (Figure 14). These smaller dimensions reduce t_{90} from 90 to 30 s. A mirrow (an aluminum-covered glass rod) improves return light intensity. The optrode was used to measure pH in the wall of a beating canine heart. The device, along with computerized signal processing and three-point calibration, was claimed to have a resolution of 0.001 pH units and was used to sense transmural pH gradients in canine myocardial ischemia.[59]

The sensor was implanted at different depths (5.5 to 8 mm) in the left ventricular wall

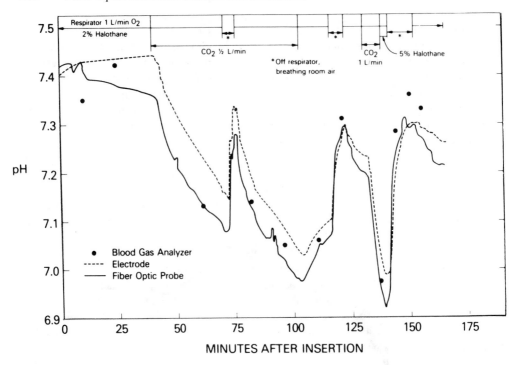

FIGURE 13. *In vivo* evaluation of a pH sensor in sheep undergoing varying respiration conditions. The blood pH analyzer data were obtained *ex vivo*, the electrode data with an inserted microelectrode. (After Peterson, J. I., Goldstein, S. R., Fitzgerald, R. V., and Buckhold, D. K., *Anal. Chem.*, 52, 864, 1980. With permission.)

(normal and ischemic subendocardium, and subepicardium, depth 3 to 4 mm). After reducing the coronary blood flow to 20 ± 5% in open-chest anesthesized dogs, and to 45 ± 5% in two dogs, pH gradients were determined after heparinization. There was a large transmural gradient, i.e., from normal pH values (7.36) in the subepicardium to severely acidotic (6.94) pHs 2 mm deeper in the subendocardium.

An even thinner catheter fitting in a 22-gauge needle and with similar working principle has been used[60] to monitor intravascular pH. The LED-based instrument that measures the ratio of reflected red and green light (see Chapter 8) as a function of pH was equipped with a commercial data acquisition and processing system capable of monitoring up to five sensors simultaneously. The system provides six analog outputs (five pH channels, one temperature channel) and a standard digital interface.

The sensor was placed in the carotid artery of anesthetized dogs, and comparisons were made between arterial pH as measured by the sensor and an electrode-based commercial blood pH meter. Tests in both normovolemic-normotensive and hypovolemic-hypotensive states over a pH range of 6.500 to 7.770 showed the mean difference between optrode and electrode measurements for 204 comparisons to be 0.060 pH ± 0.004 (SEM). The overall correlation was 0.92. The sensor performed equally well in the presence of normotension or hypotension and during respiratory or metabolic acidosis or alkalosis. Sensor drift during an experimental period of more than 6 h was no more than 0.042 ± 0.006 pH units. No dysfunction of the sensor, once inserted, was observed. Alterations in fiber optic pH values occurred if the sensor slipped from the lumen of the carotid artery, but would be correct by reinsertion of the sensor.

The same sensor was used in a study on the relationship between conjunctival pH, arterial pH, and cardiorespiratory animals in anesthetized dogs during a sequential hemorrhage protocol.[61] After three-point calibration at 37°C, the fiberoptic conjunctival pH sensor was placed in the animal's eye against the lateral, superior palpebral conjunctiva, and the eye taped

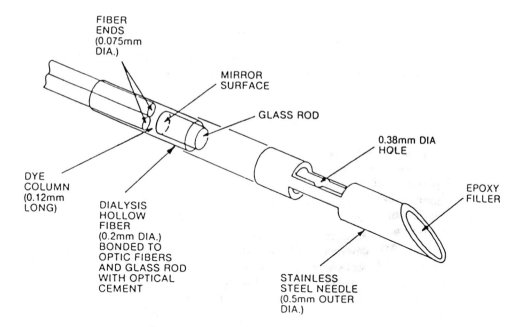

FIGURE 14.   Improved fiber optic invasive pH sensor design. (From Reference 58. With permission.)

closed. In these experiments, conjunctival pH measurements were consistently less than arterial values, with an average offset of 0.2 pH units. Such differences were previously also found with electrodes.

2. *In Vivo* pO_2 Sensing

The plain fiber sensors described in Section II.A measure *oxygen saturation* in blood. In blood gas analysis, in contrast, the *partial pressure* of molecular oxygen is determined. While oxygen partial pressure has been measured optically on the surface of the isolated guinea pig heart using a fluorescence-based oxygen sensor membrane,[49] it was only in 1984 that the first *in vivo* study using a fiber optrode was reported.[62] The sensor is based on the quenching of the fluorescence by molecular oxygen of a dye adsorbed on hydrophobic particles and kept in position at the fiber end with a porous polyethylene tubing (see Chapters 10 and 18). A diagrammatic view of the sensing unit is shown in Chapter 10, Figure 10.

The dye is excited at around 470 nm and fluoresces maximally at 515 nm with an intensity that depends on oxygen partial pressure. The ratio of green fluorescence to scattered blue excitation light (which serves as an internal reference) is the optical information. A resolution of ±1 torr up to 150 torr oxygen is reported. Figure 15 shows the trace of a continuous recording of oxygen pressure in the blood stream of a ewe sheep after the sensor had been inserted through a teflon catheter.

The experiment lasted over 3 h, and blood values for pO_2 as measured with an electrode are shown as dots. The program of variation of oxygen flow rate to the respirator is shown on top of the figure. Upon the initial insertion of the sensor, the pO_2 indication rose rapidly to a value somewhat above the level established by blood samples, over 250 torr. This level was substantially above the calibration point of 150 torr, so the agreement is as good as expected. Then the oxygen flow rate was halved and the blood pO_2 dropped into the calibration range, going below the values of the blood samples. At 66 min the oxygen flow was increased and there was an immediate response of the sensor, but at about 80 min the probe performance became poor. Clot formation at around the tip was suspected and it was removed from the artery at 116 min. After cleaning and re-insertion, whereupon it indicated correct values, the oxygen flow was decreased to give a very low oxygen partial pressure in blood, and increased again. The fiber sensor agreed reasonably well with the blood sample values.

It was not clear why clotting occurred in the heparinzed animal because Celgard® (the exterior material of the sensor) is used for other *in vivo* applications too. In conclusion, the work demonstrated the suitability of the pO_2 sensor for *in vivo* blood measurements where halocarbon anesthetics are not present. Their effect may be minimized by covering the sensing material with a perm-selective PTFE membrane.

3. *In Vivo* pCO_2 Sensing

Aside from pCO_2 measurements on the surface of the isolated guinea pig heart,[49] there are only a few reports on the *in vivo* determination of CO_2 partial pressure. Optical pCO_2 sensors are mostly based on the measurement of spectral changes of a pH-sensitive dye dissolved in buffer solution and entrapped in a proton-impermeable polymer at the fiber end (see Chapter 3, Section III.3, and Chapter 11). Vurek et al.[63] described theory, construction of such a sensor called Opticap. One plastic fiber carries light to the sensitive tip which is a silicone rubber tube 0.6 mm in diameter, and 1.0 mm long, and filled with a solution of phenol red in 35 mM bicarbonate. Ambient pCO_2 controls the pH of the solution which changes the optical absorption of phenol red (which is a weak acid). A second fiber carries the transmitted signal to a photoreceiver. The resulting electrical signal is linearly related to pCO_2 over the range of 2.7 to 10.7 kPa. The probe was tested as a tissue pCO_2 sensor on the cerebral cortex of the cat and as an arterial pCO_2 sensor. Drift over 1 d use was 0.6 Pa or less, and individual optrodes have been used as long as 12 weeks.

Fluorescence-based sensors are a promising alternative to reflectometric sensors in view of the sensitivity of fluorescence and, consequently, the potential of having quite minute volumes of sensing chemistries. Thus, a pCO_2 optrode with a nanoliter size droplet of a concentrate dichlorofluorescein (which fluoresces as a function of the solvent pH) in a buffered solution was described.[64] The pH of the droplet equilibrates with the volatile components if blood across an air bubble trapped at the tip of the capillary surrounding the fiber. The 488-nm line, which matches the absorption of the dye, was used to excite the fluorophore whose emission changes by 25% in going from 1.97 to 7.40% CO_2 in air at controlled humidity and temperature. Substantial photobleaching was observed. While this arrangement is likely to be impractical in case of real samples and routine analysis, and also would suffer from drying-out at low humidity, it obviously offers an elegant method for remotely sensing using the single fiber approach.

D. EXTRACORPOREAL ON-LINE BLOOD GAS ANALYZERS

The continuous monitoring of blood gases during circulation is the only reliable method of making sure that the proper amount of oxygen is being delivered to tissue. Measurement of acid-base balance and venous oxygen tension is the best method of measuring the adequacy of perfusion. Unfortunately, off-line blood gas analysis (BGA)* have been the only option for quite a time. Obviously, there was an urgent demand for all kinds of on-line analyzers. Fiber optic sensors appear to contribute to solve this problem.

1. The CDI-3M Healthcare System

Following the initial concept of Lübbers and Opitz, Cardiovascular Devices, Inc. (CDI, Irvine, CA) has developed a device suitable for continuous monitoring of pH, pCO_2, and pO_2 in an extracorporeal blood circuit. The instrument was marketed in 1984 by Bentley under the

---

* By convention, BGA means analysis of pO_2, pCO_2, *and* pH.

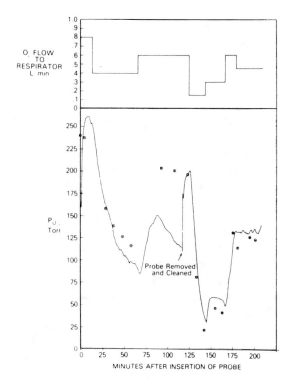

FIGURE 15. *In vivo* test of an oxygen optrode in a ewe. The breath gas composition was varied as shown on top, and the change in pO_2 followed with the inserted fiber sensor. The data points are values obtained *ex vivo* with a blood gas analyzer. (From Peterson, J. I., Fitzgerald, R. V., and Buckhold, D. K., *Anal. Chem.*, 56, 62, 1984. With permission.)

tradename GasStat 100. It is now marketed by CDI-3M Healthcare as the CDI System 400 which is the last in the series of models 200, 300, and 400. The working chemistry (fluorescent dyes, polymers, immobilization techniques) is essentially based on the research performed in the author's laboratory in the early 1980s as described in previous sections.

Since at that time sensing spots did not provide sufficient signal to allow the construction of a minute-size invasive triple sensor, it was decided to first design and develop a device coupled to an extracorporeal blood loop. Figure 16 shows an exploded view of the disposable sensor and how it is connected to the loop. Two permeable membranes separate the bloodstream from the sensor chemistry placed on the disposable sensor which is connected to the fiber optic cable and also has an electrical contact for temperature measurement using a thermistor. Light from a xenon lamp is guided through the fibers to the sensor spots (two of which are red at the sample side because of a red layer that serves as an optical isolation). Under blue light excitation, the sensor spots display green fluorescence.

The pH-sensitive chemistry consists of a cellulosic material to which hydroxypyrene trisulfonate (HPTS) is covalently bonded.[67,68] The CO_2-sensitive material[68] is a fine emulsion of a bicarbonate buffer (plus HPTS) in a two-component silicone. The oxygen-sensitive chemistry[41] simply is a solution of chemically modified decacyclene (which is strongly quenched by oxygen) in a one-component silicone. To make it insensitive toward halothane (an inhalation narcotic), it is covered with a thin layer of black PTFE which also serves as an optical isolation. The fluorescence intensities of the three sensing spots can be related to pO_2, pH, and pCO_2 via modified Stern-Volmer or Henderson-Hasselbalch algorithms.

The device has two channels, one for arterial and one for venous blood, and is capable of recording data for both the actual temperature and the corresponding 37°C data. Before use, the sensors are two-point calibrated using a calibrating device supplied with the instrument. This device contains two disposable gas cylinders with tonometered gas levels of pO_2 and pCO_2. During calibration, the gases bubble through a 300 mOsm bicarbonate buffer solution in the sensor cuvette for 7 to 10 min at room temperature to establish the two-point gas calibration curve. The pH calibration points are determined by the levels of carbon dioxide and the buffer composition in the cuvette which are, for gas 1, 9.87 kPa pO_2 and 2.47 kPa pCO_2. This gas adjusts a pH of 7.61. For gas 2, the respective numbers are 29.7, 7.87, and 7.10. The data refer to atmospheric pressure (101.325 kPa) and 22°C and have to be corrected to the actual barometric pressure for precise measurements. Both sensors are calibrated simultaneously prior to their insertion in the flow-through chamber.

The response time (t_{90}) is in the order of 3 min. The device is said to be stable for up to 12 h in operation. As in all kinds of temperature-corrected measurements, an inherent source of error results from the fact that temperature coefficients (tempcos) for pH, pCO_2, and pO_2 all vary with the hemoglobin and/or the protein concentration, which both can vary extensively during cardiopulmonary bypass (CPB) with hemodilution.[69] Typical specifications of the GasStat System 300 are the following: pH range 6.50 to 8.00; pCO_2 range 10 to 120 torr (1 to 16 kPa); pO_2 20 to 500 torr (2 to 67 kPa); base excess -25 to 25 meq/l; bicarbonate 0 to 50 meq/l; venous oxygen saturation 0 to 100%; temperature range 10 to 45°C. Figure 17 shows the instrument which, in its various versions, has been applied so far in several hundred thousand open heart operations. Figure 18 shows the parts of the parts of the disposable sensor cell and how it is aligned to the extracorporeal loop. Some 13,000 units are produced now every month.

From the user (physician) point of view, a continuous sensor for blood gases has several attractive practical features: (1) While accuracy is considered important, accurate *trending* is imperative so to enable quick countermeasures to be undertaken in critical situations. (2) When arterial and venous gases are monitored one can determine not only the true status of the patient by the venous gases but can also monitor the performance of the oxygenator by simply observing the arterial blood gas values. Should the oxygenator begin to fail, the drop in arterial pO_2 is almost immediate. (3) By continuously observing pH it is possible to determine acidosis or alkalosis and respond with the appropriate treatment. Bearing in mind that these are real-time parameters with the option of temperature correction, it is possible to observe the effects of hydrothermia on the patient. The benefit of a continuous device is obviously tremendous and offers a unique margin of safety that did not exist previously.

Early results comparing GasStat vs. intermittent samples analyzed on a blood gas analyzer showed less than ideal correlation. With continued clinical experience and further system refinements, data agreement improved substantially.[70] Main refinements of the initial system included the following: (1) A high overall tempco was identified and a correction factor incorporated into the microprocessor. (2) There was an offset between human and bovine blood (which was the material in the developmental work) which had to be compensated for. (3) A barometer was added to compensate for effects of altitude. As a result of these improvements, a good agreement was found[70] between blood gas analyzer and GasStat model 100 data. At a pO_2 of 200 torr at 37°C the electrochemical analyzer understates the tonometered value by approximately 5%, while the optical detector was within 1.5%. At 28°C, correlation between electrochemical and optical data of tonometered solutions was even worse.

Various evaluations of the performance of the GasStat have appeared. Hill et al.[71] determined blood gas values in 50 patients undergoing CPB using a conventional off-line blood gas machine (BGM) and both the GasStat and the CardioMet 4000 on-line monitors (see next chapter). The latter two achieved agreements within 7.5% for pO_2 values during the entire

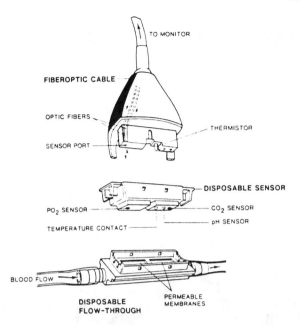

FIGURE 16. Exploded view of a disposable sensor for on-line blood gas analysis in an extracorporeal loop during cardiopulmonary by pass. (After Philbin, D. M., Inada, E., Sims, N., Misiano, D., and Schneider, R. C., *Perfusion*, 2, 127, 1987. With permission.)

range of temperatures. pH values from the GasStat appeared within 5% of the corrected values from the BGM. pCO_2 values varied within 11% during hypothermia. The GasStat agreed with the Cardiomet over varying temperatures and ranges of pO_2, while the BGM again demonstrated inaccuracy above 100 torr pO_2 and below 37°C. It was concluded that, due to the increased cost of on-line monitoring, the cost/benefit ratio should be considered, but that the advantage of on-line, real time parameters provides a much safer and precise perfusion management system giving the essential information for making rapid accurate decisions.

The GasStat was also used in another series of on-line measurements of blood gases during rewarming on hypothermic CPB.[72] In 20 patients, blood samples were obtained for 25, 30, 34, and 36°C, respectively, and analyzed with both the continuous sensor and with a BGM, for determination of oxygen content and calculation of whole body oxygen consumption (VO_2). On-line measurement of venous pO_2 provided a reasonably accurate reflection of the changing metabolic requirements during the rewarming period, more so than arterial pO_2. The data suggested that the continuous measurement of oxygen tension, particularly pvO_2, provides a simple and effective method of monitoring the adequacy of oxygen delivery, particularly during the period rewarming. Results were consistently within 5% of the values obtained with a BGM in the blood gas laboratory.

An authorative study on the performance of the GasStat has been published by Sigaard-Andersen et al.[73] The system was evaluated by reference to *in vitro* measurements using a BGM. GasStat measurements were performed at the actual temperature of the blood in the extracorporeal circuit, while reference measurements were performed at two fixed temperatures (25 and 37°C) with interpolation of the values to the actual temperature of the GasStat. Ten patients undergoing coronary artery bypass grafting during hypothermic extracorporeal circulation with hemodilution were monitored in the venous as well as the arterial line with the GasStat with 6 to 9 samplings of arterial and venous blood from each patient, a total of 136 samples.

The comparisons revealed a large scatter (albeit less than previously reported by others) which was due partly to interoptrode, partly to intraoptrode variation, and partly to a memory

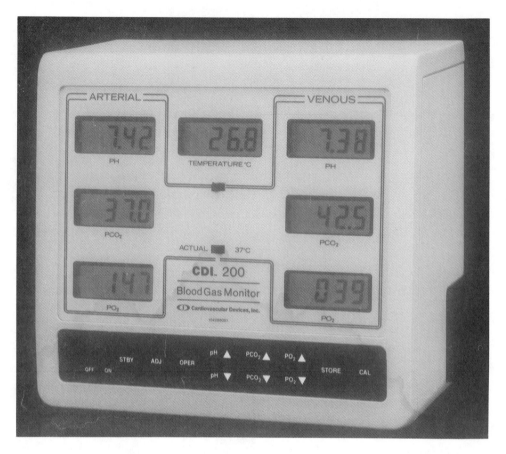

FIGURE 17.   The CDI-3M Healthcare System 300 extracorporeal blood gas monitoring system. The fibers leading to the sensors are not shown. (Courtesy CDI-3M Healthcare, Irvine, CA).

effect which rendered the initial measurements unreliable. Omitting the first samples, standard deviations for inter- and intraoptrode variations were 0.018 and 0.030 for pH, 0.026 and 0.034 for $^{10}\log pCO_2$, and 0.041 and 0.066 for $^{10}\log pO_2$. There was a minor systematic difference between the GasStat and the reference method[54] for pH, the bias being +0.017. For pCO_2 and pO_2 the bias was insignificant. No protein effect upon pH (as a result of the Donnan effect) was observed. Correlation data between GasStat 100 and blood gas machine data are given in Figures 19, 20, and 21. and demonstrate the agreement between the two methods.

While the overall accuracy of the system was deemed to be acceptable, a greater imprecision (which is still acceptable) was observed. Several causes could be identified: (1) The pH and pCO_2 calibrations are interdependent (pH is adjusted via pCO_2), and it was suggested to improve the calibration procedure. (2) Memory effects from the calibration procedure initially bias measured pHs into the direction of lower values. (3) Sampling errors may contribute to the large intrasensor variation. Consistently higher intrasensor variations were found with arterial sensors than with the venous.

The excessive data material presented indicated that the calibration procedure for the GasStat should be improved. Occasional samplings with *in vitro* measurement are necessary for quality control purposes, but an *in vivo* calibration cannot replace an accurate *in vitro* calibration. The results indicate that optrodes adequately respond to pH, pCO_2, and pO_2 with acceptable accuracy and precision for practical clinical purposes.

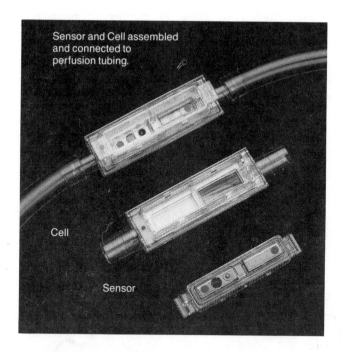

FIGURE 18. The CDI-3M Healthcare System 300 blood gas sensors and flow-through cell, and how they are assembled and connected to the perfusion tubing.

2. Other Extracorporeal Blood Gas Analyzers

Another extracorporeal blood gas system (the Cardiomet 4000®) is available from Biomedical Sensors, Ltd. (High Wycombe, Bucks, England). It provides continuous monitoring of pO_2, pCO_2, pH, and temperature, with continuously computed values for base excess, bicarbonate, and oxygen saturation. pH and pCO_2 are determined via fiber optic reflectometry using the pH-sensitive absorption indicator phenol red immobilized on a hydrophilic polymer gel. A section through the pH sensor is shown in Figure 22. The fiber cables and the electrical cable of the oxygen sensor are linked to the extracorporeal loop. The in-line connector is made from biocompatible polysulfone which is reported to exhibit greater resistance than polycarbonate. A stainless steel reinforced membrane system is designed to withstand positive and negative pressure challenges and allows the removal of the sensors during CPB without compromising patient safety.

The microprocessor-based system is calibrated without the need for precision gases, and a check routine of a few minutes is reported to suffice for a whole day, thus obviating the requirement to calibrate for each patient. A full calibration, lasting for 1 week, requires 15 min. The calibration data are retained for future patient use even when the monitor is turned off, so in an emergency the Cardiomet 4000 memory-stored calibration constants allow for clinically acceptable monitoring based upon the previous calibration.

Arterial and venous blood can be analyzed at both the actual temperature or as 37°C data. Because there is an electrical link to the body (via the oxygen electrode), the system is highly isolated (>4 kV AC). Sensor sterilization is achieved by irradiation. The connectors have a shelf life of 2 years limited by sterilization and are not sensitive to light, acidic priming solutions, or to high pCO_2 commonly found in CO_2-primed systems. The data presented in a clinical study[74] show that the system has a clinically acceptable bias and precision. The rise time and decay time of the currently available continuous blood gas monitors were evaluated by Riley et al.[75]

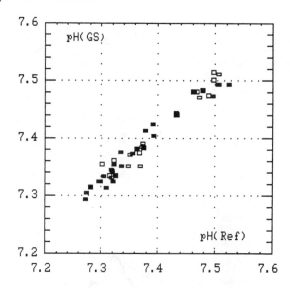

FIGURE 19. Correlation between data of the GasStat (GS) and the electrochemical blood pH reference analyzer after outlier rejection. (From Siggaard-Andersen, O., Gothgen, I. H., Wimberley, P. D., Rasmussen, J. P., and Fogh-Andersen, N., *Scand. J. Clin. Lab. Invest.*, 48 (Suppl. 189), 77, 1988. With permission.)

E. INVASIVE MEASUREMENT OF BLOOD GASES AND BLOOD pH

The ultimate goal in continuous monitoring of blood gases and blood pH is, of course, a simple device that would allow on-line monitoring without the need for extracorporeal loops. Since blood pH cannot be measured transcutaneously, measurements have to be performed using an invasive catheter. Following the development of the CDI 200 system (and its successor models) with their rather large sensing spots, the same company has developed a catheter for measurement of pH, pO_2, pCO_2, and temperature intravascularly.[76]

The main challenges in the design of such a sensor include miniaturization of sensors, in transmission, acquisition, and processing of low levels of signals, maintenance of mechanical integrity, and achieving a 99.9+ reliability when used *in vivo*. The fiber optic approach has been chosen because it offers a way for sensor miniaturization so that the fiber bundle can be introduced into the radial artery of a patient through a catheter.

The working chemistries of the three sensors (which are slightly different from those of the GasStat) are placed at the tip of three 140-μm fibers, while a thermocouple gives a direct reading of the sensor temperature at the tip. Fused silica fibers were selected for their superior transmission characteristics over the entire visible spectrum. The fragility of the sensor has been minimized by appropriate selection of materials to both reduce sensitivity to handling and assure mechanical integrity in the event of fiber breakage. Blood gas values are calculated from the measured temperature and standardized alogorithms and the known tempcos of the sensors. The issue of thrombogenicity has been addressed by designing a smooth tip surface and covalently immobilizing heparin on its surface. A schematic of the fiber tip comprising the sensor bundle and the temperature meter is shown in Figure 23.

The pH-sensitive dye in the pH sensor is hydroxypyrene trisulfonate[65-67] whose pK_a is 7.0 when immobilized on aminoethyl-cellulose. The matrix, in turn, is covered with an opaque cellulose overcoat which provides both mechanical intensity and optical isolation from ambient light and background fluorescence from blood (see Chapter 8, Figure 14). A two-wavelength approach is used for internal referencing (Chapter 3, Section VIII). In the pCO_2 sensor,[68] the same dye is used to measure changes in the pH of an isolated bicarbonate buffer

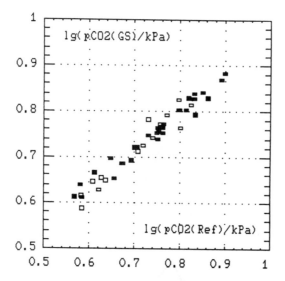

FIGURE 20. Correlation between data of the GasStat 100 (GS) and the electrochemical blood pCO_2 analyzer after outlier rejection. (From Siggaard-Andersen, O., Gothgen, I. H., Wimberley, P. D., Rasmussen, J. P., and Fogh-Andersen, N., *Scand. J. Clin. Lab. Invest.*, 48 (Suppl. 189), 77, 1988. With permission.)

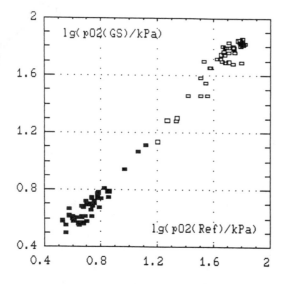

FIGURE 21. Correlation between data of the GasStat 100 (GS) and the electrochemical blood pO_2 analyzer after outlier rejection. (From Siggaard-Andersen, O., Gothgen, I. H., Wimberley, P. D., Rasmussen, J. P., and Fogh-Andersen, N., *Scand. J. Clin. Lab. Invest.*, 48 (Suppl. 189), 77, 1988. With permission.)

with changing pCO_2. The buffer is encapsulated in a silicone matrix which is coated with an overcoat similar to the one in the pH sensor. Changes in blood CO_2 tension equilibrate across the silicone causing a change in pH in the buffer droplets.

The pO_2 sensor, at present, consists of a solution of solubilized decaycyclene in silicone rubber.[41] The dye has a long lifetime and is strongly susceptible to oxygen quenching. Its excitation spectrum extends from 350 to 450 nm, and fluorescence (of the excimer band) is

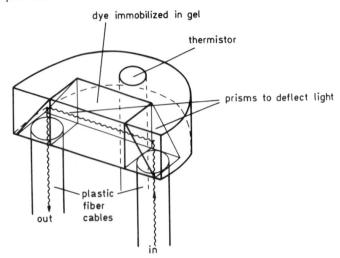

FIGURE 22.   Section through the Cardiomet 4000™ pH sensor. Light from the light source is guided, via plastic fibers, to a deflecting prism, passes the gel with immobilized phenol red, is again deflected (into the outgoing fiber), and guided to a photodiode. The gel is in contact with blood and changes its absorption with varying pH. (Redrawn from material kindly made available by Biomedical Sensors Ltd. [High Wycombe, UK].)

measured at 520 nm. To provide a reference signal, a second (longwave absorbing and fluorescing) dye is added which is unaffected by oxygen. Again, an opaque cellulose coating is applied. The oxygen sensor is cross-sensitive to halothane and indicates an apparent increase in pO_2 in its presence.

The optoelectronic system[77] is composed of a microprocessor-based monitor where excitation light originates from a xenon lamp (operated at 20 Hz) and is focused into the respective quartz fibers. Returning fluorescence signal is collected in a subunit called patient interface module (PIM) near the patient after signal demultiplexing and preamplification. The three active and reference signals are normalized against lamp intensity, filtered, and referenced, and then used to calculate actual values according to modified Henderson-Hasselbalch and Stern-Volmer equations, respectively. The calibration device contains two gas bottles and is similar to the one described for the extracorporeal CDI system (Section III.3 herein).

The heart of the instrument lies within the PIM. Each of the three channels is composed of two highly collinear quarter-pitch graded-index ("grin") lenses separated by a 5-mm cube beamsplitter, as shown in Figure 24. The grin lens on the right is part of the disposable. Above and below the cube is a photodiode plus filter stack, for purposes of fluorescence emission detection and excitation normalization, respectively. This arrangement combines the three functions of a reproducible low-less optical connector, a means of picking off some excitation light for lamp energy normalization purposes, and a means of picking off most of the fluorescence emissions returning from the fiber tips. The PIM is connected to the xenon light source via three 250-μm fibers. When a few microjoules per pulse of light are delivered into each fiber, about one nJ of fluorescence is collected by each signal photodiode.

The three sensing fibers are bonded together over their last 6 cm and then chemically sterilized, and are designed to reside within a standard 20-gauge (1 mm) radial artery catheter. The probe is thin enough to allow easy blood withdrawal and undistorted propagation of the arterial pressure signal to a pressure transducer. The PIM is clamped at the bedside and connected to the monitor located away.

Before use, the sensors are calibrated by a standard method: after attachment of the disposable to the PIM, the calibration cuvette containing the sensor tip is placed in the calibration device and an automatic cycle is started. During the calibration process the

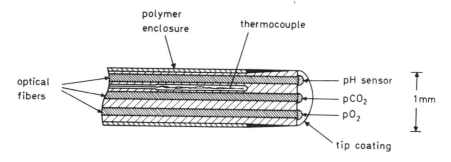

FIGURE 23. Triple fiber sensor for on-line measurement of pO_2, pCO_2, and pH in the radial artery. (After Gehrich, J. L., Lübbers, D. W., Opitz, N., Hansmann, D. R., Miller, W. W., Tusa, J. K., and Yasufo, M., *IEEE Trans. Biomed. Eng.*, 33, 117, 1986. With permission.)

temperature of the solution is maintained at 37°C while two primary standard calibration gases are sequentially bubbled through the solution until equilibrium is attained. Again, pH is adjusted via a defined pCO_2 which establishes a defined pH in a 300 mOsm sodium bicarbonate buffer. At the end of the calibration (which usually takes 20 to 30 min) the tip is removed from the calibration cuvette and threaded through the catheter into a blood circuit as described above. Figure 25 demonstrates the technique of inserting the catheter into the blood vessel.

Miller et al.[78] describe the *in vivo* performance of the probe. It displays biocompatibility for a one-time use of up to 72 h. The *in vitro* accuracy of the system has been tested against commercial blood gas measurement instruments. Comparison with tonometry and blood gas values gave r > 0.98 for all three sensors. System precision (1 SD) in tonometered bovine blood was 0.03 pH units, 2 torr pCO_2, and 4 torr pO_2. The standard error was within accuracy guidelines. The response times in animal and *in vitro* studies were less than 2 min. A small interference on all three parameters was observed with nitrous oxide. Halothane increases the pO_2 indication. Average drifts measured over a 200 h continuous study period showed 0.005 pH units, 0.9 torr pCO_2, and 2.1 torr pO_2 per 24-h period.[79] Figure 26 shows the response of each sensor to changes in pH, pO_2, and pCO_2 in bovine blood. Figures 27, 28, and 29 are typical recorder tracks obtained in animal studies. Obviously, there is good correlation with data obtained *ex vivo*, using an electrochemical blood gas analyzer.

In vivo animal studies (dog, pig, sheep, rabbit) demonstrated the need for antithrombogenic materials. Covalently bound heparin on the coating is a proper solution. Clinical trials in volunteers and in critical care and surgical patients, however, turned out to exhibit a poor correlation between sensor readings and blood gas machine readings. Two different physiological mechanisms were identified as being responsible for the error patterns: if the intravascular sensor touches the arterial wall, it senses the tissue gas value which, for oxygen, is lower than in arterial blood. The effect on pH and CO_2 is not as dramatic and, hence, the "wall effect" results in a transitory low oxygen value. This artifact can immediately be reversed if the motion of the wrist and catheter pull the probe tip away from the vessel wall.

The second pattern is consistent with the appearance of a thrombus anywhere around the sensor tip. It appears that the metabolic characteristics of most thrombi create a low oxygen, high CO_2, and low pH condition. These conditions can be duplicated in the animal model, but neither of the effects were seen as often as in humans in actual clinical trials.

As a result, the sensor/catheter configuration has been redesigned to eliminate the possibility of wall contact. Subsequent clinical studies[80] demonstrated a significant improvement in performance which approaches that of routine clinical blood gas analysis. Table 1 shows a summary of data from a clinical trial on 20 critically ill patients. Precision and accuracy are significantly improved over earlier studies. Particular emphasis was given to the accuracy and precision of pH measurements.[81]

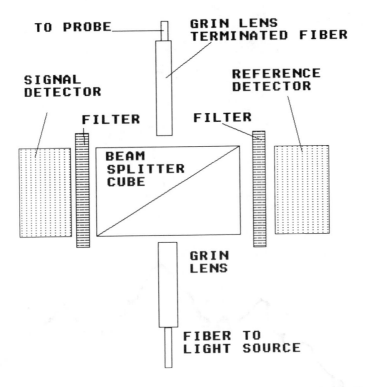

FIGURE 24.   Optical components of the patient interface module. (From Tusa, J., Hacker, T., Hansmann, D. R., Kaput, T. M., and Maxwell, T. P., *Proc. SPIE*, 713, 137, 1986. With permission.)

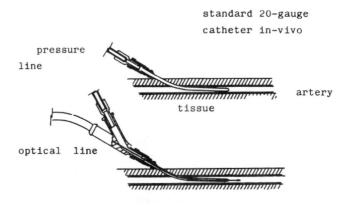

FIGURE 25.   Radial artery catheter before and after insertion of the optical sensor. (From Tusa, J., Hacker, T., Hansmann, D. R., Kaput, T. M., and Maxwell, T. P., *Proc. SPIE*, 713, 137, 1986. With permission.)

F. *EX VIVO* BLOOD GAS AND BLOOD pH ANALYSIS

Optrodes also are considered to offer advantages over electrodes in *ex vivo* blood gas analysis. Marsoner et al.[82] developed blood gas sensing devices consisting of a cylindrical glass support, 3 mm in diameter, covered with a thin layer of a working chemistry and a 12-μm black PTFE layer on top which acts as an optical isolation. The cylinders form the end of the light guides and are placed in a measuring chamber which allows the introduction of blood samples, calibration media, and washing solution.

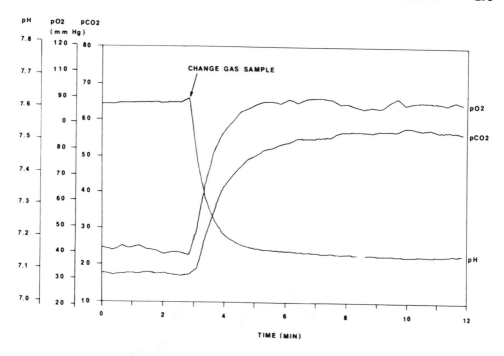

FIGURE 26. Response of the sensor of the intravascular blood gas system towards pH, pCO_2, and pO_2 of bovine blood.

pO_2 is measured via its capability of quenching the fluorescence of benzo(ghi)perylene or decacylene dissolved in a 20-μm silicone layer. There was perfect linearity between pO_2 and I/I (with I and I being the fluorescence intensities in the absence and presence of oxygen, respectively). No interferences were observed with water vapor, N_2O, enfluorane, and ethrane. Halothane acts as an efficient quencher of the fluorescence of benzo(ghi)perylene, but is excluded by the PTFE membrane.

A 10-μl sample chamber thermostatted at 37°C was used for whole blood measurements after two-point calibration with pure nitrogen and air containing 5.5% CO_2. Blood was tonometered in the usual way. An excellent agreement was found between actual and experimental data, the SD being between 0.63 and 2.90 torr in the 0 to 175 torr oxygen range. At 700 torr oxygen the SD was 7.6 torr, which still is significantly better than that of the Clark electrode.

For measurement of pH, a pH-sensitive 7-hydroxy-coumarin was immobilized on top of the porosified surface of the glass cylinder. The dye anion has excitation/emission maxima at 415/452 nm, respectively. The dye shows some dependence on ionic strength which partially was overcome by chemically modifying the environment of the sensing layer.[67] Within the ionic strength range from 120 to 200 mmol/l, the apparent pH error is ±0.003 units. However, when used in whole blood pH determination, the standard deviation still is 0.02 pH units. The sensor is reported to give a poor optical signal.

The CO_2 sensor consisted of a microemulsion of a buffered solution of hydroxypyrene trisulfonate in silicone rubber, spread as a 20-μm layer onto the glass cylinder. A 20-μm black silicone membrane on top serves as an optical isolation. Measurements performed with 10-μl whole blood samples revealed the SD to lie between 0.01 and 0.28 torr in the 14.8 to 62.9 torr pCO_2 range. The 95% response time was reported to be 22 s in the 40 to 80 torr range, but distinctly longer at higher pCO_2. The major problem with this kind of sensor results from the fact that it tends to be destabilized when stored in air with its low pCO_2. Following prolonged

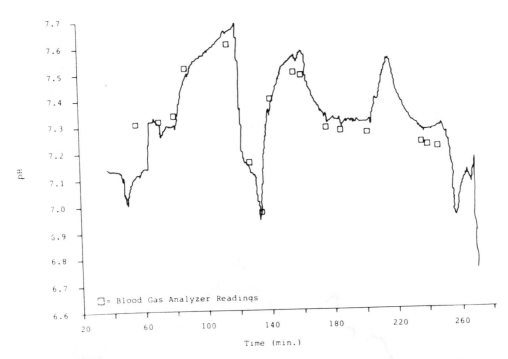

FIGURE 27.   Sensor performance in animal model. pH via fiber optics vs. laboratory electrochemical blood gas analyzer data.

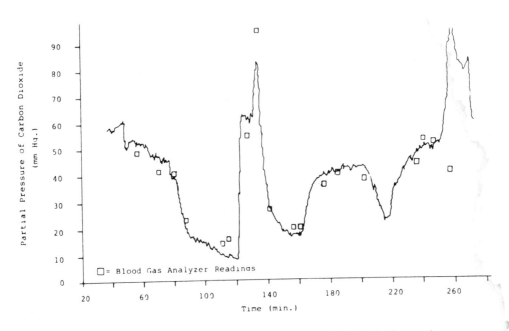

FIGURE 28.   Sensor performance in animal model. pCO_2 vs. laboratory blood gas analyzer.

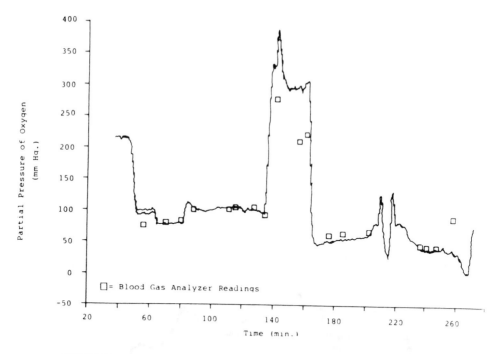

FIGURE 29.   Sensor performance in animal model. pO₂ vs. laboratory blood gas analyzer.

TABLE 1
Clinical Performance of the CDI 1000
Fiberoptic Invasive Blood Gas
and Blood pH Meter

	Accuracy	Precision (1 SD)	N
pH	0.000	0.023	158
pCO_2	−1.4 torr	4.4 torr	156
pO_2	−0.3 torr	14.6 torr	158

Note: Data from Reference 71.

storage in air, the sensor requires several hours to display a stable baseline again when exposed to physiological pCO_2.

There is evidence in the patent literature[83,84] that small and disposable sensor kits are being developed for use in single shot blood gas and blood pH analysis. For a schematic of such a device see the section on CO_2 sensors in chapter 11.

IV. *IN VIVO* APPLICATIONS OF BIOSENSORS

No published material is available so far on the *in vivo* performance of enzyme-based or immunoreaction-based fiber optic biosensors.

REFERENCES

1. **Neuman, M. R., Fleming, D. G., and Cheung, P. W., Eds.,** *Physical Sensors for Biomedical Applications,* CRC Press, Boca Raton, FL, 1980.
2. **Martin, M. J., Wickramasinghe, Y. A. B. D., Newson, T. P., and Crowe, J. A.,** Fiber optics and optical sensors in medicine, *Med. Biol. Eng. Comput.,* 25, 597, 1987.
3. **Peterson, J. I. and Vurek, G. G.,** Fiber optic sensors for biomedical applications, *Science,* 224, 123, 1984.
4. **Wolfbeis, O. S.,** Fiber optic fluorosensors in analytical and clinical chemistry, in *Molecular Luminescence Spectroscopy: Methods and Applications,* Vol. 2, Schulman, S. G., Ed., John Wiley & Sons, New York, 1988, chap. 3.
5. **Polanyi, M. L. and Hehir, R. M.,** In vivo oximeter with fast dynamic response, *Rev. Sci. Instrum.,* 33, 1050, 1962.
6. **Enson, Y., Briscoe, W. A., Polanyi, M. L., and Cournand, A.,** In vivo studies with an intravascular and intracardiac reflection oximeter, *J. Appl. Physiol.,* 17, 552, 1962.
7. **Kapany, N. S. and Silbertrust, N.,** Fiber optics spectrophotometer for in-vivo oximetry, *Nature (London),* 204, 138, 1964.
8. **Gamble, W. J., Hugenholtz, P. G., Monroe, P. G., Polanyi, M., and Nadas, A. S.,** The use of fiber optics in clinical cardiac catheterization, *Circulation,* 31, 328, 1965.
9. **Enson, Y., Jameson, A. G., and Cournand, A.,** Intracardiac oximetry in congenital heart disease, *Circulation,* 29, 499, 1964.
10. **Mook, G. A., Osypka, P., Sturm, R. E., and Wood, E. H.,** Fiber optic densitometry on blood, *Physiologist,* 7, 208, 1964.
11. **Harrison, D. C., Kapanyi, N. S., Miller, H. A., Silbertrust, N., Henry, W. L., and Drake, R. P.,** Fiber optics for continuous in-vivo monitoring of oxygen saturation, *Am. Heart J.,* 71, 766, 1966.
12. **Mook, G. A., Osypka, P., Sturm, R. E., and Wood, E. H.,** Fiber optic reflection photometry on blood, *Cardiovasc Res.,* 2, 199, 1968.
13. **Johnson, C. C., Palm, R. D., Stewart, D. C., and Martin, W. E.,** A solid state fiberoptics oximeter, *J. Assoc. Adv. Med. Instrum.,* 5, 77, 1971.
14. **Johnson, C. C.,** Fiber optic probe for oxygen saturation and dye concentration monitoring, *Biomed. Sci. Instrum.,* 10, 45, 1974.
15. **Klemp, H. W., Schmidt, E., Bender, F., Most, E., and Hewing, R.,** A solid state fiber optic system for continuous measurment of oxygen saturation and for cardiac output, *Z. Kardiol.,* 66, 257, 1977; see also References 7, 8, 11, 12, 19, 23, 36, 39, 43, and 54 herein.
16. **Landman, M. L. J., Knop, N., Kwant, G., Mook, G. A., and Zijlstra, W. G.,** A fiber optic reflection oximeter, *Eur. J. Physiol.,* 373, 273, 1978.
17. **Gattinioni, L., Kolobow, T., Damia, G., Apostoni, A., and Presenti, A.,** Extracorporeal CO_2 removal: a new form of respiratory assistance, *Int. J. Artif. Org.,* 2, 183, 1979.
18. **Mendelson, Y., Cheung, P. W., and Neuman, M. R.,** Noninvasive monitoring of arterial blood oxygen saturation, in Proc. Symp. Biosensors, Los Angeles, September 15 to 17, 1984, Potrin, A. R. and Neuman, M. R., Eds., 40.
18A. **Keller, H. and Lübbers, D. W.,** Reflection photometry of oxygen supply of skin flaps and replanted fingers, *J. Reconstr. Microsurg.,* 2, 241, 1986.
19. **Milano, M. J. and Kim, K. Y.,** Diode array spectrometer for the simultaneous determination of hemoglobins in whole blood, *Anal. Chem.,* 49, 555, 1977.
20. **Scheja, B., Cochet, F., Parriaux, O., and Depeursinge, C. D.,** In situ fiber optic spectroscopy of blood, *Proc. SPIE,* 482, 53, 1984.
21. **Steinmann, J., Böck, J., Hoeft, A., Korb, H., Wolpers, H. G., and Hellige, G.,** In-vivo measurement of oxygen saturation of blood using quartz fiber optics and an optical multichannel analyzer, *Biomed. Tech.,* 31, 246, 1986.
22. **Baele, P. L., McMihan, J. C., and Marsh, H. M.,** Continuous monitoring of mixed venous oxygen saturation in critically ill patients, *Anesth. Analg.,* 61, 513, 1982.
23. **Yoshiya, I., Shimada, Y., and Tanaka, K.,** Spectrophotometric monitoring of arterial oxygen saturation in the fingertip, *Med. Biol. Eng.,* 18, 27, 1980, and references cited therein.
24. **Schmitt, J. M., Meindl, J. D., and Mihm, F. G.,** An integrated circuit based optical sensor for in vivo measurement of blood oxygenation, *IEEE Trans. Biomed. Eng.,* 33, 98, 1986.
25. **Schweiss, J. F., Ed.,** *Continuous Measurement of Blood Oxygen Saturation in the High Risk Patient,* Oximetric Inc., Mountain View, CA.
26. **Domanski, A. W., Kostrzewa, S., Wolinski, T. R., and Dorosz, J.,** Opto-electronic system for the fiber optic adsorptive oximeter for simultaneous determination of hematocrit and oxyhemoglobin saturation, *Proc. SPIE,* 1085, xxx, 1989, (in press).

27. **Kilis, A., Wolinski, T. R., and Wermenski, K.,** Hemoglobin determination problems in fiber optic adsorptive oximeters, *Proc. SPIE*, 1085, yyy, 1989, (in press).

28. **Crowe, J. A., Rea, P. A., Wickramasinghe, Y. A. B. D., and Rolfe, P.,** Towards noninvasive monitoring of cerebral metabolism, in *Fetal and Neonatal Physiological Measurements*, Vol. 2, Rolfe, P., Ed., Butterworths, London, 1987, 150.

29. **Chance, B., Legallais, V., Sorge, J., and Graham, N.,** A versatile time-sharing multichannel spectrophotometer, reflectometer, and fluorometer, *Anal. Biochem.*, 66, 498, 1975.

30. **Mayevsky, A. and Chance, B.,** Intracellular oxidation-reduction state measured in situ by a multichannel fiberoptic surface fluorometer, *Science*, 217, 537, 1982.

31. **Renault, G., Duboc, D., and Degeorges, M.,** In situ laser fluorimetry in cardiology: preliminary results and perspectives, *J. Appl. Cardiol.*, 2, 91, 1987.

32. **Alvager, T. and Branham, M.,** Time resolved fluorescence spectroscopy for in-situ measurements, *Proc. Indiana Acad. Sci.*, 87, 365, 1977.

33. **Tromberg, B. J., Eastham, J. F., and Sepaniak, M. J.,** Optical fiber probes for biological measurements, *Appl. Spectrosc.*, 38, 38, 1984.

34. **Sepaniak, M. J.,** The clinical use of laser-excited fluorometry, *Clin. Chem.*, 31, 671, 1985, and references cited therein.

35. **Coleman, J. T., Eastham, J. F., and Sepaniak, M. J.,** Fiber optic based sensor for bioanalytical absorbance measurements, *Anal. Chem.*, 56, 2246, 1984.

36. **Labeyrie, E. and Koechlin, Y.,** Photoelectrode with a very short time constant for recording intercerebrally Ca^{2+} transients at a cellular level, *J. Neurosci. Methods*, 1, 35, 1979.

37. **Kobayashi, S., Kaede, K., Nishiki, K., and Ogata, E.,** Microfluorimetry of oxidation-reduction state of rat kidney in situ, *J. Appl. Physiol.*, 31, 693, 1971.

38. **Manuccia, T. J. and Eden, J. G.,** U.S. Patent 4,509,522, 1985.

39. **Böck, J., Gersing, E., Sundmacher, F., and Hellige, G.,** Intravascular fiberoptic detection of D_2O concentration in blood, *Proc. SPIE*, 906, 169, 1988.

40. **Kroneis, H. W.,** Unpublished results, 1983.

41. **Wolfbeis, O. S., Posch, H. E., and Kroneis, H. W.,** Fiber optical fluorosensor for determination of halothane and/or oxygen, *Anal. Chem.*, 57, 2556, 1985.

42. **Ultman, J. S. and Burzstein, S.,** Analysis of error in the determination of respiratory gas exchange, *J. Appl. Physiol. Respir. Environ. Exercise Physiol.*, 50, 210, 1981.

43. **Bacon, J. R. and Demas, J. N.,** Determination of oxygen concentrations by luminescence quenching of a polymer-immobilized transition metal complex, *Anal. Chem.*, 59, 2780, 1987.

44. **Burkhard, O. and Barnikol, W. K. R.,** A fast oxygen detector for analyzing lung function, *Prax. Klin. Pneumol.*, 37, 805, 1983.

45. **Barnikol, W. K. R., Gaertner, Th., Weiler, N., and Burkhard, O.,** Microdetector for rapid changes of oxygen partial pressure during the respiratory cycle in small laboratory animals, *Rev. Sci. Instrum.*, 59, 1204, 1988.

46. **Gaertner, Th., Barnikol, W. K. R., and Burkhard, O.,** Proof of bronchial reactions in anesthetized guinea pig by measurement of bronchial volume breath-by-breath, *Prax. Klin. Pneumol.*, 41, 539, 1987.

47. **Lübbers, D. W.,** Theory and development of transcutaneous oxygen pressure measurement, in *Adv. Oxygen Monitoring*, Vol. 25, Tremper, K. K. and Barker, S. J., Eds., Little, Brown, Boston, 1987, 31.

48. **Wolfbeis, O. S. and Lübbers, D. W.,** Unpublished experiments, 1990.

49. **Opitz, N., Weigelt, H., Barankay, T., and Lübbers, D. W.,** Application of the optode to measurement of surface pO_2 and pCO_2 of the isolated guinea pig heart, in *Oxygen Transport to Tissue*, Vol. 3, Silver, I. A., Erecinska, M., and Bicher, H. I., Eds., Plenum Press, New York, 1978, 99.

50. **Opitz, N. and Lübbers, D. W.,** Oxygen pressure measurements on moving organ surfaces by fluorescence sensor membranes using contactless signal transmission via fluorescence sensor radiation, *Adv. Exp. Med. Biol.*, 200, 367, 1987.

51. **Bates, R. G.,** *Determination of pH. Theory and Practice*, John Wiley & Sons, New York, 1973.

52. **Burnett, R. W.,** Problems associated with the definition of measured and calculated quantities in blood pH and gas analysis, NBS Spec. Publ. 450 (Proc. Workshop on Blood Gases and pH), Vol 450, National Bureau of Standards, Gaithersburg, MD, 1975, 613.

53. **Siggaard-Anderson, O.,** *The Acid-Base Status of the Blood*, Munksgaard, Copenhagen, 1976.

54. **Maas, A. H. J., Ed.,** Approved IFCC methods. Reference method (1986) for pH measurement in blood, *J. Clin. Chem. Clin. Biochem.*, 25, 281, 1987.

55. **Peterson, J. I., Goldstein, S. R., Fitzgerald, R. V., and Buckhold, D. K.,** Fiber optic pH probe for physiological use, *Anal. Chem.*, 52, 864, 1980.

56. **Goldstein, S. R., Peterson, J. I., and Fitzgerald, R. V.,** A miniature fiber optic pH sensor for physiological use, *J. Biomech. Eng.*, 102, 141, 1980.

57. **Tait, G. A., Young, R. B., Wilson, G. J., Steward, D. J., and MacGregor, D. C.,** Myocardial pH during ischemia: evaluation of a fiber-optic photometric probe, *Am. J. Physiol.*, 243, H1027, 1982.

58. **Markle, D. R., McGuire, D. A., Goldstein, S. R., Patterson, R. E., and Watson, R. M.,** A pH measurement system for use in tissue and blood, employing miniature fiber optic probes, in *1981 Adv. Bioeng.*, Viano D. C., Ed., Am. Soc. Mech. Eng., New York, 1981, 123.

59. **Watson, R. M., Markle, D. R., Ro, Y. M., Goldstein, S. R., McGuire, D. A., Peterson, J. I., and Patterson, R. E.,** Transmural pH gradient in canine myocardial ischemia, *Am. J. Physiol.*, 246, H 232, 1984.

60. **Abraham, E., Markle, D. R., Fink, S., and Ehrlich, H.,** Continuous measurement of intravascular pH with a fiberoptic sensor, *Anesth. Analg.*, 64, 731, 1985.

61. **Abraham, E., Fink, S. E., Markle, D. R., Plinholster, G., and Tsang, M.,** Continuous monitoring of tissue pH with a fiberoptic conjunctival sensor, *Am. Energ. Med.*, 14, 840, 1985.

62. **Peterson, J. I., Fitzgerald, R. V., and Buckhold, D. K.,** Fiber optic probe for in-vivo measurement of oxygen partial pressure, *Anal. Chem.*, 56, 62, 1984.

63. **Vurek, G. G., Feustel, P. J., and Severinghaus, J. W.,** A fiber optic pCO_2 sensor, *Ann. Biomed. Eng.*, 11, 499, 1983.

64. **Hirschfeld, T., Miller, F., Thomas, S., Miller, H., Milanovich, F., and Gaver, R. W.,** Laser fiber optic "optrode" for real time in-vivo blood carbon dioxide level monitoring, *J. Lightwave Tech.*, 5, 1027, 1987.

65. **Wolfbeis, O. S., Fürlinger, E., Kroneis, H. W., and Marsoner, H. J.,** A study on fluorescent indicators for near-neutral ("physiological") pH values, *Fresenius Z. Anal. Chem.*, 314, 119, 1983.

66. **Offenbacher, H., Wolfbeis, O. S., and Fürlinger, E.,** Fluorescence optical sensors for continuous determination of near-neutral pH values, *Sensors Actuators*, 9, 73, 1986.

67. **Wolfbeis, O. S. and Offenbacher, H.,** Optical sensor for continuous measurement of ionic strength and physiological pH values, *Sensors Actuators*, 9, 85, 1986.

68. **Wolfbeis, O. S.,** Analytical chemistry with optical sensors, *Fresenius Z. Anal. Chem.*, 325, 387, 1986.

69. **Siggaard-Anderson, O., Wimberley, P. D., Gothgen, I. H., and Fogh-Andersen, N.,** Variations in the anaerobic temperature coefficients for pH, pCO_2, and pO_2 with the composition of the blood, *Scand. J. Clin. Lab. Invest.*, 43, 85, 1988.

70. **Clark, C. L., O'Brien, J., McCulloch, J., Webster, M., and Gehrich, J.,** Early clinical experience with GasStat™, *J. Extra-Corporal Technol.*, 18, 185, 1986.

71. **Hill, A. G., Groom, R. C., Vinansky, R. P., and Lefrak, E. A.,** On-line or off-line blood gas analysis: cost vs time vs accuracy, *Proc. Am. Acad. Cardiovasc. Perfusion*, 6, 148, 1985.

72. **Philbin, D. M., Inada, E., Sims, N., Misiano, D., and Schneider, R. C.,** Oxygen consumption and online blood gas determinations during rewarming on cardiopulmonary bypass, *Perfusion*, 2, 127, 1987.

73. **Siggaard-Anderson, O., Gothgen, I. H., Wimberley, P. D., Rasmussen, J. P., and Fogh-Andersen, N.,** Evaluation of the GasStat® fluorescence sensors for continuous measurement of pH, pCO_2, and pO_2 during CPB and hypothermia, *Scand. J. Clin. Lab. Invest.*, 48 (Suppl. 189), 77, 1988.

74. **Basha, J. W., Sternlieb, J. J., Bjork, V. O., Bretz, P. D., and Gabrielson, T. W.,** Clinical evaluation of Cardiomet 4000 continuous on-line blood gas analyzer, in Proc. 26th Am. Soc. Extracorp. Technol. Meeting, Anaheim, CA, March 1988.

75. **Riley, J. B., Fletcher, R. W., Jeudaitis, M., Finks, S., Hoerr, H. R., Bell, C., and Crowley, J. C.,** Comparison of the response time of various sensors for continuous monitoring of blood gases, in Proc. 26th Am. Soc. Extracorp. Technol. Meet., Anaheim, CA, March 1988.

76. **Gehrich, J. L., Lübbers, D. W., Opitz, N., Hansmann, D. R., Miller, W. W., Tusa, J. K., and Yasufo, M.,** Optical fluorescence and its application to an intravascular blood gas monitoring system, *IEEE Trans. Biomed. Eng.*, 33, 117, 1986.

77. **Tusa, J., Hacker, T., Hansmann, D. R., Kaput, T. M., and Maxwell, T. P.,** Fiber optic microsensor for continuous in-vivo measurement of blood gases, *Proc. SPIE*, 713, 137, 1986.

78. **Miller, W. W., Yafuso, M., Yan, C. F., Hui, H. K., and Arick, S.,** Performance of an in-vivo, continuous blood-gas monitor with disposable probe, *Clin. Chem.*, 33, 1538, 1987.

79. **Hansmann, D. R. and Gehrich, J. L.,** Practical perspectives of the in-vitro and in-vivo evaluation of a fiber optic blood gas sensor, *Proc. SPIE*, 906, 4, 1988.

80. **Shapiro, B. A., Cane, R. D., Chomka, C. M., and Gehrich, J. L.,** Evaluation of a new intra-arterial blood gas system in dogs and humans, *Anesthesiology*, 67, A 640, 1987 (abstract only).

81. **Yafuso, M., Arick, S. A., Hansmann, D., Holody, M., Miller, W. W., Yan, C. Y., and Mahutte, K.,** Optical pH measurements in blood, *Proc. SPIE*, 1067, xxx, 1989, (in press).

82. **Marsoner, H. J., Kroneis, H. K., Offenbacher, H., and Karpf, H.,** Optical sensor for pH and blood gas analysis, in Proceed. of the IFCC Meet. Physiol. Methodol. of Blood Gases and pH, Vol. 6, Maas, A. H. J., Boink, F. B. T. J., Saris, N. E. L., Sprokholt, R., and Wimberley, P. D., Eds., Radiometer, Copenhagen, 1986.

Chapter 20

CHEMILUMINESCENCE AND BIOLUMINESCENCE BASED OPTICAL PROBES

L. J. Blum and P. R. Coulet

TABLE OF CONTENTS

I. INTRODUCTION

Luminescence is the production of light from molecules promoted to an excited state and returning to the ground state. Chemiluminescence and fluorescence are related phenomena, but the source of the energy producing molecules in an excited state is different: chemiluminescence is luminescence as the result of a chemical reaction whereas in fluorescence, incident radiation is the source of energy. Bioluminescence is simply a special form of visible chemiluminescence occurring in living organism.

Chemi- and bioluminescence reactions are powerful tools for biochemical and clinical analysis since the availability of modern instrumentation allows measurement of light emitted with a great sensitivity. Then, the determination of species involved in such reactions can be performed at a very low detection level.

Chemiluminescence reactions which require hydrogen peroxide for the light emission are of particular interest in biochemical analysis. As a matter of fact, H_2O_2 is a product of several enzymatic reactions which can be coupled to a chemiluminescent detection. Bioluminescence reactions used for analytical purposes allow the determination with an extreme sensitivity of ATP (firefly bioluminescence) and NADH (bacterial bioluminescence).

Over the past decade, immobilized bio- and chemiluminescent reagents have been gaining in popularity because they offer the economy and convenience of matrix-bound compounds in addition to the sensitivity of luminescence technique.

Recently, a strong interest arose in the design of sensors or probes based on fiber optics or optoelectronic devices associated with an immobilized reagent phase. Fiber-optic chemical sensors and biosensors are generally based on colorimetric or fluorimetric reaction, and only a few of them involve bioluminescence or chemiluminescence reactions. The emphasis of this chapter is put on the analytical use and potentialities of bio- and chemiluminescence-based fiber-optic probes.

II. LUMINESCENT REACTIONS

A. CHEMILUMINESCENCE

Several chemiluminescent reagents including luminol, lucigenin, oxalate derivatives, and lophine can be used for H_2O_2 determination, but only luminol (5-amino-2,3-dihydrophthalazine-1,4-dione) and oxalate derivatives have found widespread use and were evaluated for hydrogen peroxide detection.

The luminol chemiluminescence reaction with hydrogen peroxide occurs under alkaline conditions in the presence of a catalyst and/or a co-oxidant[1] according to the following reaction:

$$2H_2O_2 + luminol + OH^- \xrightarrow[\text{co-oxidant}]{\text{catalyst or}} 3 - \text{aminophthalate} + N_2 + 3H_2O + h\upsilon \qquad (1)$$
$$(\lambda \max = 430nm)$$

Ferricyanide is both a catalyst and a co-oxidant which can be reduced to ferrocyanide by oxidizing luminol and then is reoxidized to ferricyanide. Peroxidase can be used as a catalyst with the advantage over ferricyanide that this enzyme allows the chemiluminescence reaction to proceed at near-neutral pH values.[2] Instead of a catalyst, the generation of light can be achieved by using a positively biased electrode.[3,4] In this technique (electrogenerated chemiluminescence), luminol is electrochemically oxidized and in the presence of H_2O_2, the chemiluminescence reaction occurs.

Peroxyoxalate chemiluminescence refers to reactions which involve esters of oxalic acid and hydrogen peroxide.[5] Oxalate derivatives react with hydrogen peroxide to produce 1,2-dioxetane-dione, a high energy intermediate which transfers its energy to a fluorescer. The excited fluorescer then emits light by returning to the ground state as shown below.

$$R-O-CO-CO-O-R + H_2O_2 \longrightarrow 1,2 \text{ - dioxe tan e - dione} + 2\ R-OH \qquad (2)$$

$$1,2 \text{ - dioxetane - dione} + \text{Fluorescer} \to FL* + 2CO_2 \qquad (3)$$

$$FL^* \longrightarrow FL + h\upsilon \qquad (4)$$

Oxalate esters such as bis (2,4,6-trichlorophenyl) oxalate (TCPO) and bis (2,4,5-trichloro-6-pentoxycarbonyl) oxalate (CPPO) are often used with perylene or diphenylanthracene as a fluorescer. The peroxyoxalate chemiluminescence reaction can proceed within a greater pH range than luminol but the efficient oxalate derivatives are only soluble in organic solvents such as dimethoxyethane, dioxane, or ethyl acetate. This can be a major drawback for the use of these chemiluminescence reactions directly coupled to enzyme-catalyzed systems producing H_2O_2.

B. BIOLUMINESCENCE
1. Firefly Bioluminescence
The light-producing reaction of the American firefly *Photinus pyralis* has been the most extensively studied bioluminescent system, and since the discovery (by McElroy) of the absolute requirement of ATP for the bioluminescent process,[6] this reaction has been widely used for analytical purposes. The light emission occurs in the presence of the enzyme luciferase, Mg^{2+}, molecular oxygen, ATP, and of a specific substrate, the luciferin [D-(−)-2-(6′-hydroxy-2′-benzothiazolyl)-Δ^2- thiazoline-4-carboxylic acid], the structure of which has been elucidated and the synthesis achieved by White et al.[7] The mechanism of the firefly luciferase reaction has been thoroughly reviewed by DeLuca and McElroy[8-10] and the simplified overall reaction is shown below:

$$ATP + luciferin + O_2 \xrightarrow{\text{luciferase}} AMP + oxyluciferin + PPi + CO_2 + h\upsilon \qquad (5)$$

$$(\lambda \max = 560 nm)$$

Under appropriate experimental conditions, the light intensity is proportional to the ATP concentration.

2. Bacterial Bioluminescence
In the luminescent marine bacteria system used for analytical purposes, the light is produced by two consecutive enzymatic reactions:

$$NAD(P)H + H^+ + FMN \xrightarrow{\text{oxidoreductase}} NAD(P)^+ + FMNH_2 \qquad (6)$$

$$FMNH_2 + R-CHO + O_2 \xrightarrow{\text{luciferase}} FMN + R-COOH + H_2O + h\upsilon \qquad (7)$$

$$(\lambda \max = 490 nm)$$

In the first one, catalyzed by NAD(P)H: FMN oxidoreductase, $FMNH_2$ is produced (Reaction 6) and then utilized in the second reaction catalyzed by bacterial luciferase to produce light (Reaction 7) in the presence of molecular oxygen and of a long-chain aldehyde (R–CHO). When NAD(P)H is the limiting substrate of this bi-enzymatic system, the light intensity is proportional to NAD(P)H concentration. In the different luminescent bacteria, the same light emission mechanism is involved and the luciferases are similar.[11] The two mostly used light-emitting enzyme systems were isolated from *Vibrio harveyi* and *V. fischeri*.

C. ANALYTICAL POTENTIALITIES
1. Selectivity and Sensitivity
The bioluminescence reactions are enzyme-catalyzed processes and are specific for a particular substrate. The firefly luciferase has a remarkable specificity for ATP since other nucleotide triphosphates do not lead to light emission.[12] The bacterial luciferase reaction can be considered to be highly specific for $FMNH_2$, although related flavin derivatives can act on light emission but with a much lower efficiency.[13] The oxidoreductase from *V. fischeri* can react either with NADH or NADPH whereas distinct FMN reductases, one specific for NADH, the other specific for NADPH, have been isolated from *V. harveyi*.[14]

Chemiluminescent reactions are far less specific and oxidants other than hydrogen peroxide can produce light. However, the analyte to be measured is generally first oxidized by an enzyme-catalyzed process leading to H_2O_2 and the analytical method is then selective.

The sensitivity of these luminescent analytical methods mainly depends on the light detector and on the quantum yield of the light-emitting reaction. The quantum yield, defined as the ratio between the total number of photons emitted and the number of luminescent molecules reacting, is about 0.01 for many artificial chemiluminescence reactions, whereas it is about 0.2 for the bacterial bioluminescence and close to 1 for the firefly bioluminescence. With soluble reagents, the detection limit for the bioluminescent assay of ATP or NADH is at the picomolar level whereas it is 1 nM for the chemiluminescent assay of hydrogen peroxide. Photomultiplier tubes have lower detection limits than other types of detectors. For specific applications, however, silicon photodiodes can be used but many measurements in bioluminescence and chemiluminescence are performed below the detection limit of photodiodes and the photomultiplier tube is still the detector of choice. It must be noted that a charge-coupled device (CCD) has been recently used to detect the chemiluminescence of luminol with hydrogen peroxide and transition metal ions.[15] The detection limits were at the femtomol level for Cr^{3+}, Co^{2+}, and H_2O_2.

2. Availability of Biocatalysts

For practical purposes, it is highly desirable to have at one's disposal reliable sources of enzymes. Such characteristics as stability and specific activity must be constant from batch-to-batch preparation to guarantee the user reproducible performances of the biosensor. This is now possible with some commercially available enzymatic preparations. Crystallized and lyophilized preparations of firefly luciferase from *P. pyralis* can be obtained from different suppliers. Partially purified preparations of bacterial luciferase containing oxidoreductases are also available. Concerning the luminol chemiluminescent reaction, horseradish peroxidase preparations with different specific activities are widely available.

3. Potential Applications

Theoretically, in addition to the direct measurement of ATP in biological extracts, kinetic determination of metabolites and enzymes participating in ATP-converting reactions, and endpoint determination of metabolites participating in ATP-converting reactions can be performed using the firefly reaction.

The use of the bacterial bioluminescent system for NAD(P)H measurements can also be extended to other analytes, and the analysis of any NAD(P)H dependent enzyme as well as related substrates (Reaction 8) can be performed via this luminescent system.

$$S_{red} + NAD(P)^+ \xleftarrow{\text{dehydrogenase}} S_{ox} + NAD(P)H + H^+ \tag{8}$$

In the same manner, any enzyme-catalyzed reaction leading to hydrogen peroxide can be coupled to a chemiluminescence reaction involving H_2O_2. This represents an attractive alternative to colorimetric or electrochemical measurements with the advantage of a lower detection limit. Some examples of enzymatic systems producing H_2O_2 are given below:

$$\beta - D - glucose + O_2 \xrightarrow{\text{glucose oxidase}} gluconic\ acid + H_2O_2 \tag{9}$$

$$cholesterol + O_2 \xrightarrow{\text{cholesterol oxidase}} \Delta - 4 - cholestenone + H_2O_2 \tag{10}$$

$$D - glutamate + O_2 \xrightarrow{\text{glutamate oxidase}} NH_3 + 2 - oxoglutarate + H_2O_2 \tag{11}$$

$$L - lactate + O_2 \xrightarrow{\text{lactate oxidase}} pyruvate + H_2O_2 \tag{12}$$

$$pyruvate + HPO_4^{2-} \xrightarrow{\text{pyruvate oxidase}} CO_2 + acetyl\ phosphate + H_2O_2 \tag{13}$$

III. IMMOBILIZED REAGENTS FOR BIO- AND CHEMILUMINESCENCE ANALYSIS

The immobilization of enzymes on solid supports allows their reusability and may lead to an increase in their stability. Various supports with different immobilization procedures have been used for immobilizing biocatalysts including adsorption, entrapment or cross-linking in polymeric gels, and covalent binding to an activated support. Details on the subject can be found elsewhere[16,17] and in Chapter 7.

In the past decade, several papers dealing with the use of insoluble derivatives of bioluminescence enzymes from both the firefly and bacteria as well as of chemiluminescence reagents have been published. Generally, light measurements were performed using batchwise systems in which assay cuvettes containing the immobilized enzymes and suitable reagents were disposed in batch luminometer. Another approach which may take great advantage of bio- and chemiluminescence reactions is the use of flow analysis methods involving immobilized catalysts or luminescent reagents in association with modified analytical devices.

Some selected references concerning the use of immobilized compounds for bio- or chemiluminescence analysis in batchwise sytems are listed in Tables 1 to 3. These systems are mentioned in this study, because most of them could be adapted to a fiber-optic detection. Details on other systems can be found in extensive reviews on both batchwise and flow systems recently published by Ugarova and Lebedeva[42] and by Coulet and Blum.[43,44]

IV. FIBER-OPTIC PROBES

In contrast to UV/vis spectrophotometric or fluorimetric measurements, where a light source is needed, luminescence analysis only requires a photon-measuring system. With enzymes free in solution, this can be achieved with batch luminometers. Solid-phase systems allow to design fiber-optic probes. Characteristics and properties of such devices recently described in the literature are reported in this section.

A. CHEMILUMINESCENCE-BASED FIBER-OPTIC PROBES

Freeman and Seitz[45] have reported a new approach in chemiluminescence analysis by describing a fiber optic probe for hydrogen peroxide based on the luminol reaction with horseradish peroxidase immobilized in a polyacrylamide gel on the end of a fiber optic. When the probe was immersed in a solution of substrates, the light generated at the enzyme phase was transmitted through the fiber optic to a photomultiplier tube. The detection limit was close to 1×10^{-6} M for H_2O_2 and the response-time which depended on the hydrogen peroxide concentration was about 4 s at 1×10^{-5} M H_2O_2.

Although no fiber optic was involved, the biophotodiode described by Aizawa et al.[46] can be quoted here. This sensor was a unique combination of a silicon photodiode and matrix-bound enzymes. Hydrogen peroxide has been assayed in the range 1×10^{-3} to 1×10^{-2} M by using the chemiluminescent reaction of luminol catalyzed by peroxidase entrapped in a polyacrylamide gel. The applicability of the system was extended to glucose determination by coimmobilizing glucose oxidase with peroxidase. A standard curve for glucose could then be drawn from 0.1 to 1.5 M.

Abdel-Latif and Guilbault[47] have described the use of cetyltrimethylammonium (CTAB) as a surfactant to enhance the chemiluminescence of TCPO with hydrogen peroxide in the presence of perylene with incorporation of fiber optics. The authors showed that the use of micelles to enhance the chemiluminescence intensity of the peroxyoxalate reaction could be used for very sensitive assays of hydrogen peroxide. The dissolution problems of organic

TABLE 1
Use of Immobilized Firefly Luciferase for Bioluminescence Analysis
in Batchwise Systems

Analyte	Dynamic range	Support and activation technique	Stability	Ref.
ATP	1×10^{-8} to 1×10^{-5} M	Alkylamine glass beads	3 weeks	18
Creatine kinase	1.27 to 9.5 U	+ glutaraldehyde	(4°C)	
ATP	Detection limit: 0.1 pM	CNBr Sepharose 4B	60% (1 month)	19
ATPase		and CNBr cellophane films		
Pyruvate kinase			50% (5 d)	
ATP	0.4×10^{-9} to 4×10^{-6} M	CNBr Sepharose	Stable at	20
		4B and CL 6B	−196°C	
ATP	1×10^{-10} to 1×10^{-6} M	NaIO$_4$ and cyanuric chloride		21
		activated cellulose films		
ATP	1×10^{-11} to 1×10^{-6} M	Acylazide-activated	20% (8 months)	22
		collagen film	(dehydrated form)	
Creatine kinase	0.5 to 1000 U/l	CNBr Sepharose		23

solvents in aqueous solvents were eliminated and reproducible results were obtained. The determination of H_2O_2 with a detection limit of 2.5×10^{-9} M was linear in the range 8×10^{-4} to 8×10^{-9} M with a coefficient of variation of 0.3% for five replicates at 1×10^{-7} M. Using glucose oxidase immobilized on a polyamide membrane, the fiber-optic sensor allowed measurement of glucose concentration in a linear range of 3×10^{-6} to 3×10^{-2} M with a detection limit equal to 6×10^{-7} M. The time required for the maximum chemiluminescent intensity was comprised between 4 and 10 s for both analytes.

Electrogenerated chemiluminescence of luminol has been used by VanDyke and Cheng[48] by incorporating an electrically conductive material in the fabrication of a fiber-optic probe. The fiber-optic/electrochemical probe was designed to accommodate a flow injection analysis system for the determination of hydrogen peroxide. The detection limit was 1×10^{-6} M, but the calibration curve was nonlinear with the luminescence signal and leveled off at 1×10^{-3} M.

B. BIOLUMINESCENCE-BASED FIBER-OPTIC PROBES

The use of bioluminescence reactions to design novel fiber-optic biosensing probes have been described for the first time in our laboratory.[49,50] The probe consisted of an immobilized enzymatic phase in close contact with an optical glass fiber bundle connected to a photomultiplier tube associated with a signal processing sytem.

Based on this approach, immobilized firefly luciferase from *P. pyralis*, coimmobilized bacterial luciferase and oxidoreductase from *V. harveyi* have been used and allowed the bioluminescent analysis of both ATP and NADH with a low detection limit and in a wide dynamic range. In addition, the luminol chemiluminescence reaction catalyzed by immobilized horseradish peroxidase has also been used for hydrogen peroxide determination.

Covalent immobilization of enzymes was performed by applying 10 µl of a suitable enzymatic solution on a disk (10 mm in diameter) cut out of a polyamide membrane supplied in preactivated form according to the general procedure previously described in our laboratory.[51] The binding reaction was complete after 1 min at room temperature.

The fiber-optic biosensing probe (Figure 1) consists of a 1-m glass fiber bundle of 8 mm in diameter. One end of the bundle was connected to the photomultiplier tube of a Berthold Biolumat LB 9500 luminometer. The bioactive disk was maintained in close contact with the other end of the fiber bundle by means of a screw-cap. The probe was immersed in a 4.5 ml stirred and thermostated reaction medium. The reaction vessel was surrounded by a black

TABLE 2
Use of Immobilized Bacterial Bioluminescence Enzymes in Batchwise Systems

Analyte	Dynamic range	Support and activation technique	Stability	Ref.
NADH	1×10^{-8} — 5×10^{-4} *M*	Diazo coupling on	2 weeks	24
NADPH	1×10^{-7} — 2×10^{-3} *M*	arylamine glass beads	(4°C)	
Alcohol DH	0.015 — 3.0 pmol	Diazo coupling on		25
Ethanol	0.0004% — 0.015%	arylamine glass beads		
Glucose-6-P DH	0.0015 — 0.1 pmol			
D-glucose	0.15 — 15 nmol			
Hexokinase	0.1 — 2.0 pmol			
Lactate DH	0.003 — 0.7 pmol			
Malate DH	0.007 — 0.7 pmol			
Glucose-6-P	1 pmol — 2 nmol	Diazo coupling on		26
D-Glucose	20 pmol — 10 nmol	arylamine glass beads		
NADH	1 pmol — 2 nmol	CNBr Sepharose 4B		27
Glucose-6-P	10 pmol — 20 nmol			
Androsterone	0.3 pmol — 2 nmol			
Testosterone	0.8 pmol — 1 nmol			
D-Glucose	10 pmol — 1.5 nmol	CNBr Sepharose 4B	Several months at 0—4°C	28
L-Lactate	0.1 nmol — 100 nmol			
6-P-Gluconate	10 pmol — 100 nmol			
L-Malate	0.1 nmol — 10 nmol			
L-Malate	10 pmol — 10 nmol			
L-Alanine	50 pmol — 10 nmol			
L-Glutamate	0.1 nmol — 5 nmol			
NAD+	1 pmol — 1 nmol			
NADP+	0.2 pmol — 0.2 nmol			
NADH	0.2 pmol — 1 nmol	CNBr Sepharose 4B	2 months (−196°C)	20
Chenodiol	1×10^{-6} — 32×10^{-6} *M*	CNBr Sepharose 4B		29
3 αOH bile acids	1 pmol — 20 nmol	CNBr Sepharose 4B	4 months (4°C)	30
7αOH bile acids	1.5×10^{-6} — 50×10^{-6} *M*	CNBr Sepharose 4B	Stable at 4°C	31
12αOH bile acids	4 pmol — 2 nmol	CNBr Sepharose 4B	2 months (4°C)	32
Creatine kinase	1 — 10 U/ml	BSA gel	3 months (4°C)	33
NADH	1×10^{-9} — 2×10^{-5} *M*	Acylazide activated collagen films	3 weeks (4 °C)	34
NADH	1×10^{-12} — $1 \times 10^{-7} M$	CNBr activated		35
NAD+	1×10^{-12} — 1×10^{-7} *M*	agarose		
FMN	1×10^{-10} — 1×10^{-7} *M*			
Glucose-1-P	1×10^{-9} — 1×10^{-5} *M*			
Glucose-6-P	1×10^{-9} — 1×10^{-5} *M*			

Note: DH, dehydrogenase; OH, hydroxy.

polyvinyl chloride jacket to avoid ambient light interferences. Samples were injected with a syringe through a septum and the light signal was monitored on a single channel chart recorder.

It must be noted that some differences in the kinetics of light emission can be observed with the different light emitting systems associated with the fiber-optic sensor. The time required for reaching a steady state with immobilized firefly luciferase mainly depends on the ATP concentration and varies from 1 min at 2.8×10^{-10} *M* to 6—7 min at 3×10^{-6} *M*; then the light emission is stable for several minutes. With the bacterial bioluminescence enzymes, the time necessary to reach the plateau varies with NADH concentration from 1 to 3 min but here, the light emission is stable for only 1 to 2 min. It must be pointed out that with the luciferase-

TABLE 3
Use of Immobilized Reagents for Chemiluminescence Analysis in Batchwise Systems

Analyte	Detection limit (*) or linearity	Immobilized reagent	Support and activation technique	Chemiluminescent reagent	Ref.
α-Chymotrypsin	*0.2 μg/l	Isoluminol coupled to specific peptides	Activated crosslinked agarose beads (Affi-Gel 10)	Isoluminol/hematin	36
Trypsin	*20 μg/l				
Human thrombin	*0.5 μg/l				
Thiol compounds	*5 pmol	Mercaptoacetyl-luminol	Thiopropyl-Sepharsoe 6B	Luminol/ferricyanide	37
H₂O₂	*0.2 μM	Peroxidase	Photocrosslinkable resin prepolymer	*Cypridina* luciferin analog	38
Xanthine	*0.02 μM	Xanthine oxidase			
Hypoxanthine	*0.02 μM	Xanthine oxidase			
Cholesterol	*2 μM	Cholesterol oxidase	Ion-exchangeable cellulose beads + glutaraldehyde		
β-D-Glucose	*0.4 μM	Glucose oxidase			
Uric acid	*2 μM	Urate oxidase			
β-D-Glucose	*28 nmol	Luminol	Impregnated dialysis membrane	luminol/peroxidase	39
β-D-Glucose	*5×10⁻⁸M	Peroxidase-glucose oxidase	Impregnated filter membrane		
		Glucose oxidase	Ion exchanger or controlled pore glass + glutaraldehyde	TCPO/3-amino-fluoranthene	40
		Solid TCPO and 3-amino-fluoranthene	Controlled pore glass		
H₂O₂	1×10⁻⁸-10⁻⁴M	Peroxidase	Polyamide membrane	Luminol/peroxidase	41
Cholesterol	1×10⁻⁶-2.5×10⁻⁴M				

Note: TCPO = bis(2,4,6-trichlorophenyl) oxalate.

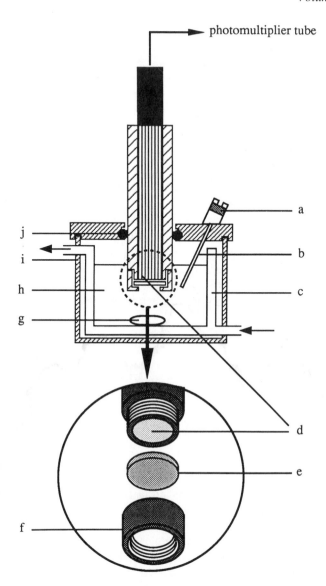

FIGURE 1. Fiber-optic biosensing probe setup. (a) Septum; (b) needle guide; (c) thermostated reaction vessel; (d) fiber bundle; (e) enzymatic membrane; (f) screw-cap; (g) stirring bar; (h) reaction medium; (i) black PVC jacket; (j) O-ring. The arrows indicate the water-flow during thermostatization. (Reprinted from Blum, L. J., Gautier, S. M., and Coulet, P. R., *Anal. Lett.*, 21, 721, 1988. With permission.)

oxidoreductase system, the shape of the bioluminescent signal depends on the relative activity of the two enzymes. A longer steady-state light emission could be obtained by decreasing the oxidoreductase activity.

With the luminol chemiluminescence reaction for H_2O_2 measurements, the maximum light intensity is reached within 1 min but the signal decreases after a 20 to 30 s stable light emission. Here also, a variable transition time phase is observed and in this case, it mainly depends on hydrogen peroxide and luminol concentrations.

The detection limit, defined as the lowest concentration yielding an increase of light above the background level for a signal/noise ratio of 2, was equal to 2.8×10^{-10} *M* ATP, and linearity of the calibration curve was observed up to 1.4×10^{-6} *M*. The detection limit for NADH was 3×10^{-10} *M*; however, the calibration graph obtained by measurements of the maximum light intensity was linear from 1×10^{-9} to 3×10^{-6} M. The amount of background light associated with the enzymatic nylon membranes was found generally proportional to the immobilized activities, and wider linear dynamic ranges were obtained with our sensor compared to other optical devices based on colorimetric or fluorimetric reactions.

For hydrogen peroxide measurements, the background light level which also imposed the detection limit depended on the luminol concentration and on the pH of the reaction medium. With the conditions selected (pH 8.5, 0.05 mM luminol), a linear relationship of peak light intensity vs. H_2O_2 concentration was observed between 2×10^{-8} and 2×10^{-5} M.

Much attention has been paid to the NADH fiber-optic probe because of the strong interest of the microdetermination of this analyte in clinical analysis,[52] and operational and storage stability have been carefully studied for proposing a reliable probe mainly devoted to NADH monitoring.[53] By coimmobilizing on preactivated polyamide membranes the bioluminescent bacterial system with suitable dehydrogenases, the determination of sorbitol, ethanol, and oxaloacetate has been performed at the nanomolar level with a good precision.[54]

For alcohol and sorbitol determinations, the dehydrogenase reactions catalyzed by alcohol dehydrogenase (ADH) and sorbitol dehydrogenase (SDH), respectively (Reactions 14 and 15), have been directly coupled to the luminescent system in the presence of NAD^+ in excess. Then the rate of NADH production was measured by the peak light intensity.

$$\text{alcohol} + \text{NAD}^+ \xrightarrow{\text{ADH}} \text{aldehyde} + \text{NADH} + \text{H}^+ \tag{14}$$

$$\text{D} - \text{sorbitol} + \text{NAD}^+ \xrightarrow{\text{SDH}} \text{D} - \text{fructose} + \text{NADH} + \text{H}^+ \tag{15}$$

The assay of oxaloacetate involves concomitant consumption of NADH by both malate dehydrogenase (MDH) (Reaction 16) and oxidoreductase (Reaction 6) and must be conducted in a sequential manner.

$$\text{NADH} + \text{H}^+ + \text{oxaloacetate} \xrightarrow{\text{MDH}} \text{NAD}^+ + \text{L} - \text{malate} \tag{16}$$

First, the injection of NADH leads to a constant light emission, then the oxaloacetate containing sample is injected leading to a light decrease. Either the rate of light decrease or the variation of light intensity can be linearly related to the oxaloacetate concentration.

Recently, we have described a continuous flow method for the bioluminescent determination of NADH using the fiber-optic biosensing probe associated with a specially designed flow cell[55] (Figure 2). The assay was linear from 2×10^{-12} to 1×10^{-9} mol NADH with a precision of 3.4 % for 1×10^{-10} mol. It was possible to measure 25 samples per hour with no carry-over. Furthermore, no loss of activity was observed after 150 assays performed within 3 d.

V. CONCLUSION AND FUTURE TRENDS

Chemiluminescence and bioluminescence optical probes represent a class of new analytical tools in the promising field of fiber-optic-based sensing devices. Emphasis can be put on several advantages. No light source is needed, since light emission is a product of the luminescence reaction itself, which allows the use of instrumentation simpler than with classical spectroscopy. Miniaturization could be more easily achieved with new generations of detectors, i.e., photodiodes and charge-coupled devices. Furthermore, the use of fiber optics coupled to such systems will allow remote sensing with signal processing distant from the measurement site. The ultrasensitivity allied to a wide dynamic range obtainable with these light-emitting reactions, suit the strong demand for monitoring analytes at a very low level in biomedical applications, food industry, as well as in the expanding environmental area. The

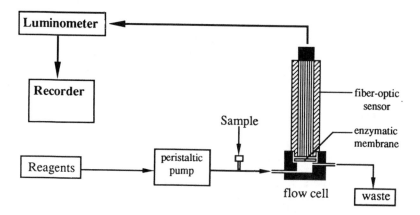

FIGURE 2. Flow diagram of the fiber-optic continuous bioluminescent system for the determination of NADH.

coupling of auxiliary enzymes, mainly oxidases and dehydrogenases, to luminescence reactions makes the approach even more versatile, with the possible measurement of the number of target organic compounds very difficult to detect at a trace level by other means.

The combination of fiber-optic biosensing probes with flow injection analysis represents a challenging goal which may lead to a decisive breakthrough in the coming years.

REFERENCES

1. **Grayeski, M. L.,** Chemiluminescence analyses in solution, in *Chemi- and Bioluminescence*, Burr, J. G., Ed., Marcel Dekker, New York, 1985, 469.
2. **Cormier, M. J. and Prichard, P. M.,** An investigation of the luminescent peroxidation of luminol by stopped flow techniques, *J. Biol. Chem.*, 243, 4706, 1968.
3. **Kuwana, T.,** Electro-oxidation followed by light emission, *J. Electroanal. Chem.*, 6, 164, 1963.
4. **Epstein, B. and Kuwana, T.,** Electrooxidation of phthalhydrazides, *J. Electroanal. Chem.*, 15, 386, 1967.
5. **Imai, K., Miyaguchi, K., and Honda, K.,** High-performance liquid chromatography-chemiluminescence reaction detection system of fluorescent compounds using TCPO and hydrogen peroxide, in *Bioluminescence and Chemiluminescence : Instruments and Applications*, Vol. 2, Van Dyke, K., Ed., CRC Press, Boca Raton, FL, 1985, 65.
6. **McElroy, W. D.,** The energy source for bioluminescence in an isolated system, *Proc. Natl. Acad. Sci. U.S.A.*, 33, 341, 1947.
7. **White, E. H., McCapra, F., Field, G. F., and McElroy, W. D.,** The structure and synthesis of firefly luciferin, *J. Am. Chem. Soc.*, 83, 2402, 1961.
8. **DeLuca, M.,** Firefly luciferase, in *Advances in Enzymology*, Vol. 44, Meister, A., Ed., John Wiley & Sons, New York, 1976, 37.
9. **DeLuca, M. and McElroy, W. D.,** Purification and properties of firefly luciferase, in *Methods in Enzymology*, Vol. 57, DeLuca, M. A., Ed., Academic Press, New York, 1978, 3.
10. **McElroy, W. D. and DeLuca, M.,** Firefly luminescence, in *Chemi- and Bioluminescence*, Burr, J. G., Ed., Marcel Dekker, New York, 1985, 387.
11. **Nealson, K. H.,** Isolation, identification, and manipulation of luminous bacteria, in *Methods in Enzymology*, Vol. 57, DeLuca, M. A., Ed., Academic Press, New York, 1978, 153.
12. **Lee, R. T., Denburg, J. L., and McElroy, W. D.,** Substrate-binding properties of firefly luciferase. II. ATP-binding site, *Arch. Biochem. Biophys.*, 141, 38, 1970.
13. **Meighen, E. A. and MacKenzie, R. E.,** Flavine specificity of enzyme-substrate intermediates in the bacterial bioluminescent reaction. Structural requirements of the flavine side chain, *Biochemistry*, 12, 1482, 1973.
14. **Gerlo, E. and Charlier, J.,** Identification of NADH-specific and NADPH-specific FMN reductases in *Beneckea harveyi, Eur. J. Biochem.*, 57, 461, 1975.

15. **Jalkian, R. D. and Denton, M. B.,** Ultra-trace-level determination of cobalt, chromium, and hydrogen peroxide by luminol chemiluminescence detected with a charge-coupled device, *Appl. Spectrosc.,* 42, 1194, 1988.

16. **Mosbach, K.,** *Immobilized Enzymes, Methods in Enzymology,* Vol. 44, Academic Press, New York, 1976.

17. **Mattiasson, B.,** *Immobilized Cells and Organelles,* Vol. 1 and Vol. 2, CRC Press, Boca Raton, FL, 1983.

18. **Lee, Y., Jablonski I., and DeLuca M.,** Immobilization of firefly luciferase on glass rods. Properties of immobilized enzyme, *Anal. Biochem.,* 80, 496, 1977.

19. **Ugarova, N. N., Brovko, L. Yu., and Berezin, I. V.,** Immobilized firefly luciferase and its use in analysis, *Anal. Lett.,* 13, 881, 1980.

20. **Wienhausen, G. K., Kricka, L. J., Hinkley, J. E., and DeLuca, M.,** Properties of bacterial luciferase/ NADH:FMN oxidoreductase and firefly luciferase immobilized onto Sepharose, *Appl. Biochem. Biotechnol.,* 7, 463, 1982.

21. **Ugarova, N. N., Brovko, L. Y., and Beliaieva, E. I.,** Immobilization of luciferase from the firefly *Luciola mingrelica:* catalytic properties and thermostability of the enzyme immobilized on cellulose films, *Enzyme Microbiol.Technol.,* 5, 60, 1983.

22. **Blum, L. J., Coulet, P. R., and Gautheron, D. C.,** Collagen strip with immobilized luciferase for ATP bioluminescent determination, *Biotechnol. Bioeng.,* 27, 232, 1985.

23. **Ugarova, N. N., Brovko, L. Yu., Ivanova, L. V., Shekhovtsova, T. N., and Dolmanova, I. F.,** Bioluminescent assay of creatine kinase activity using immobilized firefly extract , *Anal. Biochem.,* 158, 1, 1986.

24. **Jablonski, E. and DeLuca, M.,** Immobilization of bacterial luciferase and FMN reductase on glass rods, *Proc. Natl. Acad. Sci. U.S.A.,* 73, 3848, 1976.

25. **Haggerty, C., Jablonski, E., Stav, L., and DeLuca, M.,** Continuous monitoring of reactions that produce NADH and NADPH using immobilized luciferase and oxidoreductases from *Beneckea harveyi, Anal. Biochem.,* 88, 162, 1978.

26. **Jablonski, E. and DeLuca, M.,** Properties and uses of immobilized light-emitting enzyme systems from *Beneckea harveyi , Clin. Chem.,* 25, 1622, 1979.

27. **Ford, J. and DeLuca, M.,** A new assay for picomole levels of androsterone and testosterone using co-immobilized luciferase, oxidoreductase and steroid dehydrogenase, *Anal. Biochem.,* 110, 43, 1981.

28. **Wienhausen, G. and DeLuca, M.,** Bioluminescent assays of picomole levels of various metabolites using immobilized enzymes, *Anal. Biochem.,* 127, 380, 1982.

29. **Rossi, S. S., Clayton, L. M., and Hofmann, A. F.,** Determination of chenodiol bioequivalence using an immobilized multi-enzyme bioluminescence technique, *J. Pharm. Sci.,* 75, 288, 1986.

30. **Schoelmerich, J., van Berge Henegouwen, G. P., Hofmann, A. F., and DeLuca, M.,** A bioluminescence assay for total 3α-hydroxy bile acids in serum using immobilized enzymes, *Clin. Chim. Acta,* 137, 21, 1984.

31. **Roda, A., Kricka, L. J., DeLuca, M., and Hofmann, A. F.,** Bioluminescence measurement of primary bile acids using immobilized 7 α - hydroxy-steroid dehydrogenase: application to serum bile acids, *J. Lipid Res.,* 23, 1354, 1982.

32. **Schoelmerich, J., Hinkley, J. E., MacDonald, I. A., Hofmann, A. F., and DeLuca, M.,** A bioluminescent assay for 12-α-hydroxy bile acids using immobilized enzymes, *Anal. Biochem.,* 133, 244, 1983.

33. **Rodriguez, O. and Guilbault, G. G.,** Immobilized bacterial luciferase for microscale analysis of creatine kinase activity, *Enzyme Microbiol. Technol.,* 3, 69, 1981.

34. **Blum, L. J. and Coulet, P. R.,** Bioluminescent determination of reduced nicotinamide adenine dinucleotide with immobilized bacterial luciferase and flavin mononucleotide oxidoreductase on collagen film, *Anal. Chim. Acta,* 161, 355, 1984.

35. **Ugarova, N. N., Lebedeva, O. V., and Frumkina, I. G.,** Bioluminescent microassay of various metabolites using bacterial luciferase co-immobilized with multienzyme systems, *Anal. Biochem.,* 173, 221, 1988.

36. **Branchini, B. R., Salituro, F. G., Hermes, J. D., and Post, N. J.,** Highly sensitive assays for proteinases using immobilized luminogenic substrates, *Biochem. Biophys. Res. Commun.,* 97, 334, 1980.

37. **Lippman, R. D.,** Sensitive solid-phase chemiluminescence microassay of thiols, *Anal. Chim. Acta,* 116, 181, 1980.

38. **Kobayashi, T., Saga, K., Shimizu, S., and Goto, T.,** The application of chemiluminescence of a *cypridina* luciferin analog to immobilized enzyme sensors, *Agric. Biol. Chem.,* 45, 1403, 1981.

39. **Carter, T. J. N., Whitehead, T. P., and Kricka, L. J.,** Investigation of a novel solid-phase chemiluminescent analytical system incorporating photographic detection for the measurement of glucose, *Talanta,* 29, 529, 1982.

40. **van Zoonen, P., de Herder, I, Gooijer, C., Velthorst, N. H., and Frei, R. W.,** Detection of oxidase generated hydrogen peroxide by a solid state peroxyoxalate chemiluminescence detector, *Anal. Lett.,* 19, 1949, 1986.

41. **Blum, L. J., Plaza, J. M., and Coulet, P. R.,** Chemiluminescent analyte microdetection based on the luminol H_2O_2 reaction using peroxidase immobilized on new synthetic membranes, *Anal. Lett.*, 20, 317, 1987.

42. **Ugarova, N. N. and Lebedeva, O. V.,** Immobilized bacterial luciferase and its applications, *Appl. Biochem. Biotechnol.*, 15, 35, 1987.

43. **Coulet, P. R. and Blum, L. J.,** Immobilized biological compounds in bio- and chemiluminescence assays, in *Analytical Uses of Immobilized Biological Compounds for Detection, Medical and Industrial Uses,* Guilbault, G. G. and Mascini, M., Eds., D. Reidel, Dordrecht, Holland, 1988, 237.

44. **Coulet, P. R. and Blum, L. J.,** Chemiluminescence and photobiosensors, in *Practical Fluorescence,* 2nd edition, Guilbault, G. G., Ed., Marcel Dekker, New York, 1990, 543.

45. **Freeman, T. M. and Seitz, W. R.,** Chemiluminescence fiber optic probe for hydrogen peroxide based on the luminol reaction, *Anal. Chem.*, 50, 1242, 1978.

46. **Aizawa, M., Ikariyama, Y., and Kuno, H.,** Photovoltaic determination of hydrogen peroxide with a biophotodiode, *Anal. Lett.*, 17, 555, 1984.

47. **Abdel-Latif, M. S. and Guilbault, G. G.,** Fiber-optic sensor for the determination of glucose using micellar enhanced chemiluminescence of the peroxyoxalate reaction, *Anal. Chem.*, 60, 2671, 1988.

48. **VanDyke, D. A. and Cheng, H.-Y.,** Electrochemical manipulation of fluorescence and chemiluminescence signals at fiber-optic probes, *Anal. Chem.*, 61, 633, 1989.

49. **Blum, L. J., Gautier, S. M., and Coulet, P. R.,** Luminescence fiber-optic biosensor, *Anal. Lett.*, 21, 717, 1988.

50. **Blum, L. J., Gautier, S. M., and Coulet, P. R.,** Design of luminescence photobiosensors, *J. Biolum. Chemilum.*, 4, 543, 1989.

51. **Assolant-Vinet, C. H. and Coulet, P. R.,** New immobilized enzymes membranes for tailor-made biosensors, *Anal. Lett.*, 19, 875, 1986.

52. **Coulet, P. R., Blum, L. J., and Gautier, S. M.,** Luminescence optical biosensor oriented to clinical analysis, in *Automation and New Technology in the Clinical Laboratory,* Okuda, K., Ed., Blackwell Scientific Publications, 1990, 239.

53. **Gautier, S. M., Blum, L. J., and Coulet, P. R.,** Fiberoptic sensor based on luminescence and immobilized enzymes: microdetermination of sorbitol, ethanol and oxaloacetate, *J. Biolum. Chemilum.*, 5, 57, 1990.

54. **Gautier, S. M., Blum, L. J., and Coulet, P. R.,** Fiberoptic biosensor based on luminescence and immobilized enzymes: microdetermination of sorbitol, ethanol and oxaloacetate, *J. Biolumin. Chemilumin.*, 3, in press, 1989.

55. **Blum, L. J., Gautier, S. M., and Coulet, P. R.,** Continuous flow bioluminescent assay of NADH using a fiber-optic sensor, *Anal. Chim. Acta*, 226, 331, 1989.

Chapter 21

FIBER OPTIC CHEMORECEPTION

Ulrich J. Krull, R. Stephen Brown, and Elaine T. Vandenberg

TABLE OF CONTENTS

I. OPTICAL BIOSENSORS AND SELECTIVE BIOCHEMICAL RECOGNITION PROCESSES

The possibility of using classical biochemical recognition processes as the selectivity element in a fiber-optic biosensor is extremely attractive for many reasons. The biochemical interactions of enzymes with substrates, antibodies with antigens, and lectins with saccharides are selective, sensitive, and fast. Many compounds that are of potential analytical interest for rapid and/or continuous measurement by a biosensor are molecules which are often detected quickly and efficiently by natural systems. In addition, selectivities of the enzyme, antibody, or lectin systems can be altered by chemical modification or genetic engineering. The opportunity to customize the design of certain receptors for ligands of interest may be a goal realized in the near future. It is therefore understandable that there is much interest in incorporating biological molecules with fiber-optic devices to create biosensors, especially given that the relevant commercial interests of industries such as clinical medicine and food processing remain largely undeveloped.

In the biosensor field, the term "receptor" is often used with a broader definition than is given in the field of biochemistry. A receptor is defined for the purposes of this article to be any molecule which specifically binds complementary molecules by way of an arrangement of functional groups in three-dimensional space. Certain organic molecules such as chelators and crown ethers are considered receptors, as are biological compounds such as enzymes, antibodies, lectins, and molecular receptors. We will be focussing on biological receptors in this chapter.

The receptors described in the following sections have several common features. They are all proteins or glycoproteins, most are soluble in aqueous media, and all have selectivity for certain ligand types. Biological receptors usually contain a cavity conventionally known as a binding site. The steric shape and/or charge distribution within this cavity controls the selectivity of binding to various ligands. Generally as the binding site increases in size and complexity a concurrent increase in specificity for ligands is achieved, owing to a greater number of interactions that must be favorable towards a particular ligand. It should be noted that the individual binding forces are weak, consisting of electrostatic interactions, hydrogen bonds, Van der Waals interactions, and hydrophobic forces. The sum of all these interactions determines the strength and specificity of the binding.

The binding of a ligand to a cavity in natural systems may initiate secondary effects such as ligand cleavage, complement binding, molecular conformational changes, or initiation of cell processes. These secondary effects result from chemical, spatial, and/or electrostatic alterations caused by the complexation event. It is therefore possible to transduce the physical changes associated with the biochemical recognition into a signal suitable for measurement by a device. A common method of investigating these physical changes in biological systems is through the use of fluorescence spectroscopy. The sensitivity and multidimensional nature of fluorescence make it the method of choice for optical sensor development. The optical signal generated may be in the form of variations in wavelength, intensity, lifetime, or polarization.

A. ENZYME-SUBSTRATE SYSTEMS

An enzyme is defined as a catalytic protein which increases reaction rates in biological systems on the order of 10^6 by lowering the Gibbs free energy of activation.[1] The selectivity of this catalysis varies from enzyme to enzyme, as does the binding constant of 10^2 to 10^8 M, corresponding to -3 to -12 kcal .mol^{-1} binding energy, and the turnover number, from 0.5 to 600,000 s^{-1}.[1] Some enzymes demonstrate large changes in conformation on ligand binding while others show little.[1] The binding site of the enzyme contains amino acid residues known as catalytic groups that assist the formation or dissociation of covalent bonds.

The analytical attractiveness of enzymes for use in fiber optic sensors lies in their selectivity and reversibility. Only certain types of ligand are bound by the enzyme, and in some cases selectivities are well documented. Enzymes are fully regenerated immediately after the catalytic reaction, and thus response is reversible. Enzyme action is usually very fast, as demonstrated by high turnover rate. Some enzymes, such as carbonic anhydrase and acetylcholinesterase, are limited in catalytic rate only by the diffusion of substrate to enzyme.[2]

To be useful in a fiber-optic sensor, the enzyme must directly or indirectly transduce a chemical signal to an optical signal. The main method of optically detecting intact proteins is by measurement of fluorescence. Radiation of 280 nm will be absorbed and fluorescence emitted from some proteins, indicating the presence of tryptophan, tyrosine, or phenylalanine residues.[3] In order for binding of a substrate to induce a change in this fluorescence, the emitting residues must be exposed to a different environment on binding, implying a significant change in the conformation or electrostatic arrangement of the enzyme. Although some enzymes demonstrate large conformational changes for certain catalytic residues (e.g., in carboxypeptidase A, tyrosine 248 rotates by 12 Å),[4] this avenue of transduction is unlikely to be sufficiently sensitive in most cases due to the small conformational changes associated with most binding processes and limited concentrations of the fluorescent species.

Fluorescent labels can be placed on the enzyme, and a signal produced if the environment of the label changes when substrate binding occurs. The label should be near the binding site in order to be affected by ligand binding, but must not obscure the binding site and alter the activity of the enzyme. Also, the label should be placed on the same residue(s) of each enzyme in the sensor to give maximum reproducibility of response. Unfortunately, only empirical testing will determine the feasibility of this scheme.

An alternate philosophy of signal transduction is to use an indirect method of producing an optical signal. The products resulting from an enzymatic reaction may exhibit characteristic absorption or fluorescence properties, in which case the reaction could be monitored by an optical fiber system. An example of absorption measurement is the catalysis of p-nitrophenyl-phosphate to p-nitrophenoxide by alkaline phosphatase, where the product absorbs at 404 nm.[5] The limitations of this approach lie in the limitations of the Beer-Lambert Law, illustrated in this case by linearity in analyte concentration over only one order of magnitude. An example of a fluorescent product is the production of NADH from NAD$^+$ during the oxidation of ethanol by alcohol dehydrogenase.[6] A modification of these schemes involves a fluorescently tagged substrate which is immobilized in the sensor system. An analyte enzyme cleaves the substrate, releasing a fluorescent material which may be quantitatively measured.[7]

B. ANTIBODY-ANTIGEN SYSTEMS

An antibody is a protein which specifically binds a defined region (antigenic determinant) on a second molecule, called an antigen.[8] Antibodies, or immunoglobulins, are produced by animals in response to the presence of foreign substances, and they have a specific affinity for the foreign material that elicited their synthesis. The antigenic determinant normally occurs in only one molecule in the sample of interest; thus antibodies are usually considered specific for the target antigen. A schematic of the antibody protein immunoglobulin G is shown in Figure 1.

The binding site of the antibody is typically 6 to 34 Å deep; thus the antibody is capable of wrapping around the antigen. In some classes of antibody, binding creates a change in the F_c portion of the antibody, causing the complement system to bind to the C_H2 domain of the F_c. Typical antibody-antigen binding constants are 10^3 to 10^{10} M, corresponding to free energies of binding from -6 to -15 kcal/mol.

The antibody binding site is often a nonpolar niche in the protein, and the antigen is bound to this site through many weak interactions to provide high complexation affinity (binding energy) and little rotational freedom. The binding is diffusion-controlled, with a rate constant of 10^8 $M^{-1} s^{-1}$. The existence of two binding sites on the antibody allows for increased affinity of the antibody-antigen complex by a factor of 10^4 in the case of a multivalent antigen.

One major drawback of using an antibody-antigen system is that this binding is normally irreversible, and therefore restricted to single-use applications. Some complexes may be dissociated, such as when increasing ionic strength to extract material in affinity chromatography.

Catalytic antibodies have recently been developed to combine the extreme specificity of

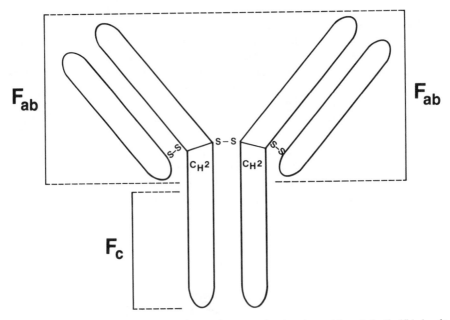

FIGURE 1. Schematic diagram of immunoglobulin G, showing domains and interchain disulfide bonds.

antibody recognition with a reaction rate increase to within a few orders of magnitude of the rate of typical enzymes.[9,10] These antibodies have a high affinity for the transition state conformation of the antigen/substrate, so that the act of binding to the antibody catalyzes breaking of bonds. This catalyzed reaction goes in the forward direction only, unlike enzymatic reactions which generally catalyze forward and reverse reactions.

The main use of fluorescence measurements with antibody-antigen systems is in conventional fluoroimmunoassay. Analysis is generally performed after some washing procedure, and therefore does not operate in a sensing mode. Adaptations of this system for use with fiber-optic technology have been reported.[11]

Since no large-scale conformational rearrangements are seen, detection of antibody-antigen complexation is through simple intensity increases on accumulation of fluorescent material, implying fluorescently labeled analyte, or decreases due to loss of fluorescent material where competitive binding is possible.

A more generic approach to detection through changes in fluorescence may be possible if the fluorophore is placed near to the binding site, so it is perturbed by the complexation but does not significantly alter the selectivity of the antibody. For some systems, a conformational change seems to be transmitted to the C_H2 region of the F_c portion (see Figure 1), triggering complement binding. A fluorophore attached in this region may be able to report the initial complexation.

C. LECTIN-SACCHARIDE SYSTEMS

Lectins are proteins which selectively bind one or more sugar residues. Derived from plants, their function is unknown, but they have been observed to stimulate mitosis. Lectins are selective for certain types of sugar residues. For example, concanavalin A prefers internal and nonreducing α-mannosyl residues, whereas wheat germ agglutinin prefers terminal N-acetylglucosamines.[12] Binding constants are typically 10^2 to 10^8 M;[13] thus, some interactions are reversible and others are relatively irreversible. Dissociation may be induced by increasing ion concentration or by adding higher concentrations of saccharides which competitively bind with the protein.

Lectins have been used in the development of a fiber-optic sensor using a competitive displacement reaction.[14] A fluorescently labeled saccharide was allowed to complex with lectin bound to the surface of a semipermeable membrane compartment at the end of an optical fiber. Unlabeled glucose entered the compartment from solution through the membrane and competed for binding sites on the lectin. The subsequently released fluorescent saccharide entered into the optical fiber light path, where it was detected as a fluorescence increase.

As is the case for antibody-antigen complexation, the binding of lectin-saccharide does not induce large scale conformational changes, nor is there any *in vivo* method of chemical signal transduction observed. This indicates that generic signal transduction will be possible only if the fluorescent label is located near the binding site, with the same considerations as discussed previously.

D. MOLECULAR RECEPTOR-LIGAND SYSTEMS

Molecular receptors are a membrane-embedded class of proteins which transduce chemical signals from the extracellular to the intracellular environment. These proteins are very attractive for use in biosensors for several reasons: they exhibit good selectivity, tight ligand binding (e.g., $K_d = \mu M$ to nM for the acetylcholine receptor[15]), reversibility, and intrinsic transduction of the binding to a second messenger (e.g., Na^+ influx or activation of cyclic AMP). The difficulty in working with these systems is that the receptors must be kept in a lipid membrane which is difficult to maintain in a practical device assembly. Accordingly, molecular receptors have never been used in fiber optic devices, although attempts are being made at incorporating these proteins into electrochemical devices.[16-18]

The intrinsic signal transduction of molecular receptors provides unique characteristics to this class of selective binding proteins. This transduction implies that a conformational and/ or electrostatic change occurs within the receptor due to ligand binding, in a manner which also perturbs the surrounding lipid membrane (see Section IV) or other intrinsic proteins (e.g., the G-protein). Fluorescent probes on the receptor protein or in the lipid may therefore provide a direct signal sensitive to the physical alterations caused by receptor-ligand interaction. For example, acetylcholine receptor labeled with 4-[N-(iodoacetoxy)ethyl-N-methyl)amino-7-nitrobenz-2-oxa-1,3-diazole (IANBD) exhibits fluorescence enhancement upon selective binding to carbamylcholine.[19-21]

E. ENERGETICS OF SELECTIVE BINDING — INTRINSIC ANALYTICAL INFORMATION

The energetics of enzyme-substrate reactions have been studied extensively by Jencks and others.[22] The concepts can be applied directly to catalytic antibody reactions, and in many cases to antibody-antigen, lectin-saccharide, and molecular receptor-ligand interactions.

The most popular theories in enzyme-substrate binding that account for the maximized reaction rates, but not maximal binding constants, are those of induced fit, nonproductive binding, and rate acceleration by the induction of strain. Elements of all three may contribute to the reaction. The induced fit theory states that the catalytic groups in the enzyme are not in a position to catalyze a reaction when a substrate of low binding constant (poor substrate) is bound. However, when a good substrate is bound, it forces the enzyme into an energetically less favorable, but catalytically more favorable, conformation. Thus there are structural differences between the free and bound enzyme. Optimal catalysis in both directions occurs when the enzyme best fits the transition state conformation of the reaction from substrate to product, and optimal catalysis in the forward direction only occurs when the enzyme best fits the product, thus forcing the reaction to reduce strain in the enzyme. The energies of repulsion of functional groups on the enzyme and substrate in the initial conformation are used to distort the active site to provide a better fit for the substrate.

The nonproductive binding theory refers to the situation in which a substrate binds tightly to the enzyme in a conformation in which the catalytic groups are not in a favorable arrangement for catalysis. In this case, the energy of binding prevents its reaction, thus contributing to the specificity of the enzyme. Nonproductive binding of substrates appears kinetically as though competitive inhibition is taking place.

The induction of strain is a complex subject. Some manifestations of strain include geometric destabilization, desolvation, electrostatic destabilization, and induced destabilization. With geometric destabilization, models usually consider the enzyme to be rigid and the substrate deformable, or vice versa. In reality, probably both the enzyme and the substrate are deformable to some extent. In all cases, the observed binding energy is the difference between the binding energy of the substrate and the energy required to distort the enzyme and/or substrate. In other words, the binding energy is used to destabilize the starting materials, thus decreasing the free energy of activation of the reaction, as well as providing specificity at high substrate concentrations. If binding of the substrate involves desolvating polar groups or causes electrostatic repulsions between charged groups, this energy can be released by resolvating or moving the charges apart, contributing to the driving force for the reaction to proceed. In the case of desolvation, polar groups may become more reactive when moved into a medium of lower dielectric than water. Induced destabilization refers to conformational effects being translated over a distance in the protein by a three-dimensional conformation. This is seen especially clearly in the case of elastase, where lengthening the substrate increases the free energy of binding, even though the added residues are far from the catalytic groups.

The concepts of induced fit and nonproductive binding can be applied to the antibody-antigen and lectin-saccharide cases to explain the varying degrees of selectivity of receptor for ligand. Induction of strain is unlikely to be important due to the general lack of conformational changes in these systems. Molecular receptor behavior is energetically similar to the enzyme system, but with greater complexity since the receptors are known to change conformation reversibly with kinetics that are dependent on many extrinsic variables (see Figure 2 for an energy diagram of the ternary complex model of molecular receptor binding).[23]

II. CHEMORECEPTION — NATURE'S ANALYTICAL CHEMISTRY

It is interesting to note that the same factor which is directing the present development of sensor technology has also shaped the evolution of the specialized sensors which can be found in natural organisms. The necessity of rapid selective chemical analysis for communication purposes in nature has prompted the invention of a "natural analytical chemistry" which may offer guidelines for the development of a new sensor technology.[24]

The sensitivity of animal life to chemicals is multifaceted. Biological species respond internally to messages in the form of transmitters and hormones, as well as externally through receiving sites (receptor cells) for gleaning information about the external environment. In this respect, there is a tremendous amount of variety in the design and location of such cells. For example, insects sense chemicals with their antennae, certain invertebrates have chemical sensors in their feet, and selected land vertebrates such as snakes, rabbits, and dogs have a special structure called the vomeronasal organ. Despite these variations there is a great similarity in the application of these senses. Chemical sensitivity is used for assessment of the proximity of food or other environmental objects, for communication purposes, and through release of pheremones, for mating, alarm, trail marking, aggregation and dispersion, etc.

The olfactory epithelium is made up of an array of 10 million to 20 million receptor cells.[25] A mucous layer (10 to 55 μm thick)[26] formed from secretions emanating from Bowman's glands covers the surface. Each olfactory receptor cell is a bipolar neuron which is composed of the cell body and axon. The dendritic portion (1 to 3 μm in diameter),[26] which generates electrical potentials as a result of chemical stimulation, terminates at the olfactory rod. These

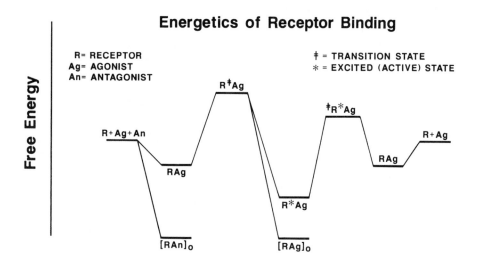

FIGURE 2. An energy diagram of the ternary complex model describing the behavior of the β-adrenergic system.

structures are widely believed to bear the molecular receptors for odorous molecules. Based on the number and size of cilia per cell (10 to 20 for vertebrates[25]) and the total number of receptor cells (e.g., 10 million to 20 million for humans vs. 100 million for rabbits), the superiority of different species for odor discrimination has been correlated with overall receptive area, viz., humans 250 cm² and rabbits 600 cm².[27]

Receptor cells are regarded as biological transducers which convert the energy of a given stimulus into an electrical signal responsible for initiation of neural impulses. The transduction process involves the modulation of the flow of ion current through electrochemically resistive cell membranes. These membranes are fundamental assemblies of two layers of lipid molecules interspersed with a multitude of different proteins. The proteins provide structural stabilization and cellular differentiation with respect to biological activity, while the underlying bilayer lipid membrane (BLM) structure provides a relatively ion-impermeable barrier between the cytoplasm and the external environment of the cell.

Natural organisms employ specialized chemically sensitive binding tissue which can interact with particular stimuli at the cell membrane surface. The operation of this molecular recognition process is based on the availability of membrane-embedded receptors, where three-dimensional molecular shape and electronic distribution govern the degree of binding. Interaction with a stimulant at the receptor can generate a sudden measurable inorganic ion flux increase across the cell membrane, which is equivalent to the establishment of a signal containing chemical information. The controllable resistance of the cell membrane to the passage of ions provides the natural system with the ability to transduce the chemical binding event into an electrical signal with a sensitivity so great that it is possible to recognize a few molecules of stimulant as is the case for pheromones.

It is possible that some odorants interact directly with the lipid bilayer causing changes in permeability by a variety of processes including phase transitions, fluidity, packing, and lipid-protein interactions. Odorants have been found to change the surface pressure of lipid monolayer films and these changes have been correlated with the threshold concentrations necessary for odor perception.[28] The role of lipids has been found to be critical in the activity of certain proteins contained in olfactory preparations.[29] Also, odorants have been observed to change the ion conductivity of artificial bilayer membranes made from lecithin (a major component of natural membranes), but a correlation with the olfactory properties of the odorants was not established.

The significance of the "natural analytical chemistry" stems from an intrinsic property of signal amplification combined with selectivity derived from tertiary protein structure. Conventional electrochemical devices which are in widespread use by the analytical community operate on the basis of selective complexation which creates a product which is either potentiometrically or coulometrically measured by conventional electrodes. The sensitivity is determined in part by the quantity of products produced. For example, consider an enzymatic reaction which generates ions; if one complexation reaction provides one product ion, ideally one unit of analytical signal is available for further analysis. An important advantage of the lipid membrane system is that one complexation can modulate the conduction of thousands of ions (or more) implying thousands of units of analytical signal and therefore increased inherent sensitivity. This is a nonequilibrium process, which speeds response times and is not sensitive to interferences commonly encountered in equilibrium processes such as mixed potentials.

A. ARTIFICIAL LIPID MEMBRANE STRUCTURE

The ability of lipids to self-assemble into organized structures such as micelles, vesicles, monolayers, and bilayers is well known and has been extensively investigated.[30,31] Molecular charge, steric profile, and relative acyl chain lengths determine the shapes of the final lipid assemblies. Molecular associations are driven by electrostatic interactions between functional groups in the headgroup region, and by sequestering of acyl chains to form a hydrophobic phase of low dielectric constant. Artificial bilayer lipid membranes were first introduced as experimental models appropriate for studies of biochemical processes at cell membranes by Ti Tien et al.[32] The basic premise of electroanalytical signal generation was first demonstrated by del Castillo.[33] Detailed studies of BLMs have subsequently revealed that transmembrane structure, and therefore transmembrane electrochemical ion current, is dependent upon four major physical features.[34]

First, the phospholipid acyl chain length and/or residual hydrocarbon solvent content within a BLM can control the thickness of the membrane, and therefore the proportion of the hydrophobic zone of low dielectric constant located in the interior of the membrane. An ideal Born interaction energy can be calculated for ions in the nonpolar interior of the membrane. The integration of the Born energy barrier across the hydrophobic width of the membrane provides an indication of the magnitude of the opposition to ionic permeability through the zone of low dielectric constant. The ideal image potential would also be altered as a function of the thickness of the membrane.

Second, the extent of unsaturation of the phospholipid chains has been related to the control of fluidity and thickness of BLMs. The choice of the term fluidity is ambiguous as the molecular motion of the lipids can be axial or lateral, and the term does not account for the proximity of neighboring lipid molecules. It will be assumed that the greater lipid mobility also includes a decreased lipid packing density, and the effects of the unsaturated acyl chains will therefore relate to packing/fluidity alterations. An increase in the extent of acyl chain unsaturation causes an increase in the rate constant of ionic translocation, resulting in the observed higher ionic currents, through increased molecular fluidity and decreased packing density.

Third, the chemical alteration of the headgroup (polar) region can result in substantial differences in the extent of hydrogen bonding and the magnitude of the dipolar potential. The dipolar potential results from the anisotropic structure of the membrane. Dipolar moieties in the polar headgroup zone have a time-averaged alignment which results in the establishment of a net electrostatic polarization. The dipolar potential or Volta potential at gas/liquid or liquid/liquid interfaces, ΔV, can be regarded as the average dipole-based electrostatic voltage perpendicular to the BLM surface. The average molecular dipole moment perpendicular to the plane of the membrane, μ, can be represented as $\mu = \Delta V/4\pi n$, where n is the number of molecules in a defined area. If the molecular area, A, is considered, then the expression can be rewritten as $\mu = A\Delta V/12\pi$.

Fourth, the preparation of BLMs from mixed lipid systems such as phospholipid and steroid can lead to the concurrent alterations of dipolar potential, fluidity, and packing. The partition coefficients evaluated for a number of different ions have been found to be relatively independent of the presence of certain common steroids, though large permeability changes due to dipolar potential and packing/fluidity alterations have been observed.

Experimental work has indicated that the properties of charge, size, and hydrophobicity of the permeating ion; the ion adsorption plane; the density and mobility of the lipid; the dielectric distribution; and the dipolar potential of a BLM are the most significant features which combine to establish the magnitude of the transmembrane ion current. The binding processes attributed to receptors in BLMs are governed mainly by electrostatic fields and dipolar vectors, and local electron density and delocalization. These properties combine to produce the conventional macroscopic features of molecular shape, size, polarity, and hydrophobicity commonly used to describe molecular interactions. These latter features are only poor representations of the underlying electronic structure of any molecule, and serve to disguise the true nature of molecular interactions. However, they are indicative of the synergy between protein-based receptors and the lipid matrix since the same forms of chemical interactions are responsible for their mutual structures and functions. Alterations of the structure, and subsequently the conductivity, of a lipid membrane by selective protein mediated mechanisms can be attributed to one or more of the following processes:

1. Contribution of the stimulant dipole to the existing transmembrane electric field reduces the dipolar potential ion energy barrier.
2. Complexation of stimulant and receptor alters the original electric field contribution of the receptor alone.
3. The electric field changes of items (1) or (2) above may locally change the ion conductivity of the lipid membrane matrix, or may locally alter the conductance of a threshold voltage dependent protein channel or pore. The channel and receptor need not be physically coupled in such a process.
4. The electric field variations may cause a synergic variation of intermolecular association in the membrane.

An overview of these processes reveals that the origin of the electroanalytical signal may be derived from a physical or electrostatic perturbation of the receptor, or the lipid matrix in which the receptor is situated. The observed alteration of transmembrane ion current is therefore a secondary process driven by a primary change in the properties of the membrane.

B. AN OPTICAL ANALOGY OF THE ELECTROCHEMICAL PROCESS

It has been well documented that optical techniques, particularly those using fluorescence probes, may be employed to quantitatively evaluate the electrical potentials, microviscosities, and dielectric characteristics of lipid membranes.[35] Fluorescence techniques therefore provide an alternative method for the generation of secondary signals of analytical utility based on a primary perturbation of lipid membranes by a selective binding event. The fluorescence strategy offers an analytically powerful method suitable for membrane analysis in contrast to the electrochemical system, which can only detect whether or not a membrane perturbation has occurred. Intensity measurements to determine the extent of analyte binding and wavelength shifts to determine structural order can readily be pooled to extensively define the quantitative analytical signal as well as interferences.

Previous work in the area of biosensors introduced the manipulation of properties of ordered BLMs for implementation of a generic transduction mechanism suitable for electrochemical sensing of selective receptor binding events taking place within the sensing membrane. The change in ion current is related to the degree of perturbation of the membrane by receptor complexation, and has provided an analytical signal for analyte concentrations as low as 10^{-13} M in some cases.

Even though the BLM is ideal for sensitive response, and assists in maintenance of an environment suitable for maximizing receptor-binding activity, the electrochemical transduction is limited by extreme sensitivity to electrical noise and requires the complete structural integrity of the lipid membrane. It is possible to avoid these practical limitations by nonelectrochemical analytical exploitation of the membrane perturbation using the high sensitivity of fluorescence spectrophotometry and the ability of membrane-embedded fluorophores to respond to perturbation of the membrane potential and molecular packing/fluidity.

The availability of selective and reversible molecular receptors does not necessarily provide inherent optical transduction capability associated with suitable sensitivity. A mechanism must be established whereby the selective binding process is physically transduced into some form of optical signal. Previous experimentation with antibody-antigen systems has indicated that the optical signal can be derived by extrinsically labeling either reactive constituent with a fluorescent probe, by observing intrinsic protein fluorescence or absorption, or by relying on special optical properties of the receptor-analyte complexation such as excimer formation. Each of these scenarios has limited applicability, and the majority of selective receptor-analyte systems require preliminary covalent attachment of a suitable optical probe. A general sensing strategy capable of detecting the binding event, but eliminating any customized pretreatment of each unique receptor system, is required.

The concept of using membrane perturbation to establish an analytical signal is in principle very different from those proposed recently by Wolfbeis and Schaffar[36] or Zhujun and Seitz,[37] who have introduced the use of lipid membranes in analytical systems. In one instance, the ion complexing agent valinomycin is incorporated into a lipid multilayer with a potential sensitive dye. Variations in cation concentrations are indicated in alterations of fluorescence intensity due to partitioning of ions into the lipophilic complexing agent. In another unrelated report, lipid vesicles are used as reservoirs of agents in a fluorescence experiment. The lysis of the vesicles by an immunochemical reaction releases the indicator and results in fluorescence intensity variation. Neither of these strategies addresses the analytical potential of lipid membranes as generic transducers in an optical mode, and both are unable to provide signals for the broad range of selective biochemical interactions that are presently available as described in Section I.

A generic mechanism for fiber-optic sensor design would include protein receptor incorporation into a membrane, and fluorescence monitoring as a function of perturbation of the lipid membrane due to analyte-receptor binding. Signal transduction of proteins in biological systems is in part a result of their ability to perturb the surrounding cell membrane by affecting membrane structure or electrostatic fields when a binding event occurs at the protein,[22] which should cause a change in fluorescence of certain fluorophores in the membrane. The result would be a generic optrode series, where each sensor was of the same basic structure, but incorporated a different receptor protein according to the analyte to be measured. The generic sensor would exploit the natural selectivity, sensitivity, and reversibility demonstrated by some biological membrane systems, and also has the potential of affecting many fluorescent molecules with a single binding event which constitutes chemical signal amplification. In addition, such experiments may provide the opportunity to further understand the function of biological membranes.

Monolayer investigations at the air-water interface by comparison methods, dipolar potential measurements, and particularly by fluorescence microscopy, have amply demonstrated the tremendous sensitivity of the lipid membrane structure to most of the selective complementary protein binding systems that are known.[38] Epifluorescence results[39] have confirmed that monolayer phase and domain structure are directly modulated by selective protein-based binding events. The results have confirmed that fluorescently labeled receptors could be directly observed, as could fluorescence variations from membrane soluble probes (e.g., fluorescently labeled phospholipids). Such results indicate another advantage of the optical sensing strategy over the electrochemical system. The use of multidimensional information, such as concurrent fluorescence measurements of components at different wavelengths from labeled receptors and lipids, combined with measurements of optical polarization and fluorescence lifetime, provides sufficient information to ascertain whether selective binding has occurred, and what action should be taken to control nonselective interference.

One aspect of the electrochemical system which may be deemed superior to fluorescence experiment is sensitivity. A single ion conductive event can be observed electrochemically, and may result from a single receptor binding event. However, the intrinsic sensitivity of the fluorescence experiment, coupled with the ability to perturb membrane structures so as to "switch" between structural phases or domain distribution,[40] provides better potential for success of such a generic optical transduction strategy.

III. LIPID MEMBRANE STABILIZATION ONTO OPTICAL DEVICE STRUCTURES

The optical implementation of organized planar assemblies of lipid membranes associated with active chemically selective binding agents presents numerous technical difficulties. The major problems which must be addressed concern structural stabilization of extremely thin lipid layers, development of "receptors" which can perturb the lipid matrix where selective binding takes place, and the incorporation of such "receptors" into membranes in such a manner which provides reproducibility and extended lifetime. Of these areas, lipid membrane stabilization and protein incorporation have been extensively investigated.

A. LANGMUIR-BLODGETT FILMS: MONOLAYERS AND MULTILAYERS

One of the most successful methods of preparing surface-deposited lipid monolayers and multilayers for spectroscopic investigation is that of dip-casting prepared lipid films from an air-water interface onto a substrate. Insoluble surfactants in either pure or mixed forms can be compressed as monolayers on a Langmuir trough to establish characteristic pressure-area curves, which provide information about phase/domain structures within the membranes.[41] These films can be transferred onto substrates such as quartz, silicon, glass, or metal surfaces by forcing the substrate through the monolayer. Adhesion of the film, and orientation of the polar and nonpolar areas of molecules comprising the film, is a function of the surface free energy of the substrate and the monolayer. For example, the length of time of immersion of a glass substrate in the trough subphase has an effect on the transfer of a phospholipid monolayer to the glass surface. As the glass is allowed to soak for longer periods in the trough, the free silanols at the solution interface become deprotonated and provide a net negative surface charge.[42] This provides a more polar substrate surface which is suitable for interaction with the polar moieties of the lipid. The lipid monolayer is deposited onto polar surfaces upon withdrawal of the substrate from the subphase, resulting in a monolayer orientation with hydrophilic lipid headgroups adjacent to the support surface and hydrophobic acyl chains extending towards the gas phase. Prior experimental results have indicated that a hydration layer may be trapped between the lipid headgroups and the polar substrate surface. It has been suggested that the monolayer is stabilized on the support by electrostatic interactions of the zwitterionic phospholipid headgroups and the negatively charged support surface.[42]

Orientation and monolayer structure may be controlled in part by electrolyte content and pH of the subphase solution and by the casting speed. Slow and extremely regular casting speeds of 1 mm/min will provide a reproducible transfer of the structure of a monolayer from an air-water interface to a solid substrate.[38] Bilayers and multilayers can be prepared by the casting method, but this requires modifications of the basic technique. For example, multiple deposition of monolayers of stearic acid can be achieved by incorporating divalent cations such as cadmium ion into the aqueous subphase to compensate for the electrostatic headgroup repulsion of the fatty acids. Multilayers of stearic acid can also be produced by extremely rapid repetitive casting (>1 cm.sec^{-1}) of the monolayer from 0.1 M aqueous salt solution onto a surface, at the expense of maintaining the physical structure of the original monolayer.

The deposition of lipid from an air-water interface onto a solid support can be qualitatively monitored by observation of the compensating movement of the compression barrier when the monolayer surface pressure is maintained at a fixed magnitude. Phospholipid deposition onto glass, quartz, and oxidized lipid surfaces are all similar with respect to observed casting properties. Phospholipid casting onto glass surfaces shows no change in monolayer surface area during the immersion sequence, but during substrate withdrawal the compression barrier reduces the surface area of the monolayer corresponding to transfer to the substrate. The transfer function, representing the ratio of the loss of trough monolayer surface area to the total substrate surface area, has been measured as 1.0 ± 0.1 at slow casting speed.[43]

Surface wettability estimates can clearly indicate the presence of phospholipid monolayers after the withdrawal of the support from the aqueous trough subphase. Wettability angle estimates ranged from $10 \pm 3°$ for cleaned glass surfaces to $60 \pm 5°$ for glass wafers coated with monolayers at pressures maintained at 35 mN.m^{-1}. These results confirm that the orientation of the phospholipid is such that the acyl chains are exposed at the air-monolayer interface. Since very hydrophobic surface provides contact angle measurements close to $90°$, the results for the phospholipid monolayers indicate a partial disordering of the monolayer structure, as the acyl chain region of a monolayer is in principle extremely hydrophobic. The structural rearrangement of the phospholipid monolayer to a more favorable thermodynamic equilibrium does not preclude the use of phospholipid monolayers as analytical transducers, as shown by previous fluorescence response results acheived for membrane perturbation.[44]

The ability of a BLM to transduce a receptor binding event is derived from perturbation within the lipid headgroup zone. The orientation of the lipid monolayers on cleaned glass and quartz causes difficulty in the facility with which headgroup perturbations may be achieved. Ideally the polar headgroup region must be readily accessible and therefore should face the external environment. A further limitation stems from the fact that a number of receptors have physical dimensions which span a BLM, and the adsorbed monolayer structure cannot provide physical accomodation for such species. These factors necessitate implementation of a modified support surface based on alteration from a polar hydrophilic nature to an extremely hydrophobic interface.

A relatively smooth and hydrophobic surface was prepared on glass and quartz by prior alkylation with octadecyltrichlorosilane (OTS). A dense monolayer coverage of OTS on the surface was readily achieved. The degree of hydrophobicity was established by means of surface wettability measurements, which indicated values of 90 ± 5 for alkylated substrates. The alkylated slides were chemically stable as judged by subsequent contact angle measurements, and could be stored for weeks before further use. Casting of the planar alkylated surfaces through a phospholipid monolayer resulted in a substrate coating with a transfer ratio of 1.0 ± 0.1 during the subphase immersion sequence. Withdrawal of the alkyated support through the monolayer produced no additional lipid deposition. Consecutive casting cycles produced no further compression barrier movement after the first subphase immersion sequence was completed. Wettability measurements provided values of $80 \pm 5°$, indicating that the phospholipid membrane structure was not well-aligned with polar headgroups exposed at the gas-surface interface. However, a stable phospholipid monolayer does form if the monolayer-coated wafer is not drawn into the atmosphere from the trough subphase. The resulting lipid structure assumes a BLM polar interface configuration, and also provides membrane thicknesses (including phospholipid and alkylation) similar to those observed from BLMs.

B. COVALENT STABILIZATION OF LIPID MONOLAYERS

Lipid monolayers and multilayers prepared by casting films from an air-water interface are easily prepared in the laboratory, but remain sensitive to destruction by relatively minor mechanical, electrical, or osmotic shock. The preparation of multilayers of phospholipid on quartz exemplifies these difficulties, as such monolayers can easily mechanically wash off the substrate, and can rearrange after deposition if moved between aqueous solutions through the gas phase. Further structural stabilization by means of covalent cross-linking between lipid molecules within organized membranes has provided a significant improvement in membrane stability.[45,46] This can be accomplished by photoinitiated polymerization reactions involving functional groups such as diacetylenic residues located on the acyl chains or headgroup region of the lipid monomers. The degree of cross-linking controls the mechanical and chemical stability of the monolayer. The deposition of such intermolecularly stabilized monolayers and multilayers onto solid substrates provides for excellent stabilization and efficient interfacing of membranes with transduction devices. Unfortunately, this method of stabilization eliminates the mobility of the lipid acyl chain regions, and therefore impedes and may eliminate optical signal generation caused by "receptor" perturbation of membrane structure during selective binding processes.

A second route of covalent membrane stabilization involves the self-assembly of lipid monolayers and multilayers on substrates by means of formal chemical reactions. Sagiv introduced the formation of self-assembled close-packed monolayers by use of silane binding to hydroxylated surfaces, and the formation of multilayers by monolayer-to-monolayer binding through siloxane functions.[47] The resulting structures are densely packed and stable, but again are highly immobile with respect to perturbations of structure by selective binding agents. Recent work has extended the implementation of siloxane chemistry to deposit phospholipid monolayers onto quartz and silicon surfaces in a manner which retains significant acyl chain mobility while maintaining high lipid density and order.[48] In this work, one end of the phospholipid acyl chain is covalently attached to a smooth mechanical support which must initially have a high density of surface –OH groups. The other acyl chain is not bound, and may rotate freely to provide fluidity in the membrane. Oxidized silica surfaces were independently treated with chlorosilylated phospholipids in a one-step process under neutral and base catalyzed conditions, providing maximum surface coverage of only 30%. The low surface density was attributed to the higher rate of hydrolysis of the chlorosilylated lipids by surface-bound water and subsequent condensation of silanols to siloxanes, compared to the anticipated condensation of the chlorosilyl moieties with the surface hydroxyls. Improvement of surface coverage was achieved by employing a two-step procedure, which initially made use of a base catalyzed reaction with aminopropyltriethoxysilane (APTES) to place a functionalized high density monolayer on a silica surface, and was followed by attachment of one acyl chain of a phospholipid to the free amino group by means of amide bond formation. The resulting stabilized lipid membrane provided surface coverage of up to 90% as determined by X-ray photoelectron spectroscopy.[49]

Chemical deposition can occur relatively irreversibly through other types of functional groups, as demonstrated by the recent attention to interactions of sulfur and gold. Spontaneous organization of monolayers of stable oriented alkyl monolayers terminated with disulfides or thiol moieties has provided densely packed membranes on fresh vacuum deposited gold films.[50,51] A variety of alkyl disulfides with different terminal functional groups such as hydroxyls, amines, and carboxylic acids have been adsorbed from bulk water and alcohol solutions. Work in our group has placed a synthetic disulfide terminated acyl chain phospholipid onto a gold support and additional tests have demonstrated long term stability in aqueous solution at neutral pH. Further investigations using fluorescently labeled disulfide terminated phospholipids are continuing to determine membrane phase structure and permeability.

C. MEMBRANE INCORPORATION OF SELECTIVE BINDING AGENTS

Membrane modification for the purpose of creating selectivity involves incorporation of binding agents, usually of natural origin, into the surface-stabilized lipid matrix. This can be done by a number of well-established procedures:[52]

1. The binding agent can be introduced into the lipid mixture used to form the membrane.
2. The binding agent can be absorbed into a prepared stable membrane from the aqueous supporting electrolyte.
3. The binding agent can be first mixed with the lipid solution, and the mixture then dispersed mechanically to form vesicles. The vesicles are then introduced to an aqueous solution supporting a stable membrane, where they coalesce with the BLM to deliver receptor to the target membrane. Alternatively, vesicles can be prepared directly from natural membranes and can be used as described here. The principle of vesicle fusion to membranes is based on differences in surface tension between the structures and indicates that control of surface pressures, ion availability, and phase structure are critical.
4. Binding agents can be prepared in a surfactant, then delivered to a stable membrane or the appropriate lipid solution. Removal of the surfactant results in receptor incorporation into the BLM.
5. Membranes from natural sources, carrying their natural receptors, can be directly isolated from tissues and chemically bonded or adsorbed to the appropriate device. This technique would also be suitable for immobilization of whole cells as previously indicated for electrochemical experiments.

Protein incorporation into lipid membranes often occurs spontaneously, and is assisted by the availability of hydrophobic interactions with the headgroup region of the membrane. Certain problems are encountered when incorporation is attempted into surface-stabilized membranes that are evident during incorporation into free BLMs. Stabilized membranes are constrained in terms of transmembrane thickness, transmembrane distribution of hydrophobic/hydrophilic functional groups, the distribution of phase domain structures associated with fluid or gel regions, and packing fluidity within fluid regions due to reduced lateral molecular mobility. Large macromolecular binding agents are often intrinsic proteins which are longer than the width of the membrane. For example, the neural receptor for acetylcholine extends 30 Å intracellularly and 90 Å extracellularly to the natural membrane. Incorporation of acetylcholine receptor into artificial surface deposited monolayers or bilayers cannot provide a natural orientation for the protein, though such an orientation may not be necessary for biological binding activity to be retained. Incorporation of slightly water-soluble proteins or lipids implies that a dynamic equilibrium exists, based on partitioning between the aqueous support solution and the membrane, and can be shifted simply by changing the surface tension (pressure) of the membrane. This process may limit the quantity of protein that can be loaded into the membrane, and therefore can limit the sensitivity of the analytical system. Sensitivity limits can also result from protein aggregation within fluid domains of the membrane, and subsequent reduction in activity, or by adhesion to gel-fluid phase boundaries as a result of interfacial tension.

Binding agent incorporation into membranes is an area fraught with technical problems. Perhaps one of the best experimental techniques disclosed to date was introduced by Heckl as demonstrated by fluorescence and scanning tunneling electron microscopic investigations of monolayers at an air-water interface.[38] Selective binding proteins were incorporated relatively homogeneously into monolayers by directly spraying an aqueous solution of the binding agents very slowly onto a monolayer surface. Incorporation was again driven by a difference in surface tension between the added droplets and the hydrophobic surface of the monolayer, and assisted by the availability of a hydrophobic region on the protein. The resulting monolayer structure provided homogeneous incorporation of a number of different proteins, some of sizes in excess of 100 kDa, with retention of biological activity as detected by fluorescence studies. It was shown that resulting monolayers could then be transferred intact by the dip casting procedure onto various substrates.

IV. RECEPTORS FOR OPTICAL TRANSDUCTION

A prospective biosensor incorporating a molecular receptor in a lipid membrane environment stabilized onto an optical device structure follows naturally from the considerations of the three previous sections. Barriers to the fabrication of such a sensor are, however, extensive. Few integral membrane receptor proteins have been isolated or characterized to a significant degree, resulting in a lack of receptor-analyte systems which may be considered. The nature of receptor-membrane interactions is still not well understood, making membrane-based transduction of receptor activity difficult to define. Stabilized lipid membrane structures exhibit poor reproducibility, stability and durability. While efforts continue to overcome these difficulties, much related work supports the eventual viability of a biosensor based on optical measurements from a membrane-based receptor system.

A. FLUORESCENTLY LABELED RECEPTOR SCHEMES

The simplest method of using fluorescence to detect an analyte uses a nonfluorescent system and a fluorescent analyte. This has been applied to AChR systems through the use of fluorescent analogues of acetylcholine.[53] This type of assay is of little use in a sensing strategy as most analytes do not fluoresce. A more desirable method uses a signal which is derived from secondary fluorescence changes induced by receptor binding of nonfluorescent analytes.

Some characterization of receptor-ligand complexation is possible where there is a change in the intrinsic fluorescence properties of the receptor, generally associated with fluorescence of tryptophan residues of the receptor. A decrease in this signal was used in investigations of the kinetics of toxin binding to AChR,[54] though low signal levels and a general lack of sensitivity of this signal to ligand binding prohibit the use of intrinsic fluorescence for most applications. While the change in signal has been attributed to a conformational change in the protein, the exact nature of that change and the mechanism of fluorescence response are as yet unknown. The intrinsic fluorescence was shown to yield no response to the binding of cholinergic ligands.[55]

A more useful signal may be obtained through the covalent attachment of extrinsic fluorophores to protein molecules. This is best exemplified by the work of Raftery and coworkers.[19-21] Here, AChR was fluorescently labeled using iodoacetoxy nitrobenzoxadiazole (IANBD) which relatively nonspecifically labels nucleophilic sites within the overall receptor system. An enhancement of the fluorescence intensity was observed on addition of agonists: 90% with 8 mM carbamylcholine and 60% with 0.5 mM acetylcholine. Through these measurements, binding kinetics and the dissociation constant of the AChR complexes were studied.

Reproduction of these results in our lab yielded greater sensitivity (e.g., 100% at 5 μmol carbamylcholine), and further spectral investigation was performed on the system. The nonspecific labeling with the IANBD is expected to result in a mixture of label sites on the AChR protein, as well as on other proteins present and on the negative phosphate sites of the lipid headgroups. When the label was added to native (unpurified) receptor vesicles, the fluorescence spectrum obtained was a broad peak centered at 570 nm. Agonist-induced enhancement was observed to affect mainly the shorter wavelength side of the spectrum, indicating that only some of the NBD-labeled components were affected by the receptor binding event. Greater signal enhancement was obtained where the fluorescence intensity was measured as the integrated area of only the shorter wavelength component (500 to 550 nm) of the spectrum.

Reproducibility in enhancement between experiments was not within 50% even for the same sample. Since the long-wavelength portion of the spectrum did not respond to the receptor binding event, the ratio of the short to long wavelength portions was taken, and yielded much better reproducibility (5%). This internal calibration is useful for compensating for many instrumental variables such as source intensity, optical efficiency, and detector sensitivity.

Labeling of purified, reconstituted receptor yielded a slightly narrower peak at 550 nm which increased uniformly on agonist binding. The narrower peak, suggesting fewer fluorescing components, prohibited the use of peak ratios suggesting that the intentional addition of other components may allow for greater reproducibility.

B. FLUORESCENT PROBES FOR SAMPLING LIPID MEMBRANE STRUCTURE

1. Vesicle Investigations

Fluorescence has long been used in the study and characterization of membrane systems. While a detailed description of the mechanisms of fluorescence response to membrane structure will not be presented here [see 56], it should be noted that all of the measurable parameters of fluorescence signals have been utilized in these studies. The membrane systems examined include unilamellar and multilamellar vesicles, monolayers, surface stabilized monolayers, and multilayers. There are many different fluorescent probes, or fluorophores, available for membrane studies. The choice of fluorophore is generally based on spectroscopic considerations and on its fluorescence response to the characteristic of interest.

Incorporation of fluorophores into vesicles has allowed for characterization of membrane phase behavior and local lipid structure.[57] Local structural studies are often aided by using a fluorophore which aligns within the membrane such that the position of the fluorescent moiety may be determined. It is important to ensure that conclusions drawn from fluorescence results are not neglecting processes such as lateral migration and changes in the transmembrane position of the fluorescent molecules. It is equally important to consider possible structural alterations of a system caused by incorporation of the fluorophore.

Fluorophores which change emission efficiency (quantum yield) or lifetime depending on microviscosity (e.g., stilbenes, quinolines) are routinely used for the detection of bulk changes in membrane structure, such as phase behavior. Other probes are used in the characterization of polarity or potential of hydrocarbon and headgroup regions by quantum yield changes and wavelength shifts (e.g., coumarins and anilinonaphthalene sulfonates).[58,59] Polarization of fluorescence can be used to identify molecular orientation where the excitation or emission dipoles of the fluorescent moiety are known, as in the identification of the orientation of chlorophyll-protein complexes within a membrane.[60]

More extensive characterization involves the study of phase structure domains in "phase coexistence" situations, which are prevalent in biological and synthetic BLMs.[61] While these domains are microscopic in nature, they may be studied by using a fluorophore which changes emission characteristics between phases. Deconvolution of multiple components is possible from fluorescence lifetime and rotational correlation measurements, yielding information as to the distribution of domains and possibly some characterization of the domains themselves. The detection of structural domains induced in red blood corpuscle membranes on addition of cholesterol was possible through the observation of biphasic rotational anisotropy of the probe 12-anthroyloxy stearic acid (12-AS).[62] Relative domain distribution has also been studied using self-quenching of fluorescence, which occurs with many probes with an increase in local fluorophore concentration. Perylene, which is known to increase local concentration by partitioning preferentially into less dense phases, was used to detect changes in relative phase distribution during a transition through fluorescence intensity measurements.[35]

Vesicle lipid exchange and vesicle fusion processes are of interest in membrane-based sensor investigations firstly as vesicle exchange systems may eventually be used to deliver receptor molecules to a stabilized membrane system, and secondly since the overall structural changes which occur on fusion or exchange produce similar alterations in local fluorophore environment to those associated with phase behavior in an individual membrane, meaning that the same fluorescence measurements may be applied. Intermolecular quenching has been used in these studies, including self-quenching, dimer and excimer formation, and resonance energy transfer. One study followed the exchange of phospholipids between vesicles by observing the increase in fluorescence of a nitrobenzoxadiazole-phosphatidylethanolamine (NBD-PE) probe. The fluorophore self-quenched at the initial high concentration, but intensity increased as it was "diluted" upon lipid exchange.[63] The relative emission intensities of pyrene from its monomer and excimer forms, which emit at 400 and 480 nm, respectively, allows for increases in the monomer/excimer ratio to be used as an indicator of lipid component dilution on membrane fusion.[64] Fluorescence resonance energy transfer between NBD-PE and rhodamine-PE pairs, which results in concentration dependent quenching of the NBD emission by the rhodamine, can be used to observe changes in probe pair proximity and concentration during fusion.[65] It has been noted that membrane fusion assays must be done carefully and results interpreted cautiously as other factors, such as lipid exchange and vesicle aggregation, may result in similar changes in fluorescence signal.[66]

2. Monolayer Investigations

The study of monolayers at an air-water interface as a model for BLM behavior (see Section III.A) has led to the extension of membrane fluorescence experiments to these systems, with the associated advantages and assumptions. The major advantage of monolayers over vesicle systems, for present considerations, is the ability to transfer the monolayer to a solid substrate. This implies that fluorescence can also be used to characterize the deposited system, as would be necessary in a sensor.

It is generally accepted that the overall phase behavior of monolayers is similar to that of BLMs (with the BLMs at a fixed lateral pressure),[41] as phase structure is dominated by acyl chain interactions which are present in both systems. Initially, the pressure-area isotherm from a Langmuir-Blodgett film balance was the principal means of characterizing monolayer phase behavior.[67] Perhaps the most interesting current developments in the study of monolayers have come through the establishment of the technique of fluorescence microscopy.[68,69] This technique has made it possible to visualize the complex mixed-phase structure which actually exists in many monolayers. Visualization is accomplished either by using fluorophores which align with phase structure and can be located by analysis of fluorescence as a function of polarization of the excitation radiation, or by using fluorophores which partition between phases leaving some domains bright and others dark.

While fluorescence microscopy is not presently a viable means of obtaining a signal from a biosensor, it is a good method for evaluation of the steps in fabricating such a sensor. Fluorescence micrographs have demonstrated that monolayer transfer to a solid substrate at slow deposition rates (~1 cm.min^{-1}) results in maintenance of the phase domain structure.[70] Conversely, deposition at a high rate (~1 cm.s^{-1}) results in distinct "streaking" of the domain structure in the deposition direction. Final deposited structure may be critical to stability and function of a sensor system.

A signal of some analytical utility may be derived from modification of phase domain structure in monolayers using changes in overall intensity similar to those of the vesicle fusion experiments. Fluorescence self-quenching has been observed to correspond to expected phase domain structure at the air-water interface by using an NBD-lipid fluorophore[71] and in a deposited monolayer on an alkylated quartz substrate by using an anthroyloxy stearic acid fluorophore.[72]

C. MEMBRANE PERTURBATION: AN EXAMPLE

The mystery surrounding the conformational change of molecular receptors on agonist binding, coupled with the complexity of microscopic membrane structure evident from monolayer experiments, is indicative of how difficult it will be to exactly determine the nature and mechanism of protein-lipid interactions.

It is known that the AChR is extremely sensitive to supporting membrane composition,[73,74] functioning best at 45 mol% cholesterol, with specific ratios of various phospholipid types. Evidence from spin-labeled lipids shows preferential interaction of the receptor with steroids, fatty acids, and phosphatidic acid. Quenching of intrinsic protein fluorescence by lipid components labeled with bromine was used to study the partitioning of some membrane components, notably cholesterol, into a concentrated annular region around an AChR molecule.[75] Fluorescence lifetime measurements of pyrene-derivatized AChR before and after agonist binding indicated a greater exposure of the receptor to external quenching agents,[76] anions (I$^-$), cations (Tl^{+2}), and neutrals (nitromethane), implying an increase in local membrane permeability.

The local membrane perturbation induced by receptor conformational changes on agonist binding may provide an indirect means of monitoring the binding process. A generic sensing scheme may then be developed which utilizes a standard host membrane structure yielding a fluorescence response to structural perturbation, and then inserting a receptor of choice into the host structure.

To demonstrate generic transduction of receptor induced membrane perturbation by fluorescence, NBD–PE was added at a 2 mol% lipid concentration to vesicles containing AChR by vesicular lipid exchange. Addition of carbamylcholine to this system resulted in an enhancement of the fluorescence intensity, similar to that described for the IANBD-labeled AChR system. While a complete description of the mechanism of enhancement is not yet possible, it is reasonable to propose mechanisms involving membrane structural changes which would be expected to cause an increase in fluorescence intensity.

In general, fluorescence increases may be described for the NBD probe in the headgroup region of the membrane as a quantum yield increase due either to a decrease in local mobility, reducing collisional quenching, or to a decrease in local probe concentration, reducing self-quenching. An increase in rigidity of the headgroup region with a conformational change of the receptor protein may be attributed to changes in lipid packing/fluidity and headgroup electrostatics of the protein-associated lipid boundary region, as well as to the extension of that boundary region to incorporate more lipid molecules. A decrease in local probe concentration depends on changes in the distribution of phase domains in the bulk lipid system. An increase in area of fluid regions over gel regions, as in an induced local phase transition, will cause dilution of the probe which is known to partition into fluid regions from gel phases. These vague mechanisms for fluorescence enhancement are further complicated by the possible contribution of lateral protein mobility. Processes such as protein aggregation, if induced by agonist binding, may also account for changes both in the structure of the headgroup region and in the overall phase domain distribution. It must be emphasized that the process is receptor protein mediated, as the same experiment using vesicles of the same lipid composition but containing the protein bovine serum albumin, or no protein at all, yielded no enhancement. Thus there remains justification for pursuing a generic mechanism based on fluorescence transduction of membrane structural changes.

V. OPTICAL DEVICE STRUCTURES

Optrodes for spectroscopic measurement of membrane-based chemoreception may be of an extrinsic or intrinsic configuration. An extrinsic optrode would incorporate a vesicle-receptor system, as described in Section IV, into a small solution compartment situated at the distal end of a fiber or fiber bundle. Excitation and emission radiation would be carried by the same fiber bundle, allowing for probe-like operation. The solution compartment would have to be isolated from bulk solution (or gas phase) by a semipermeable membrane such as a dialysis membrane. The chief difficulty associated with this system is the slow rate associated with the diffusion of any analyte through a membrane such as a dialysis membrane. While reversible, the response time would likely be on the order of minutes or hours. Further, there are potential problems with membrane clogging and with adsorption to internal and external optrode surfaces. The principal advantage of an extrinsic optrode configuration is that the large signals obtained from vesicular solutions would allow for simple instrumentation, including conventional light sources and relatively inexpensive detectors.

Stabilization of lipid membrane structures onto surfaces, as described in Section III, allows for the development of intrinsic-mode optrodes. Using an intrinsic configuration, analyte-receptor interaction would require only the diffusion of analyte through bulk solution, which is generally quite rapid (microseconds to milliseconds). The small signals associated with a monolayer or multilayer of fluorescently labeled lipid membrane will likely restrict intrinsic-mode instruments to laser excitation and photomultiplier detection of fluorescence signals. Future developments in shorter wavelength semiconductor lasers and high sensitivity semiconductor detectors may alleviate some of the problems associated with weak fluorescence signals.

A. WAVEGUIDES AND OPTICAL FIBERS

Total internal reflection fluorescence (TIRF), the process utilized in intrinsic-mode optrodes, is ideal for interacting with lipid membranes owing to their thickness (~3 nm) and the surface specificity of the technique. The first choices to be made in TIRF optrode design are centered on the substrate or waveguide composition and geometry. While most fluorescence studies tend to use quartz substrates of some geometry, spectroscopic interactions using TIRF conventionally use planar waveguides while intrinsic optrodes generally incorporate cylindrical optical fibers.

Intrinsic-mode optrodes under consideration to this point have been almost exclusively multimode optical fibers, owing to their relative ease of use. This is a result of the strength, flexibility, and manipulability of these fibers. Quartz fibers of hundreds of micrometers diameter are inexpensive, ideal for fluorescence spectroscopy, fairly rugged, and easily surface-modified through silanization chemistry. They are limited, however, to fluorescence measurements of lifetime, wavelength, and intensity.

Where the flexibility and remote capability of an optical fiber are not essential, the use of a planar waveguide in fabricating a fluorescence optrode may be advantageous for several reasons:

1. It allows strict control over incident reflection angle, permitting efficient coupling of source radiation into the waveguide and control over the penetration depth of the evanescent wave into the surface coating. Multimode fibers result in a distribution of incident angles and little control over evanescent depth.
2. Langmuir-Blodgett deposition of membranes onto planar substrates is generally well characterized. Deposition onto cylindrical substrates has not been sufficiently studied.
3. Polarization of excitation and emission radiation may be used to observe the orientation of the fluorophore excitation and emission dipoles, respectively, as the polarization direction is maintained within the waveguide. Polarization direction is lost during multimode propagation in a fiber.
4. Interferrometric measurements are possible where two beams are intersected after propagating on paths of different but well characterized pathlength or index of refraction, such as is possible with planar waveguides but is difficult using multimode optical fibers.
5. Planar waveguide configurations are ideal for adaptation to optoelectronic technology, which may play a key role in eventual device development.

Single-mode optical fibers allow for the use of such measurements as polarization and interferometry, but have not been developed for the shorter wavelengths associated with fluorescence in terms of fiber diameter or fiber material (e.g., quartz filaments of <0.5 μm diameter). Deposition and characterization of membranes on such fibers by the L-B technique would also be difficult, simply because of difficulties in physical manipulation of ultra-thin fibers.

Future developments may include hybrid waveguide systems as optrodes, perhaps fibers fused onto planar waveguides, to take advantage of the flexibility of a fiber while maintaining the optical control of the planar waveguide.

B. IS THE EVANESCENT WAVE NECESSARY?

This section could also perhaps be titled, "Is the Evanescent Wave Present in the Fluorescence Excitation Process?" This question arises as a result of a very cursory examination of a model of the complete optical system, as depicted in Figure 3. Given that a membrane may be regarded as a close-packed, organized layer, some key assumptions may be applied to the model. The first is that the substrate n_0, the coating n_1, and the ambient n_2 are all isotropic in index of refraction. The second is that all interfaces are discrete and parallel, meaning that angles and points of reflection are well defined. Further assumptions include $d_2 \rightarrow \infty$, or that the ambient thickness may be considered infinite, and $d_0 \gg d_1$, which is equivalent to stating that loss of propagated radiation from the ends of the coated region is negligible as most remains within the substrate.

On coupling of an incident laser beam into a multimode fiber, there will generally be a distribution of modes from $\theta^{0,2} = 90°$ to $\theta^{0,2} = \theta_c^{0,2}$, where $\sin \theta_c^{0,2} = n_2/n_0$. This immediately requires that $n_0 > n_2$. Given these conditions, three general schemes are possible for the intrinsic optrode configuration.

$$n_o > n_1 = n_2 \left(\text{or } n_1 < n_2 \right) \qquad \text{(S. 1)}$$

In this case, $\theta_c^{0,2} \geq \theta_c^{0,1}$ yielding the classic evanescent configuration, where only the evanescent component of the propagated radiation enters the coating phase n_1. For fluorescence

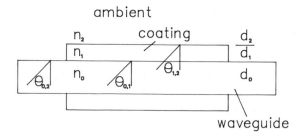

FIGURE 3. A schematic showing a coated waveguide, indicating optical parameters pertaining to internal reflection of radiation .

recovery considerations, however, this means that no fluorescence is coupled back into the waveguide, as $\theta_c^{1,2} \geq 90°$. Collection and detection of fluorescence must be done external to the waveguide.

$$n_o > n_1 > n_2 \qquad\qquad (S.\ 2)$$

In this scheme, $q_c^{0,2} < q_c^{0,1}$. All incoming modes where $q > q_c^{0,1}$ will be totally reflected as in S. 1. Modes such that $q_c^{0,2} < q^{0,1} < q_c^{0,1}$ will propagate through the coating phase n_1 and will be totally reflected at the coating-ambient interface. Thus a mixture of evanescent and direct excitation will occur, with its distribution dependent on the distribution of incident mode angles in the waveguide. Fluorescence modes such that $q^{1,2} > q_c^{1,2}$ will be incorporated into the waveguide for detection.

$$n_o = n_1 > n_2 \left(or\ n_o < n_1\right) \qquad\qquad (S.\ 3)$$

In this scheme, $q_c^{0,1} \leq q_c^{0,2}$ so all incident modes will propagate into the coating layer. All of the evanescent component will be from the coating-ambient interface, and will not participate in the excitation of fluorescence in n_1. All fluorescence such that $q \geq q_c^{0,2}$ will be propagated into the waveguide.

It is interesting to note that the index of refraction of close-packed biological layers is usually taken as $n_1 = 1.5$. The index of refraction of a quartz fiber is $n_0 = 1.47$. This means that these systems are generally operating under S. 3, and it is not appropriate to refer to them as evanescently excited, though they are still intrinsic. It is important that although S. 3 offers the greatest possible contribution of solution interference, owing to the presence of the evanescent component in the ambient, it also offers the greatest incorporation of fluorescence back into the waveguide as is desirable in an intrinsic optrode.

C. LIGHT SCATTERING TECHNIQUES

An evanescent wave originating from the optical transmission of radiation by total internal reflection within a waveguide and propagating into a chemically selective lipid membrane can cause excitation of chromophores or fluorophores. This can result in alteration of intensity, polarization, wavelength, and/or lifetime of the radiation propagated within the waveguide, and can be sampled either by collecting photons at the terminus of the waveguide, or by placing a photomultiplier in an orientation which is perpendicular to the optical train. The evanescent wave intensity near the surface of the waveguide can be limited due to restrictions of wavelength, incident angle, and refractive index similarities between the coating and the guide. Recapture of fluorescence by the waveguide is also limited by these latter criteria, combining to reduce the sensitivity of such a device. This situation has led to the development of a new strategy for amplification of fluorescence by avoiding the difficulties associated with the evanescent wave phenomenon, particularly when quartz optics are interfaced with extremely thin organic films of very similar refractive index. The evanescent arrangement does effectively incorporate the thin film as the outer surface of the waveguide, but no signal amplification and limited fluorescent efficiency could be anticipated.

An optical fiber and its union with an analytically viable, chemically selective fluorescent lipid membrane constitute the primary components of the evanescent technology and the new light scattering strategy. A quartz fiber is used so that the entire optical and near UV wavelength range is suitable for excitation purposes. The fiber is designed to contain structural imperfections such as crystallites and/or trapped gas bubbles which can scatter the optical radiation in all directions from the interior of the fiber. Hand drawn quartz fibers prepared from quartz rods are suitable for these experiments. Light scatter from these imperfections enhance leakage of radiation from the fiber and provides efficient illumination of a surface coating of lipid membrane. Fluorescent radiation is emitted in all directions, and the intensity reduction of source radiation after one pass through a membrane only 5 nm in thickness limits the sensitivity of the system. Therefore the system is enclosed in a second mirrored capillary tube which serves to eliminate escape of scattered source and fluorescent radiation. The multiple passage of excitation radiation through the membrane provides amplification of the fluorescence signal which is captured at the fiber terminus. Present research developments include the passage of a flowing solution through the interior of the capillary tube, providing a sensitive on-line sensor for a flow injection analysis system. Langmuir-Blodgett multilayer preparations of stearic acid have been shown to be stable during such experiments which transfer the lipid layers through the atmosphere to different solutions.

D. LIPID MULTILAYER WAVEGUIDES

The use of lipid membrane technology to create selective and sensitive biomembranes and prepare physical and chemical models of chemoreceptive processes has demonstrated the advantages which can be achieved by combining the selective chemistry and transduction process in one structure. Investigations studying optical transmission have demonstrated that carefully constructed monolayers of lipid can act as optical waveguides.[77,78] Structural manipulation of lipid membranes can lead to development of an intrinsic optical sensor consisting of a selective biomembrane which acts as a light guide. This physical arrangement maximizes membrane selectivity parameters as well as optical emission and collection conditions, and integrates the chemistry and transducer into a single entity.

Suitable multilayers have been prepared and examined on silicon semiconductor wafers, which have sufficient reflectivity to allow monitoring of deposition by ellipsometry, and are opaque with respect to waveguiding.[79] Stearic acid monolayers at an air-water interface held at pressures of 30 mN.m^{-1} over a 0.1M KCl subphase at pH 7.0 have been reproducibly transferred by rapid dip casting (0.1 to 1.0 cm.s^{-1}) onto a silicon semiconductor wafer to form multilayer assemblies of over 250 monolayers.

The first layer deposited is aligned by polar bond formation between the hydrophilic sites of the molecule and the silicon oxide surface, and occurred during removal from the subphase. The dense packing of the transferred monolayer caused the new surface to be hydrophobic as demonstrated by contact angle measurements. Subsequently, casting caused deposition to occur during both immersion and withdrawal through the monolayer. These layers deposit due to Van der Waals interactions between acyl chains, and polar interactions between the carboxylic acid headgroups and retained water of hydration. Barrier movement was constant for each monolayer deposition and allowed a transfer ratio of approximately 0.98 ± 0.03 to be established (estimated due to the irregular shape of the wafer). Surface wettability measurements indicated that the orientation of the surface layer of stearic acid molecules always assumed a configuration in which the acyl chains were directed away from the silicone surface.

The long-term stability of the multilayers when stored in air or aqueous solution was excellent, and no apparent thickness/refractive index alterations were observed ellipsometrically over periods of weeks.

The sequential deposition of 200 monolayers of stearic provided the base for a fluorescent waveguide. Stearic acid has a chain length of approximately 4 nm, which would indicate an approximate thickness of 800 nm for the multilayer system, if the system is assumed to have no angular dependence and no interdigitation. This thickness would allow for visible light from an argon ion laser to be guided by total internal reflection through the lipid matrix in a monomode capacity. To determine if waveguiding was occuring, 20 layers of stearic acid containing the fluorescent probe NBD hexadecyl amine were deposited onto the 200 stearic acid layers. The angle of illumination of the wafer was varied by moving an optical fiber providing excitation radiation to monitor the change in fluorescence intensity of the probe, which could only be related to the event of waveguiding. The results of this angular dependence of fluorescence indicated an angular threshold of between 42° and 45° for initiation of the waveguiding event. The critical angle was calculated to be 42° on the basis of refractive index of 1.5 for stearic acid and 1.0 for air. A decrease in signal at 70° was consistent with a decrease in the quantity of light entering the waveguide due to increased scattering.

The production of chemically selective multilayers formed by depositing a receptor-doped phospholipid monolayer matrix onto an underlying stearic acid waveguide has been initiated. The hydrophobic nature of the surface of the multilayer asssembly provides an ideal surface for deposition of a phospholipid monolayer in a biologically significant orientation. Transfer of phospholipid monolayers to such surfaces by Langmuir-Blodgett casting techniques has been done, and monolayers and multilayers have been prepared. Gas phase interactions with chloroform, hexane, and the quenching agent N,N-dimethylaniline have demonstrated the ability of the system to fluorescently indicate gas concentration quantitatively. Furthermore, the investigators have provided both fluorescent intensity enhancement and signal reduction for different gases, indicating that the lipid phase domain structure within surface deposited multilayers can be controlled *in situ* after deposition.

VI. FURTHER RESEARCH DIRECTIONS

Further development of fiber optic sensors that make use of a generic lipid membrane perturbation for transduction of chemical stimuli will lead to advances in a number of areas. Membrane stabilization to provide fluid yet dense organized films is progressing well, but requires an efficient method to allow reproducible incorporation of active binding agents. Extremely sensitive threshold concentration sensors can be made by manipulation of the phase structure of membranes. This is done by "switching" between distinct phase states by the physical perturbations caused by selective binding processes. Multidimensional fluorescence maps combining intensity, wavelength, polarization, and/or lifetime data will provide analytical surfaces of great utility in quantitatively and qualitatively defining the presence of selective binding processes as well as interference effects. Optimization of fluorescence labeling of lipid membranes and binding agents must be achieved, and will likely involve implementation of new fluorescent probes. Replacement of naturally derived receptors with artificial receptors will provide enhancement of membrane perturbations, and increased stability and functional lifetime.

Other areas of optical sensor design are being investigated and may provide complementary information to the fluorescence strategy described herein. Infrared evanescent wave processes can use the same membrane-coated optical fibers as used for fluorescence studies, greatly expanding the optical information available for characterization of binding events. Surface enhanced Raman spectroscopy and plasmon resonance spectroscopy may also be suitable for combination with chemically selective lipid membranes deposited on waveguides.

ACKNOWLEDGMENTS

We are grateful to the Natural Sciences and Engineering Research Council of Canada, the Canadian Defense Research Establishment, the Ontario Ministry of the Environment, and Imperial Oil Canada for support of this work. We would also like to acknowledge experimental assistance and useful discussions with Drs. M. Thompson, V. Ghaemmaghami, K. Kallury, and also J. Brennan, C. Cadas, R. DeBono, B. Hougham, G. McGibbon, and K. Stewart.

REFERENCES

1. **Stryer, L.,** *Biochemistry*, 2nd ed., W. H. Freeman, San Francisco, 1981, 109.
2. **Stryer, L.,** *Biochemistry*, 2nd ed., W. H. Freeman, San Francisco, 1981, 115.
3. **Udenfriend, S.,** *Fluorescence Assay in Biology and Medicine*, Academic Press, New York, 1962, chap. 5.
4. **Lipscomb, W. N.,** Structures and mechanisms of enzymes, *Proc. Robert A. Welch Found. Conf. Chem. Res.*, 15, 131, 1971.
5. **Arnold, M. A.,** Enzyme-based optical sensor, *Anal. Chem.*, 57, 565, 1985.
6. **Wolfbeis, O. S.,** Fibre-optic probe for kinetic determination of enzyme activities, *Anal. Chem.*, 58, 2874, 1986.
7. **Walters, B. S., Nielsen, T. J., and Arnold, M. A.,** Fibre-optic biosensor for ethanol based on an internal enzyme concept, *Talanta*, 35, 151, 1988.
8. **Stryer, L.,** *Biochemistry*, 2nd ed., W. H. Freeman, San Francisco, 1981, 792.
9. **Massey, R. J.,** Catalytic antibodies catching on, *Nature (London)*, 328, 457, 1987.
10. **Schultz, P. G.,** The interplay between chemistry and biology in the design of enzymatic catalysts, *Science*, 240, 426, 1988.
11. **Dahne, C., Sutherland, R. M., Place, J. F., and Ringrose, A. S.,** Detection of antibody-antigen reactions at a glass-liquid interface, in *International Conference on Optical Fibre Sensors*, Kersten, R. Th. and Kist, R., Eds., VDE-Verlag GmbH, Berlin, Offenbach, 1984.
12. **Stryer, L.,** *Biochemistry*, 2nd ed., W. H. Freeman, San Francisco, 1981, 221.
13. **MacKenzie, A. E.,** A Study of the Binding of Oligosaccharides and Glycopeptides to Concanavalin A, Ph.D. thesis, University of Toronto, Toronto, Canada, 1985.
14. **Schultz, J. S., Mansouri, S., and Goldstein, I. J.,** Affinity sensor: a new technique for developing implantable sensors for glucose and other metabolites, *Diabetes Care*, 5, 245, 1982.
15. **Changeux, J.-P., Devillers-Thiery, A., and Chemouilli, P.,** Acetylcholine receptor: an allosteric protein, *Science*, 225, 1335, 1984.
16. **Dalziel, A. W., Georger, J., Price, R. R., Singh, A., and Yager, P.,** Progress report on the fabrication of an acetylcholine receptor-based biosensor, in *Membrane Proteins: Proceedings of the Membrane Protein Symposium, August 3-6, 1986, San Diego, California*, Hjelmeland, L. M., Gennis, R., McNamee, M. G., and Goheen, S. C., Eds., Bio-Rad Publishing Co., 1986.
17. **Gotoh, M., Tamiya, E., Momoi, M., Kagawa, Y., and Karube, I.,** Sensor based on ion selective field effect transistor and acetylcholine receptor, *Anal. Lett.*, 20, 857, 1987.
18. **Valdes, J. J., Wall, J. R., Jr., Chambers, J. P., and Eldefrawi, M. E.,** A receptor-based capacitive biosensor, *Johns Hopkins APL Tech. Dig.*, 9(1), 4, 1988.
19. **Dunn, S. M. J., Blanchard, S. G., and Raftery, M. A.,** Kinetics of carbamylcholine binding to membrane-bound acetylcholine receptor monitored by fluorescence changes of a covalently bound probe, *Biochemistry*, 19, 5645, 1980.
20. **Dunn, S. M. J. and Raftery, M. A.,** Activation and sensitization of *Torpedo* acetylcholine receptor: evidence for separate binding sites, *Proc. Natl. Acad. Sci. U.S.A.*, 79, 6757, 1982.
21. **Dunn, S. M. J., Conti-Tronconi, B. M., and Raftery, M. A.,** Separate sites of low and high affinity for agonists on *Torpedo californica* acetylcholine receptor, *Biochemistry*, 22, 2512, 1983.
22. **Jencks, W. P.,** Binding energy, specificity, and enzymic catalysis: the circe effect, in *Adv. Enzymol. Relat. Areas Mol. Biol.*, Meister, A., Ed., 43, 219, 1975.
23. **De Lean, A., Stadel, J. M., and Lefkowitz, R. J.,** A ternary complex model explains the agonist-specific binding properties of the adenylate cyclase-coupled β-adrenergic receptor, *J. Biol. Chem.*, 255, 7108, 1980.
24. **Thompson, M., Dorn, W. H., Krull, U. J., Tauskela, J. S., Vandenberg, E. T., and Wong, H. E.,** The primary events in chemical sensory perception, *Anal. Chim. Acta*, 180, 251, 1986.

25. **Ganong, W. F.,** *The Nervous System*, 2nd ed., Lange, Los Altos, CA, 1979, chap. 10.
26. **Graziadei, P. P. C.,** The ultrastructure of vertebrates olfactory mucosa, in *The Ultrastructure of Sensory Organs*, Friedman, I., Ed., Elsevier, New York, 1984, chap. 4.
27. **Brown, E. L. and Deffenbacher, K.,** *Perception and the Senses*, Oxford University Press, New York, 1978, 118.
28. **Koyama, N. and Kurihara, K.,** Modification by chemical reagents of proteins in the gustatory and olfactory regions of the fleshfly and cockroach, *Nature (London)*, 236, 402, 1972.
29. **Dressen, T. D. and Koch, R. B.,** Odorous perturbants of $(Na^+ + K^+)$, *Biochem. J.*, 203, 69, 1982.
30. **Tien, H. Ti,** *Bilayer Lipid Membranes*, Marcel Dekker, New York, 1974.
31. **White, S. H.,** The physical nature of planar bilayer membranes, in *Ion Channel Reconstitution*, Miller, C., Ed., Plenum Press, New York, 1986, 3.
32. **Mueller, P., Rudin, D. O., Tien, H. Ti, and Wescott, W. C.,** Formation and properties of bimolecular lipid membranes, *Nature (London)*, 194, 979, 1962.
33. **del Castillo, J., Rodriguez, A., Romero, C. A., and Sanchez, V.,** Lipid films as transducers for detection of antibody-antigen and enzyme-substrate reactions, *Science*, 153, 185, 1966.
34. **Lauger, P., Benz, R., Stark, G., Bamberg, E., Jordan, P. C., Fahr, A., and Brock, W.,** Relaxation studies of ion transport systems in lipid bilayer membranes, *Q. Rev. Biophys.*, 14, 513, 1981.
35. **Yguerabide, J. and Foster, M. C.,** Fluorescence spectroscopy of biological membranes, in *Molecular Biology, Biochemistry and Biophysics*, Vol. 31, Grell, E., Ed., Springer-Verlag, New York, 1981, 200.
36. **Wolfbeis, O. S. and Schaffar, B. P. H.,** Optical sensors: an ion-selective optrode for potassium, *Anal. Chim. Acta*, 198, 1, 1987.
37. **Zhujun, Z. and Seitz, W. R.,** Ion-selective sensing based on potential sensitive dyes, *Proc. Optical Fibers in Medicine III*, Katzir, A., Ed., SPIE, Bellingham, WA, 1988, 74.
38. **Heckl, W. M.,** Laterale Organisation von Lipidmonoschichten bei Einbau von Amphiphilen Fremdstoffen und Proteinen, Ph.D. thesis, Technischen Universitat Munchen, Munich, 1988.
39. **Weis, R. M. and McConnell, H. M.,** Cholesterol stabilizes the crystal-liquid interface in phospholipid monolayers, *J. Phys. Chem.*, 89, 4453, 1985.
40. **Thompson, M., Krull, U. J., and Bendell-Young, L. I.,** Surface aggregate modulation of lipid membrane ion permeability, *Bioelectrochem. Bioenerg.*, 13, 255, 1984.
41. **Nagle, J. F.,** Theory of the main lipid bilayer phase transition, *Annu. Rev. Phys. Chem.*, 31, 157, 1980.
42. **Tamm, L. K. and McConnell, H. M.,** Supported phospholipid bilayers, *Biophys. J.*, 47, 105, 1985.
43. **Krull, U. J., Brown, R. S., and Safarzadeh-Amiri, A.,** Optical transduction of chemoreceptive events: towards a fiber-optic biosensor, in *Proc. Optical Fibers in Medicine III*, Katzir, A., Ed., SPIE, Bellingham, WA, 1988, 49.
44. **Krull, U. J., Bloore, C., and Gumbs, G.,** Supported chemoreceptive lipid membrane transduction by fluorescence modulation: the basis of an intrinsic fiber-optic biosensor, *Analyst*, 111, 259, 1986.
45. **Albrecht, O., Johnston, D. S., Villaverde, C., and Chapman, D.,** Stable biomembrane surfaces formed by phospholipid polymers, *Biochim. Biophys. Acta*, 687, 165, 1982.
46. **Kajar, F. and Messier, J.,** Solid State polymerization and optical properties of diacetylene Langmuir-Blodgett multilayers, *Thin Solid Films*, 99, 109, 1983.
47. **Netzer, L., Iscovici, R., and Sagiv, J.,** Adsorbed monolayers versus Langmuir-Blodgett monolayers. I. From monolayer to multilayer by adsorption, *Thin Solid Films*, 99, 235, 1983.
48. **Kallury, R. K., Krull, U. J., and Thompson, M.,** Synthesis of phospholipids suitable for covalent binding to surfaces, *J. Org. Chem.*, 52, 5478, 1987.
49. **Kallury, R. K. M., Krull, U. J., and Thompson, M.,** X-ray photoelectron spectroscopy of silica surfaces treated with polyfunctional silanes, *Anal. Chem.*, 60, 169, 1988.
50. **Nuzzo, R. G., Fuso, F. A., and Allara, D. L.,** Spontaneously organized molecular assemblies. 3. Preparation and properties of solution adsorbed monolayers of organic disulfides form solution onto gold surfaces, *J. Am. Chem. Soc.*, 109, 2358, 1987.
51. **Troughton, E. B., Bain, C. D., and Whitesides, G. M.,** Monolayer films prepared by the spontaneous self-assembly of symmetrical and unsymmetrical dialkyl sulfides from solution onto gold substrates: structure, properties, and reactivity of constituent functional groups, *Langmuir*, 4, 365, 1988.
52. **Darszon, A.,** Strategies in the reassembly of membrane proteins into lipid bilayer systems and their functional assay, *J. Bioenerg. Biomembr.*, 15, 321, 1983.
53. **Meyers, H.-W., Jurss, R., Brenner, H. R., Fels, G., Prinz, H., Watzke, H., and Maelicke, A.,** Synthesis and properties of NBD-*n*-acylcholines, fluorescent analogs of acetylcholine, *Eur. J. Biochem.*, 137, 399, 1983.
54. **Endo, T., Nakanishi, M., Furukawa, S., Joubert, F. J., Tamiya, N., and Hayashi, K.,** Stopped-flow fluorescence studies on binding kinetics of neurotoxins with acetylcholine receptor, *Biochemistry*, 25, 395, 1986.
55. **Eldefrawi, M. E., Eldefrawi, A. T., and Wilson, D. B.,** Tryptophan and cystein residues of the acetylcholine receptors of *Torpedo* species, *Biochemistry*, 14, 4304, 1975.

56. **Pesce, A. J., Rosen, C. G., and Pasby, T. L., Eds.,** *Fluorescence Spectroscopy*, Marcel Dekker, New York, 1971.
57. **Wehry, E. L., Ed.,** *Modern Fluorescence Spectroscopy 2*, Plenum Press, New York, 1976.
58. **Pal, R., Petri, W. A., Ben-Yashar, V., Wagner, R. R., and Barenholz, Y.,** Characterization of the fluophore 4-heptadecyl-7-hydroxycoumarine: a probe for the head-group region of lipid bilayers and biological membranes, *Biochemistry*, 24, 573, 1985.
59. **Winiski, A. P., Eisenberg, M., Langner, M., and McLaughlin, S.,** Fluorescent probes of electrostatic potential 1 nm from the membrane surface, *Biochemistry*, 27, 386, 1988.
60. **Faludi-Daniel, A., Szito, T., Kiss, J. G., and Garab, G. I.,** The organization of thylakoid membranes as shown by linear dichroism and fluorescence polarization of aligned membranes, *Photobiochem. Photobiophys.*, 12, 1, 1986.
61. **Brown, G. H. and Wolken, J. J.,** *Liquid Crystals and Biological Structures*, Academic Press, New York, 1979.
62. **Vanderkooi, J., Fischkoff, S., Chance, B., and Cooper, R. A.,** Fluorescent probe analysis of the lipid architecture of natural and experimental cholesterol-rich membranes, *Biochemistry*, 13, 1589, 1974.
63. **Nichols, J. W. and Pagano, R. E.,** Kinetics of soluble lipid monomer diffusion between vesicles, *Biochemistry*, 20, 2783, 1981.
64. **Pal, R., Barenholz, Y., and Wagner, R. R.,** Pyrene phospholipid as a biological fluorescent probe for studying fusion of virus membrane with liposomes, *Biochemistry*, 27, 30, 1988.
65. **Wilshut, J. and Hoekstra, D.,** Membrane fusion: lipid vesicles as a model system, *Chem. Phys. Lipids*, 40, 145, 1986.
66. **Duzgunes, N., Allen, T. M., Fedor, J., and Papahadjopoulos, D.,** Lipid mixing during membrane aggregation and fusion: why fusion assays disagree, *Biochemistry*, 26, 8435, 1987.
67. **Gaines, G. L.,** *Insoluble Monolayers at Liquid-Gas Interfaces*, Interscience, New York, 1966.
68. **Losche, M. and Mohwald, H.,** Fluorescence microscope to observe dynamical processes in monomolecular layers at the air/water interface, *Rev. Sci. Instrum.*, 55, 1968, 1984.
69. **Seul, M. and McConnell, H. M.,** Automated langmuir trough with epifluorescence attachment, *J. Phys. E*, 18, 1, 1985.
70. **McConnell, H. M., Tamm, L. K., and Weis, R. K.,** Periodic structure in lipid monolayer phase transitions, *Proc. Natl. Acad. Sci.*, 81, 3249, 1984.
71. **Owen, C. S.,** A membrane bound fluorescent probe to detect phospholipid vesicle-cell fusion, *J. Membr. Biol.*, 54, 13, 1980.
72. **Krull, U. J., Brown, R. S., DeBono, R. F., and Hougham, B. D.,** Towards a fluorescent chemoreceptive membrane-based optode, *Talanta*, 35, 129, 1988.
73. **Zabrecky, J. R. and Raftery, M. A.,** The role of lipids in the function of the acetylcholine receptor, *J. Recept. Res.*, 5, 397, 1985.
74. **Fong, T. M. and McNamee, M. G.,** Correlation between acetylcholine receptor function and structural properties of membranes, *Biochemistry*, 25, 830, 1986.
75. **Jones, O. T. and McNamee, M. G.,** Annular and nonannular binding sites for cholesterol associated with the nicotinic acetylcholine receptor, *Biochemistry*, 27, 2364, 1988.
76. **Gonzales-Ros, J. M., Farach, M. C., and Martinez-Carrion, M.,** Ligand-induced effects of regions of acetylcholine receptor accessible to membrane lipids, *Biochemistry*, 22, 3807, 1983.
77. **Pitt, C. W. and Walpita, L. M.,** Lightguiding in Langmuir-Blodgett films, *Thin Solid Films*, 68, 101, 1980.
78. **Walpita, L. M. and Pitt, C. W.,** Measurement of Langmuir film properties by optical waveguide probe, *Electron. Lett.*, 13, 210, 1977.
79. **Krull, U. J., Brown, R. S., and Stewart, K.,** An intrinsic chemically selective lipid-based waveguide for organic vapour sensing, in *Proc. 1987 Technology Transfer Conference — Analytical Methods*, Province of Ontario, 1987, D14.

INDEX

U